CSR, Sustainability, Ethics & Governance

Series editors
Samuel O. Idowu, London Metropolitan University, London, United Kingdom
René Schmidpeter, Cologne Business School, Germany

More information about this series at http://www.springer.com/series/11565

Reinhard Altenburger

Editor

Innovation Management and Corporate Social Responsibility

Social Responsibility as Competitive Advantage

 Springer

Editor
Reinhard Altenburger
Department Business
IMC University of Applied Sciences Krems
Krems, Austria

ISSN 2196-7075   ISSN 2196-7083   (electronic)
CSR, Sustainability, Ethics & Governance
ISBN 978-3-319-93628-4   ISBN 978-3-319-93629-1   (eBook)
https://doi.org/10.1007/978-3-319-93629-1

Library of Congress Control Number: 2018956809

This Springer imprint is published by the registered company Springer Nature Switzerland AG
The registered company address is: Gewerbestrasse 11, 6330 Cham, Switzerland

Responsibility and Innovation: Two Sides of the Same Coin?

In order to successful address the current societal and environmental challenges—resource scarcity, climate change, social and political tensions, and demographic changes—we need increased levels of ecological and social innovations. Companies play a crucial role in this process, as they will decisively influence both contemporary and future society through "corporate creativity" and sustainable value creation.

If one takes a look at history, for example, industrialization, it is clear that visionary entrepreneurs, executives, and economic thinkers recognized their time period's respective issues and developed new solutions, which specifically attempted to solve these societal problems. Today is no different. We will only overcome the current challenges when companies represent a part of the solution and not part of the problem, which is unfortunately not the view of many of critics who view businesses and capitalism as the problem.

The approach of sustainable value creation connected with sustainable innovation is essentially a rediscovery of an "old" management paradigm. Due to this rediscovery, the current discussion in business circles surrounding corporate social responsibility has shifted from profits and ethics being mutually exclusive to that one should consider both in equal regard.

In contrast to critics of capitalism, the common good should not be dictatorially placed above economic rationality. Rather, economic expertise should be used to reconcile the interests of companies (to generate added value) with respect to the legitimate interests of society (to generate societal added value).

Sustainable innovation bridges this gap. If one views the current social and ecological challenges, in this light, they represent a tremendous opportunity for all entrepreneurs. It is now evident that sustainable innovations create completely new products and markets that both enhance the future viability of our society and increase the competitiveness and profitability of companies.

If we succeed in understanding social and ecological issues as part of the entrepreneurial innovation process, the solutions to our current problems will be a natural result of it. That is why one can and should consider responsibility and innovation as two sides of the same coin. After all, a lack of innovative strength ultimately also means that the responsibility rightly demanded by society cannot be borne by companies.

By definition, entrepreneurs are innovators, as they are always seeking out better solutions and competitive advantages. Companies will be given the social license to operate, if they generate profits and actively help to solve the current ecological and social challenges in a constructive manner. This is not only an ethical necessity, but rather is a manner of economic necessity and ultimately lucrative for companies.

This publication breaks the mold of the public black-and-white discussion on sustainability: on the one hand through new business management approaches and on the other hand through numerous positive practical examples from very different sectors. The book provides further concepts regarding the implementation of CSR in companies. All readers are invited to take up the ideas presented in the series and to use them for their own professional challenges and to discuss them intensively with the editors, authors, and supporters of this series. Last but not least, I would like to thank the editor, Prof. Dr. Reinhard Altenburger, for his commitment and continued support and Christian Rauscher from Springer-Verlag for his excellent work and all the supporters of the series and hope that you, dear readers, have an interesting, enjoyable, and informative read.

Cologne Business School, Germany   René Schmidpeter

Preface

Enterprises play a key role in addressing environmental and social challenges. Globally, in an increasing number of companies—not only hand multinationals, but also family businesses and SMEs—an increasing involvement of innovation management with issues of sustainability and social responsibility can be identified. On the one hand, this development is driven by the perceptible pressure of numerous stakeholder groups such as customers, suppliers, NGOs, and consumer protection organizations and, on the other hand, by the growing awareness of globally active companies of taking on responsibility toward society. Increasing international interconnectedness in the value chain requires intensive consideration of aspects of sustainability and also a different understanding of "sustainability" and "responsibility" in different regions of the world. New technologies such as artificial intelligence and blockchain often enable a rapid increase in productivity but often also require an intensive examination of the social consequences. This requires executives to deal with increasing complexity in managing innovation. CEOs around the world are becoming increasingly aware that responsibility and sustainability issues require not just a change in their product mix, but often a reorientation of their business model. How far social responsibility goes is seen differently by managers in different cultures. It is not about right or wrong, but to develop suitable approaches to the challenges of the industry but also the respective culture. The international discussion of the UN Sustainable Development Goals is expected to make a significant contribution.

Interacting with stakeholders once again gains significance for sustainable innovations. Especially the cooperation with customers and suppliers but also universities and NGOs requires a professionalization of the open-innovation process and investments in a "culture of trust.". It is also required to have a more comprehensive understanding of innovation, as corporate social responsibility and sustainability will impact on innovation in processes, products, and services as well as on the organization and business model of companies.

This volume presents examples from the perspective of globally operating companies as well as SMEs and family businesses, as well as integrating contributions from the current scientific discussion on innovation and social responsibility/sustainability. Likewise, it is important to point out the challenges in various industries—from the textile, electronics, and logistics industries—to selected companies and exemplify solution approaches for sustainable innovation.

The different perspectives and approaches of the individual authors should offer suggestions for dealing with the connection between social responsibility, sustainability, and innovation management. My special thanks go to the Springer publishing house team—and in particular to Christian Rauscher and Barbara Bethke for their support in the production process of this publication.

Krems, Austria   Reinhard Altenburger
March 2018

Contents

Corporate Social Responsibility as a Driver of Innovation Processes

Reinhard Altenburger

1 Global Challenges for Sustainable and Responsible Innovation

Social and environmental challenges such as climate change, resource constraints, urbanization, loss of trust in companies and institutions, loss of biodiversity, access to clean drinking water, quality of working conditions and demographic shifts are often seen as risks by companies, but by a growing number of companies also regarded as an opportunity for innovation. In science, civil society (especially in globally active NGOs), and business (multinationals as well as medium-sized companies) but also in politics, there has been a comprehensive and varied discussion about responsibility of companies during the last years. This discussion has intensified considerably after numerous scandals such as Enron, Worldcom, Siemens (corruption), Nike (child labor), Volkswagen or TEPCO (Fukushima) and the economic and financial crisis 2008/9.

The worldwide intensive discussion of societal and environmental challenges has triggered a change in the understanding of CSR in the academic community and in numerous companies. We can see a shifting focus from risk reduction and the increase in reputation to the perception of innovation opportunities. By addressing the above mentioned issues companies from all sectors are looking for new solutions and to question their current business model—in some cases radically.

Social as well as ecological criteria have become particularly relevant in the last few years as they are increasingly taken into account in the purchasing decisions of consumers, but also play an increasingly important role in investment by companies and the public sector (Schaltegger et al. 2009). Numerous activities for the

R. Altenburger (✉)
Department of Business, IMC University of Applied Sciences Krems, Krems, Austria
e-mail: reinhard.altenburger@fh-krems.ac.at

© Springer International Publishing AG, part of Springer Nature 2018   1
R. Altenburger (ed.), *Innovation Management and Corporate Social Responsibility*,
CSR, Sustainability, Ethics & Governance,
https://doi.org/10.1007/978-3-319-93629-1_1

development of environmentally-friendly products, fair trade, social projects with NGOs have been launched by companies, but these are often uncoordinated activities which are lacking in strategic direction (Porter and Kramer 2011). Critical stakeholders increasingly focus on the impact of corporate CSR activities, and critically analyze whether CSR practices have not only a "PR-focus", or can be described as "greenwashing" or "window dressing".

For many companies, the discussion of sustainable and socially responsible products and services focuses on the question of the costs caused by this (Schreck 2012). The discussion about the so-called "business case for CSR" has intensified in recent years. In most cases, questions concerning the CSR's return on investment and whether a positive relationship between corporate social performance (CSP) and Corporate Financial Performance (CFP) in numerous contributions and studies were at the center of the discussion (Carroll and Shabana 2010). Kurucz et al. (2008) have identified four main arguments and discussion strands around the "Business Case for CSR":

- Cost and risk reduction
- Achievement of competitive advantages
- Legitimacy and reputation as well as
- Win-win situations of companies and society (joint value creation).

Schaltegger (2012) presents the discussion about the "Business Case OF Sustainability" in which the use of a trend often without "sustainability" is opposed to "Business Case FOR Sustainability" and social activities. In a recent study (Accenture & United Nations Global Compact 2016) CEOs believe that Agend 2030 provides an essential window of opportunity to rethink and reset approaches to sustainability:

- 87% believe the Sustainable Development Goals (SDGs) provide an opportunity to rethink approaches to sustainable value creation
- 80% believe that demonstrating a commitment to societal purpose is a differentiator in their industry
- 88% believe that greater integration of sustainability issues in financial markets will be essential to making progress

The European Union has made a major contribution to the global debate on CSR with the view that CSR is "the responsibility of businesses for their impact on society" (European Commission 2011). As a result of this, CSR has a considerable influence on how well a company adapts to social change and therefore plays a decisive role in securing the competitiveness of a company. Although the EU's interpretation of CSR is consistent with the results of the scientific literature, there is no clear, generally accepted definition or a comprehensive theoretical framework for CSR (Perrini 2006). Many companies now view CSR initiatives as opportunities for more efficient management of their human resources and the supply chain in order to achieve a competitive advantage. Studies show that companies are pursuing CSR approaches to increase efficiency, stimulate innovation, and create organizational growth (Stigson 2002).

The ISO 26000 (2010) is a common internationally recognized standard of principles and areas of responsibility in the exercise of social responsibility for all types of organizations. ISO 26000 defines social responsibility as the "responsibility of an organization for the impact of its decisions and activities on society and the environment through transparent and ethical conduct,

- contributes to sustainable development, health and well-being,
- takes account of the expectations of stakeholders,
- adheres to applicable law and is in accordance with international standards of conduct,
- is integrated throughout the organization;
- is lived in their relationships".

Visser (2011) describes the particular challenge for companies in the development of CSR 1.0 to CSR 2.0. CSR 1.0 is therefore strongly influenced by a philanthropic orientation, product orientation, standards and guidelines, and the focus is on multinational companies. CSR 2.0 is characterized by innovative partnership models and intensive networking with stakeholders, real-time transparency, social entrepreneurship, decentralization and shared value. CSR 2.0 according to Visser (2011) is characterized by the following principles:

- Creativity—the focus is on solving societal and environmental problems
- Scalability—Rapid implementation rather than long-term pilot projects in face-to-face international challenges
- Responsiveness-transforming approaches, the existing business models sometimes also radically change the whole industry and significantly higher transparency of knowledge when it comes to solving global challenges
- Glocality—when solving global problems, it is important to better understand local specificities in order to develop more appropriate solutions in individual countries.
- Circularity—radically alter product design, implementation of the principles of circulatory management, e.g. cradle-to-cradle solutions

Important impulses for sustainable innovation can be achieved by addressing the Sustainable Development Goals (SDGs). The SDGs (http://www.un.org/sustainabledevelopment/) are the worldwide target for sustainable development. The 17 specific objectives (and 169 sub-goals) address the major challenges faced by society, business and politics. The 193 United Nations states are committed to the achievement of these goals by 2030. The social actors from the worlds of business, politics, civil society and the media are called upon to play their part in achieving the goals. The contribution of the companies plays an important role. These SDGs also provide a valuable contribution to family businesses in the direction of their future focus on sustainability and corporate responsibility. The opportunity is here following the global sustainability discussion, the identification of innovation opportunities as well as possible weaknesses in the value chain and the targeted communication of CSR activities with regard to global targets. For the long-term orientation and the positive contribution to a social development which has been lived by many companies especially family businesses for generations, the SDGs now offer a good possibility to link to the global discussion for a global sustainable development.

2 The Potential of Corporate Social Responsibility for Innovation

Innovations depend on a multitude of factors influencing the process from idea generation through development to implementation. Sundbo and Miles (2000) describe it as an interactive process, depending on both internal and external factors. Increasingly, it is being recognised that a single organisation cannot innovate in isolation, but that it has to engage with different types of stakeholders to acquire ideas and resources from the external environment to stay abreast of competition. This is what Chesbrough (2003) calls openness or open innovation. He defines it as the *"use of purposive inflows and outflows of knowledge to accelerate internal innovation, and to expand the markets for external use of innovation, respectively"* and argues that *"open innovation is a paradigm that assumes that firms can and should use external ideas as well as internal ideas, and internal and external paths to market, as firms look to advance their technology"* (Chesbrough 2006, p. 1). Chesbrough's rather broad definition of openness is the most commonly used in the literature, as it underscores that valuable ideas emerge and can be commercialised from inside or outside the firm. Since he coined the term in 2003, the concept of open innovation has been growing rapidly (Dahlander and Gann 2010).

The roots of open innovation, however, go far back in history. Already in 1986, von Hippel (1986) worked in his 'lead user'-concept on integrating customers in the innovation process. He describes the ability of user communities to initiate and develop complex products, at times even without any specific manufacturer involvement. Around the same time, Katz and Allen (1982) focused on the 'not invented here (NIH) syndrome' in their open innovation culture studies. Prahalad and Ramaswamy (2000) popularised the concept of co-creation, as an active, creative and social process, based on collaboration between producers and users that is initiated by the firm to generate value for customers.

The basic premise of open innovation is opening up the innovation process and integrating both internal and external components. In this context, Chesbrough (2011) argues that *"in the open innovation model, there are two complementary kinds of openness. One is outside in, where a company makes great use of external ideas and technologies in its own business ... [whereas] the other kind of openness is inside out, in which a firm allows some of its own ideas and technologies outside its own business"* (p. 83).

While open innovation is usually contrasted with closed innovation, as its supposed predecessor, in reality not many firms follow a fully closed innovation approach (Huizingh 2011). In this context, Dahlander and Gann (2010) also argue that open innovation reflects much less a dichotomy, as in open versus closed, than a continuum with varying degrees of openness.

Another much quoted definition of open innovation comes from Lichtenthaler (2011), who considers the various knowledge flows in open innovation. According to him, *"open innovation is defined as systematically performing knowledge exploration, retention, and exploitation inside and outside an organization's boundaries throughout the innovation process"* (Lichtenthaler 2011, p. 77). In order for open innovation to be effective, however, Nonaka (2007) argues that organisations must have sufficient capability to successfully integrate the information obtained from the external sources into internal processes and structures.

Benefits for firms interacting extensively with their environment in the innovation process are manifold, including significant amount of external knowledge exploration and exploitation (Vanhaverbeke et al. 2008), cost reduction, leveraging complementarities by accessing unique resources; shorter time-to-market as well as gaining stronger credibility and access to partners' network (Martovoy 2014). However, the benefit of involving customers and users in the development process of a new product or service has not been without critique. Open innovation can be challenging as it requires the assimilation of knowledge and expectation management at the same time (Magnusson et al. 2003). Companies also fear the risk that involved stakeholders share knowledge with competitors (Dyer and Hatch 2006).

In many studies, CSR has repeatedly emphasized the potential for innovation and growth (e.g. Hansen et al. 2009; Schaltegger and Wagner 2011). For example, Nidumolu et al. (2013) argue that company social responsibility is one of the most important drivers for innovation, while Yoon and Tello (2009) argue that a consistent commitment to CSR helps the organization continue to innovate and to be competitive. The aim of sustainable innovation is to minimize the negative impacts of production, but at the same time to create benefits and added value for customers and other stakeholders (Klewitz and Hansen 2014).

Hansen and Grosse-Dunker (2013) show, in their concept of sustainability-oriented innovation, that sustainable development can be advanced by spreading more sustainable product and service offerings by focusing from a close focus on the direct customer value to an emphasis on the whole physical life cycle of a product. MacGregor and Fontrodona (2008) emphasize that companies must consider the social and environmental impact of their operational processes in order to be successful and to work with their customers, suppliers and other business partners to design and develop new products and services. So far, however, only a minority of CSR companies have actually used them as a means to drive innovation, most see it as a tool to reduce risks and operating costs (Hockerts 2007). MacGregor and Fontrodona (2008) also examine the relationship between innovation and CSR and suggest that there is an virtuous circle (see Fig. 1). In some companies, driven primarily by their values, CSR leads to innovation (CSR-centered motivation). CSR-driven innovation is about "doing the right things" and focuses on products and services that have a social purpose. These companies thoroughly examine the impact of their activities on the environment and on the community without jeopardizing the importance of profit. In companies that primarily focus on the creation of

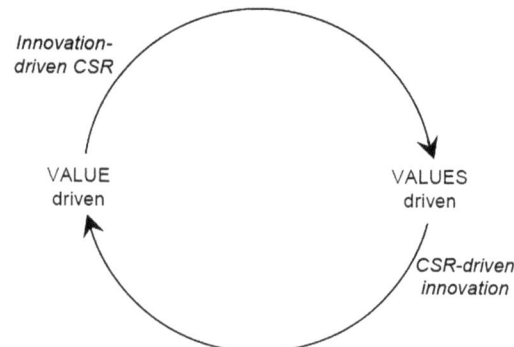

Fig. 1 Innovation-driven CSR and CSR-driven Innovation. Source MacGregor and Fontrodana (2008)

values, innovation leads to CSR (innovation-oriented approach). They value their workforce, their production chain and their customers. The process of identifying overlaps between CSR and innovation reveals the relationship (MacGregor and Fontrodona 2008).

The challenge of a CSR-driven innovation process is that social responsibility and sustainability should be integrated not only into individual phases (e.g. idea generation) but in all phases of the innovation process (see Fig. 2). The potential for the involvement of stakeholders should be explored from the idea finding, the discussion of evaluation criteria, the development and the testing of prototypes up to the launch of the market.

Values and abilities are the decisive factor in determining how innovative a company is and/or what role is played by sustainability. A critical discussion and reflection on values, abilities, but also dominant industry logic or the understanding of sustainability in the company is usually the starting point of a CSR-driven innovation process. Subsequently, it is necessary to promote appropriate values and skills and establish a coherent corporate culture.

Through this opening-up of the innovation process (open innovation), potentials from other companies, organizations as well as creative individuals can be used for the development of sustainable/socially responsible products, services and business models. An intensive exchange with stakeholders can help to get to know company-specific perspectives as well as to identify opportunities for innovation and new business segments. Stakeholder relationships are based on trustworthy exchange, binding communication, and willingness to make changes on both sides. The forms of collaboration between companies and their stakeholders can range from goal-oriented individual discussions to strategic partnerships.

This applies not only to multinational companies, but especially to SMEs, often through cooperation with regional partners, which offers numerous opportunities to secure their competitiveness in the long term. A particular challenge lies in the identification of lead stakeholders—these are particularly innovative stakeholders who deal with new social and environmental challenges at an early stage, formulate their needs for new problem solutions earlier than others, have a very high level of knowledge on these topics, and are often globally connected.

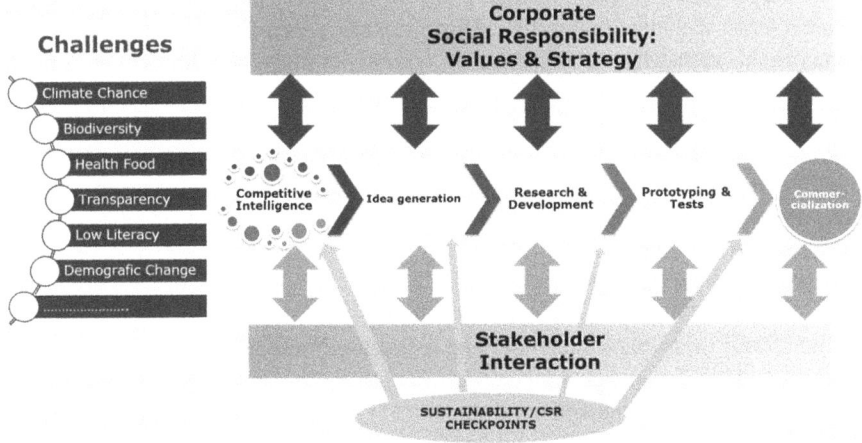

Fig. 2 Sustainability/CSR-driven innovation process

The development of sustainable solutions and business models also requires the exploration of new forms of innovation management. For example, the approach of "design thinking" (Young 2010), which is based on the way the designers develop when they develop new ideas or solutions, or "Theory U" (Scharmer 2009) offers a high potential as this approach is well suited to deal with complex problems. These approaches usually use new forms of interaction with key stakeholders and focus on a deeper understanding of "real" challenges or problems. This makes more sustainable and systemic solutions possible. A process of understanding, observation, idea-finding, refinement, execution and learning is usually carried out several times. Also important is the composition of the innovation team, which is to consist of members of different disciplines.

Internationally active companies face different challenges social and environmental issues in the individual countries or regions. The intensive examination of these topics and corresponding joint reflections with the respective stakeholders should lead to learning processes which in the medium term enable companies to react faster and better to changed framework conditions and to recognize the opportunities that arise at an early stage.

3 Stakeholder-Engagement for Sustainable Innovation

In the twentieth century, the academic and business community has started to explore the wider impacts of business, driven by the mounting challenges of doing business as well as corporate scandals. With his landmark book "Strategic Management: A Stakeholder Approach", Freeman (1984) identified the importance of the role of stakeholders in relationship to the organisation and stated that multiple players interact with the organisation. Those stakeholders must be taken into account by the organisation when facing complex conditions in the operating environment. Stakeholder theory offered a new form of managerial understanding and action in a strategic planning context and a new way to organise thinking about firms' responsibilities by suggesting that the needs of shareholders cannot be met without satisfying the needs of other stakeholders (Jonker and Foster 2002; Foster and Jonker 2005). A central tenet of stakeholder theory is that the interests of all legitimate stakeholders have value and it is critical that firms' decisions consider these interests (Donaldson and Preston 1995).

The benefits of engaging stakeholders cited in empirical literature are manifold. Sharma (2005) finds that stakeholder engagement allows the organisation to access information from its stakeholders which can then be used help understand and respond to emerging social and environmental issues. Katsoulakos and Katsoulacos (2007) argue that the existence of advantage-creating stakeholder relations supports organisational knowledge development. The value of stakeholder engagement is amplified when businesses face broad and rapidly changing conditions. Identifying and engaging key stakeholders becomes increasingly critical to long-term corporate viability (Brown and Flynn 2006).

According to AccountAbility (2011), engaging with stakeholders also

- enables better management of risk and reputation;
- allows for the pooling of resources (knowledge, people, money and technology) to solve problems and reach objectives that cannot be reached by single organisations;
- enables understanding of the complex operating environments, including market developments and cultural dynamics;
- enables learning from stakeholders, resulting in product and process improvements;
- informs, educates and influences stakeholders to improve their decisions and actions that will have an impact on the organisation and on society;
- and contributes to the development of trust-based and transparent stakeholder relationships.

Freeman et al. (2010) also highlight that motivated by values, businesspeople continuously create new sources of value by cooperating with stakeholders. Thus, stakeholder engagement can be a tremendous source of innovation. More and more companies are discovering that a growing percentage of innovation is coming from outside the organisation and not from within.

Research emphasizes the role of the stakeholder as an opportunity for facilitating innovation (Hart and Sharma 2004; Kanter 1999). Indeed, Gould (2012) argues that open innovation and stakeholder engagement describe similar organisational processes, as in both cases, the focal organisation reaches outside its boundaries making an explicit effort to access essential information. Nevertheless, the two concepts, and their associated languages and discussions, have remained isolated from each other and research linking open innovation and stakeholder engagement remains scarce.

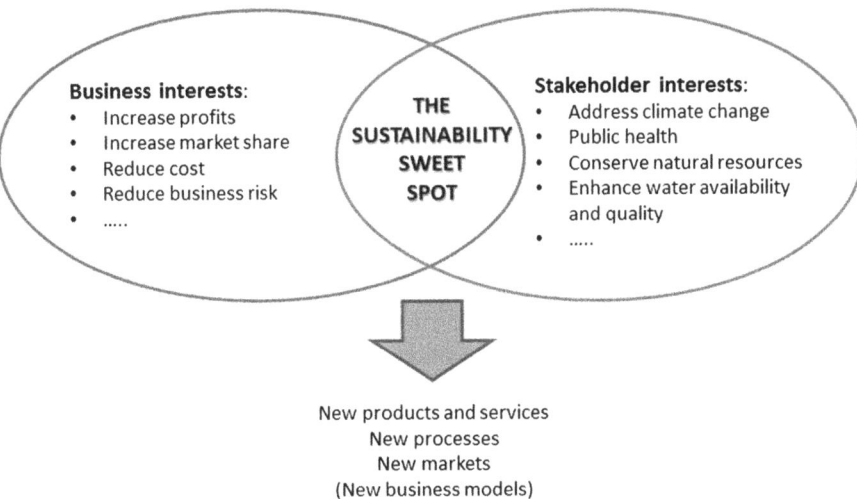

Fig. 3 Sustainability Sweet Spot (Savitz and Weber 2007)

Therefore, Gould (2012) claims, there is a need for more detailed identification and analysis of specific processes involved in both open innovation and stakeholder engagement, such as the role of intellectual property and other organisational protections, the impact of specific network configurations, as well as the roles of power and legitimacy. As stakeholders in the open innovation process are not always easily categorised as primary or secondary, further attention should be paid to defining and mapping the specifics of stakeholder relationships and engagement processes in the context of open innovation processes.

In the so-called "Sustainability Sweet Spot", first described by Savitz and Weber (2007), stakeholder interests and business goals overlap. Successful companies are continuously trying to identify and move into their sweet spots. In order to get there and stay there, they are creating sustainable innovations and developing new ways of doing business with their stakeholders (Fig. 3).

If companies want to play a leading role in the sustainability debate and not only react to external pressure, but also see sustainability and the associated social responsibility of the company as a core element of their strategy and therefore as a competitive advantage, a professional stakeholder management with adequate resources is indispensable. The challenge is not to implement a standard model of stakeholder management but to work out the unique cultural, organizational and historical development of a company (Kakabadze et al. 2005) and to develop a unique approach.

References

Accenture & United Nations Global Compact. (2016). *The UN Global Compact-Accenture strategy CEO study 2016.*

AccountAbility. (2011). *AA1000 stakeholder engagement standard 2011.* Accessed October 11, 2017, from http://www.accountability.org/images/content/3/6/362/AA1000SES%202010%20PRINT.PDF

Brown, B., & Flynn, M. (2006). The meta-trend stakeholder profile: Impacts on the finance and banking industry. *Greener Management International, 54,* 53–56.

Carroll A. B., & Shabana K. M. (2010). The business case for corporate social responsibility: A review of concepts, research and practice. *International Journal of Management Reviews.* doi: https://doi.org/10.1111/j.1468-2370.2009.00275.x

Chesbrough, H. W. (2003). *Open innovation: The new imperative for creating and profiting from technology.* Boston, MA: Harvard Business School Press.

Chesbrough, H. W. (2006). Open innovation: A new paradigm for understanding industrial innovation. In H. W. Chesbrough, W. Vanhaverbeke, & J. West (Eds.), *Open innovation: Researching a new paradigm* (pp. 1–14). Oxford: Oxford University Press.

Chesbrough, H. W. (2011). *Open services innovation rethinking your business to grow and compete in a new era.* San Francisco, CA: Jossey-Bass. Retrieved from http://www.books24x7.com/marc.asp?bookid=41007

Dahlander, L., & Gann, D. M. (2010). How open is innovation? *Research Policy, 39*(6), 699–709.

Donaldson, T., & Preston, L. E. (1995). The stakeholder theory of the corporation: Concepts, evidence, and implications. *The Academy of Management Review, 20*(1), 65–91.

Dyer, J. H., & Hatch, N. W. (2006). Relation-specific capabilities and barriers to knowledge transfers: Creating advantage through network relationships. *Strategic Management Journal, 27*(8), 701–719 https://doi.org/10.1002/smj.543

European Commission. (2011). *Communication from the Commission to the Council and the European Parliament – A renewed EU strategy 2011-14 for Corporate Social Responsibility.* Brüssel

Foster, D., & Jonker, J. (2005). Stakeholder relationships: The dialogue of engagement. *Corporate Governance: The International Journal of Business in Society, 5*(5), 51–57. https://doi.org/10.1108/14720700510630059

Freeman, R. E. (1984). *Strategic management: A stakeholder approach.* Boston: Pitman.

Freeman, R. E., Harrison, J. S., Wicks, A. S., Parmar, B. L., & de Colle, S. (2010). *Stakeholder theory: The state of the art.* Cambridge, New York: Cambridge University Press.

Gould, R. (2012). Open innovation and stakeholder engagement. *Journal of Technology Management & Innovation, 7*(3), 1–11. https://doi.org/10.4067/S0718-27242012000300001

Hansen, E. G., & Grosse-Dunker, F. (2013). Sustainability-oriented innovation. In S. O. Idowu, N. Capaldi, L. Zu, & L. Das Gupta (Eds.), *Encyclopedia of corporate social responsibility* (pp. 2407–2417). Heidelberg: Springer.

Hansen, E. G., Grosse-Dunker, F., & Reichwald, R. (2009). Sustainability innovation cube – A framework to evaluate sustainability-oriented innovations (2009). *International Journal of Innovation Management, 13*(4), 683–713.

Hart, S. L., & Sharma, S. (2004). Engaging fringe stakeholders for competitive imagination. *IEEE Engineering Management Review, 32*(3), 28–28. https://doi.org/10.1109/EMR.2004.25105

Hippel, E. (1986). Lead users: A source of novel product concepts. *Management Science, 32*(7), 791–805.

Hockerts, K. (2007). *Managerial perceptions of the business case for corporate social responsibility center for corporate social responsibility.* Frederiksberg: CBS.

Huizingh, E. K. R. E. (2011). Open innovation: State of the art and future perspectives. *Technovation, 31*(1), 2–9. https://doi.org/10.1016/j.technovation.2010.10.002

ISO 26000. (2010). *Guidance on Social Responsibility ISO 26000 (2010) (E).*

Jonker, J., & Foster, D. (2002). Stakeholder excellence: Framing the evolution and complexity of a stakeholder perspective of the firm. *Corporate Social Responsibility and Environmental Management, 9,* 187–195.

Kakabadze, N. K., Rozuel, C., & Lee-Davies, L. (2005). Corporate social responsibility and stakeholder approach: A conceptual review. *International Journal of Business Governance and Ethics, 1(4), 277–302.*

Kanter, R. M. (1999). From spare change to real change: The social sector as beta site for business innovation. *Harvard Business Review, 77*(3), 122–132.

Katsoulakos, T., & Katsoulacos, Y. (2007). Integrating corporate responsibility principles and stakeholder approaches into mainstream strategy: A stakeholder-oriented and integrative strategic management framework. *Corporate Governance: The International Journal of Business in Society, 7*(4), 355–369. https://doi.org/10.1108/14720700710820443

Katz, R., & Allen, T. J. (1982). Investigating the Not Invented Here (NIH) syndrome: A look at the performance, tenure, and communication patterns of 50 R&D Project Groups. *R&D Management, 12*(1), 7–20. https://doi.org/10.1111/j.1467-9310.1982.tb00478.x

Klewitz, J., & Hansen, E. (2014). Sustainability-oriented innovation of SMEs: A systematic review. *Journal of Cleaner Production, 65,* 57–75.

Kurucz, E., Colbert, B., & Wheeler, D. (2008). The business case for corporate social responsibility. In A. Crane, A. McWilliams, D. Matten, J. Moon, & D. Siegel (Eds.), *The Oxford handbook of corporate social responsibility* (pp. S. 83–S.112). Oxford: Oxford University Press.

Lichtenthaler, U. (2011). Open innovation: Past research, current debates, and future directions. *Academy of Management Perspectives, 25*(1), 75–93. https://doi.org/10.5465/AMP.2011. 59198451

MacGregor, S., & Fontrodona, J. (2008). *Exploring the fit between CSR and innovation* (Working Paper No. 759). Navarra, Spain: IESE Business School.

Magnusson, P. R., Matthing, J., & Kristensson, P. (2003). Managing user involvement in service innovation: Experiments with innovating end users. *Journal of Service Research, 6*(2), 111–124. https://doi.org/10.1177/1094670503257028

Martovoy, A. (2014). Advantages and disadvantages of open innovation: Evidence from financial services. In A.-L. Mention & M. Torkkeli (Eds.), *Innovation in financial services a dual ambiguity* (pp. 259–294). Newcastle upon Tyne: Cambridge Scholars Publishing. Retrieved from http://public.eblib.com/choice/PublicFullRecord.aspx?p=1821875

Nidumolu, R., Prahalad, C. K., & Rangaswami, M. R. (2013). Why sustainability is now the key driver of innovation. *Engineering Management Review, IEEE, 41,* 30–37.

Nonaka, I. (2007). The knowledge-creating company. *Harvard Business Review, 85*(7/8), 162–171.

Perrini, F. (2006). SMEs and CSR theory: Evidence and implications from an Italian perspective. *Journal of Business Ethics, 67*(3), 305–316.

Porter, M., & Kramer, M. (2011). Creating shared value. *Harvard Business Review, 89*(1/2), 62–77.

Prahalad, C. K., & Ramaswamy, V. (2000). Co-opting customer competence. *Harvard Business Review, 78*(1), 79–87.

Savitz, A. W., & Weber, K. (2007). The sustainability sweet spot. How to achieve long-term business success. *Environmental Quality Management,* Winter. https://doi.org/10.1002/tqem

Schaltegger, S., & Wagner, M. (2011). Sustainable entrepreneurship and sustainability innovation: Categories and interactions. *Business Strategy and the Environment, 20*(4), 222–237.

Schaltegger, S., et al. (2009). *Nachhaltigkeitsmanagement in der öffentlichen Verwaltung. Herausforderungen, Handlungsfelder und Methoden.* Studie im Auftrag des Rats für Nachhaltige Entwicklung der Bundesregierung. Lüneburg: Centre for Sustainability Management.

Schaltegger, S. (2012). Die Beziehung zwischen CSR und corporate sustainability. In A. Schneider & R. Schmidpeter (Hrsg.), *Corporate social responsibility. Verantwortungsvolle Unternehmens- führung in Theorie und Praxis* (S. 165–176). Berlin: Springer-Gabler.

Scharmer, C. O. (2009). *Theory U. Leading from the future as it emerges.* San Francisco: Berrett Koehler Publishers.

Schreck, P. (2012). Der business case for corporate social responsibility. In A. Schneider & R. Schmidpeter (Eds.), *Corporate Social Responsibility. Verantwortungsvolle Unternehmensführung in Theorie und Praxis* (pp. S 67–S 86). Berlin: Springer – Gabler.

Sharma, S. (2005). Through the lens of managerial interpretations: Stakeholder engagement, organizational knowledge and innovation. In S. Sharma & J. A. Aragón Correa (Eds.), *Corporate environmental strategy and competitive advantage* (pp. 49–70). Cheltenham: Edward Elgar.

Stigson, B. (2002). Pillars of change: Business is finally learning that taking care of the environment and meeting social responsibilities makes good business sense. *Forum for Applied Research and Public Policy, 16*(4), 23.

Sundbo, J., & Miles, I. (2000). Innovation as a loosely coupled system in services. In J. S. Metcalfe & I. Miles (Eds.), *Innovation systems in the service economy measurement and case study analysis* (pp. 43–68). Boston: Springer US. Retrieved from https://doi.org/10.1007/978-1-4615-4425-8

Vanhaverbeke, W., Van de Vrande, V., & Chesbrough, H. (2008). Understanding the advantages of open innovation practices in corporate venturing in terms of real options. *Creativity and Innovation Management, 17*(4), 251–258. https://doi.org/10.1111/j.1467-8691.2008.00499.x

Visser, W. (2011). *The age of responsibility: CSR 2.0 and the new DNA of business.* London: Wiley.

Yoon, E., & Tello, S. (2009). Corporate social responsibility as a driver of sustainable innovation: Greening initiatives of leading global brands. *Competition Forum, 7*(2), 290–294.

Young, G. (2010). *Design thinking and sustainability.* Accessed July 24, 2017, from http://zum.io/wp-content/uploads/2010/06/Design-thinking-and-sustainability.pdf

Reinhard Altenburger is Professor for Strategic Management, Sustainable Management / CSR and Innovation in the Department of Business of the IMC University of Applied Sciences Krems. The focus of his research is "CSR and Innovation" as well as "Innovations in Family Businesses" and the connection of social responsibility and corporate strategy. Reinhard Altenburger has 15 years of work experience in the banking industry, as a project manager and senior expert in the fields of strategic planning, retail strategy, in the Austrian Savings Banks Association, the Raiffeisen regional bank Lower Austria-Vienna and the Competence-Centre for IT-Solutions in the Banking Industry. He has held numerous presentations at scientific conferences and also published and edited books in his research field. At Springer he edited "CSR und Innovationsmanagement" and "CSR und Stakeholdermanagement".

Managing Sustainable Innovation

Annabeth Aagaard

1 Introduction

In 1983 the UN decided to establish a World Commission on Environment and Development, and in 1987 the Commission submitted its report to the general assembly of the United Nations with the title "Our Common Future". The Brundtland Report, as it was later called, gave the kick-off to a worldwide interest for the environment, which subsequently has set the agenda for both governments' and consumers' environmental considerations and thus the terms and conditions of industries and businesses. In the rise of globalization, sustainability has become a new premise for implementing business and innovation, moving it from political discourses into company strategy (Faber et al. 2005; Aagaard 2016). According to Hall and Vrendenburg (2003) it is no longer enough for businesses to create new radical innovations, if leading customer groups perceive these innovations as unsustainable. Developments in innovation management reveal that solving persistent environmental and social problems is considered a new source of inspiration for businesses in pursuing innovative opportunities (Hansen et al. 2009).

The constant need for businesses to stay at the innovative forefront has led them towards the most unexpected areas, where leading businesses increasingly realize that complex social and environmental problems can convert into new business opportunities (Kanter 1999; Venn and Berg 2013; Lodsgård and Aagaard 2017). However, studying the concept of sustainability is challenged by the fact, that it is fragmented, where some researchers even question whether sustainability is a concept or a political discourse (Dryzek 2005) and/or an artifact, as suggested by Faber et al. (2005). The pressure on corporations to demonstrate their responsibility

A. Aagaard (✉)
Department of Business Development and Technology, Aarhus University, Herning, Denmark
e-mail: aaa@btech.au.dk

© Springer International Publishing AG, part of Springer Nature 2018   13
R. Altenburger (ed.), *Innovation Management and Corporate Social Responsibility*,
CSR, Sustainability, Ethics & Governance,
https://doi.org/10.1007/978-3-319-93629-1_2

toward society has intensified over the past years, and the notion of CSR has institutionalized the ideal of an ethically alert organization able to balance its financial interests with its concern for society as a whole (Christensen et al. 2011).

The theoretical concept of sustainable innovation was originally introduced by professor Rosabeth Moss Kanter, who identified and explained a new behavior among companies. In her article from 1999, *From spare change to real change: The social sector as beta site for business innovation*. Kanter describes how companies are showing a growing interest in meeting the needs of the social sector and in creating innovation based on these needs. Her point being that companies had begun to view social problems as economic opportunities and that the solutions to these problems were attractive for the social services sector and for companies as well (Kanter 1999). In further exploration of the concept, Nidumolu et al. (2009, 57) argue that: "*sustainability is a mother lode of organizational and technological innovations that yield both bottom line and top line returns*". As such sustainable innovation encourages companies to internationalize given their own characteristics and needs, such as uniqueness of product, technological competences, economies of scale, competitive pressure, small domestic market and lack of domestic demand, overproduction, unconsolidated foreign orders, possibility to extend sales of seasonal products, and proximity to international customers (Hollensen 2008).

2 Sustainable Innovation

Some streams of the SI literature have emphasized the integration of environmental requirements into all stages of the innovation management and product development process, emphasizing the environmental impact of the product's life cycle stages in relations to raw materials, manufacturing, distribution, product use, and end of life (Choi et al. 2008). This stream of SI literature has primarily addressed the concept of eco-innovations. However, other researchers have emphasized the need to go beyond these compliances and extend the focus to more long-term innovative processes (Boons et al. 2013). In their optic, the concept of eco-innovation is inadequate as it only addresses the environmental and economic dimensions of sustainable innovation and leaves out the social and ethical dimensions (Charter and Clark 2007). Consequently, Charter and Clark (2007) offer a definition that embraces all of these elements, which is the SI definition applied in this chapter: "*Sustainable innovation is a process where sustainability considerations (environmental, social and financial) are integrated into company systems from idea generation and development (R&D) and commercialization. This applies to products, services and technologies, as well as to new business and organizational models* (Charter and Clark 2007: 9)".

Another stream of literature defines sustainable innovation by emphasizing the importance of sustainable innovative entrepreneurs with an implicit focus on creativity through sharing, creation, and use of new knowledge in collaboration with different stakeholder groups (de Sousa 2006; Schaltegger and Wagner 2011).

Despite a growing body of literature analyzing and discussing sustainability and sustainable development at the political and social level (Dryzek 2005), the operationalization of the concept in relation to business and innovation is rather still weak (Bansal 2005; Stubbs and Cocklin 2008; Zink et al. 2008; Carroll and Shabana 2010). What is really needed is an improved understanding and operationalization of sustainable business and innovation management into corporate practices and performances in creating and testing new sustainable innovations and sustainable business models on corporate level (Aagaard 2017).

Over the last decade more and more innovation research has addressed the concept and potentials of corporate social innovation (CSI), also referred to as sustainable innovation. Sustainable innovation is characterized by the fact that social responsibility is the focal point of a company's innovation process. Kaebernick et al. (2003) argued that the introduction of environmental requirements into the product development process at all stages of a product's life leads to a new paradigm of sustainability, which is reflected in a new way of thinking, new applications of tools and methodologies in every single step of product development (p. 468). Through sustainable innovation, sustainability creates new opportunities and new values for the company (Pavelin and Porter 2008). This implies that the companies can help solve some of the world's social and environmental challenges, while at the same time develop new products, markets, and business areas. Furthermore, by working strategically with CSR-driven innovations, companies can increase their growth and competitiveness. This correlation between company performance and CSI has been explored and witnessed by a number of researchers (e.g. Hull and Rothenberg 2008; Wagner 2010).

Several researchers emphasize the disruptive circumstances of external stake-holder pressure as a key motivation for developing radical, sustainable innovations. Whereas sustaining circumstances, e.g. that customers are willing to accept minor product adjustment, typically lead to incremental sustainable innovations (Christensen 1997; Steketee 2010). It appears, that companies have moved beyond eco-efficiency compliances and extended their focus to the adaption of disruptive innovative processes, where businesses respond with new game-changing business models (Schaltegger and Wagner 2011; Loorbach et al. 2009; Boons et al. 2013). This emphasizes the potentials of companies pursuing incremental innovations through the perspective of eco-efficiency in products and processes and at the same time pursuing more radical innovations through sustainable business model innovations.

2.1 Sustainable Innovation Through Circular Economy

Circular economy has gained increasing theoretical and practical emphasis. This is particularly due to the fact that EU has adopted an ambitious Circular Economy Package, including revised legislative proposals on waste to stimulate Europe's transition towards a circular economy that will boost global competitiveness, foster sustainable economic growth and generate new jobs. Furthermore, the package also consists of an EU Action Plan for the Circular Economy that establishes a concrete and ambitious programme of action, with measures covering the whole cycle: from production and consumption to waste management and the market for secondary raw materials. The proposed actions will contribute to "closing the loop" of product lifecycles through greater recycling and re-use, and bring benefits for both the environment and the economy. Waste management is an interesting trend, as waste is no longer just something that a company has to find a sustainable way to reduce or get rid off. Waste also represent new business opportunities and sustainable innovations when combined with the circular economy mindset industrial symbiosis can be stimulated and help turn one industry's by-product into another industry's raw material (Aagaard 2016).

Circular economy may be viewed as a development strategy based on restorative thinking that aims to maximize resource efficiency and minimize waste production within the framework of economic and social sustainability (Hishop and Hill 2011). The key components of a circular economy are product design, new business models, reverse-cycle networks, and conducive conditions (WEF 2010). The concept of circular economy is gaining increasing theoretical and practical emphasis. Particularly, due to the focus put by EU, which has adopted an ambitious Circular Economy Package and action plan. This package includes revised legislative proposals on waste to stimulate Europe's transition towards a circular economy that will boost global competitiveness, foster sustainable economic growth and generate new jobs. The action plan for Circular Economy constitutes an ambitious program of actions with measures covering the whole cycle: from production and consumption to waste management and the market for secondary raw materials (Aagaard 2016).

Many researchers have sought to understand and explain how long-established systems of production and consumption could be influenced to transform through innovation, in order to achieve the goal of dramatically increased environmentally sustainability (Shove 2003; Berkhout 2011; Tukker et al. 2008). Large existing businesses are seen as being trapped in systemic interdependencies (Tukker et al. 2008). This is especially so in consumer businesses with short-term profit focus, such as retailers (Charter et al. 2008). The circular economy is a direct challenge to the textile industry and marketers. Alternative strategies and business models have to be devised to change and re-design the fundamental idea that fashion products are designed to have limited useful lives so that consumers have to make further purchases once they wear out or are out of fashion.

3 Managing Sustainable Innovation

Traditional innovation management (IM) comprises the systematic processes that organizations apply to develop new and improved products, services, and business processes. This involves harnessing the creative ideas of the organization's employees and utilizing them in creating a steady pipeline of profitable new innovations to the marketplace, quickly and efficiently (Aagaard 2011). Thus, the ability of an organization to grow is dependent upon its ability to generate new ideas and to exploit them effectively for their long-term benefit of the organization (Flynn et al. 2003). The company's capacity to innovate is determined by multiple factors that relate both to the internal organization and to the market environment of the company where the task of generating and then converting ideas into usable and marketable products requires high levels of inter-functional coordination and integration (Adams et al. 2006).

Igartua et al. (2010: 42–43) propose that innovation management may be defined as the creation of preconditions to promote human creativity, including strategic commitment and context management or can be seen as a process to foster the application of knowledge. In their review of literature, they found that innovation management involves a number of different components that require management including: the strategy of innovation, portfolio management, project management, leadership and organizational culture, HRM, external relations in fostering innovation, organizational design for innovation, innovation processes, performance measures, marketing as a source of idea generation, knowledge & intellectual property management, and technology. Summarized and according to Tidd and Bessant (2013) successful innovation management requires that a company:

1. Designs innovation strategy
2. Implements the strategy effectively by synchronizing the new technology and market needs
3. Develops a supportive organizational context for innovation, and
4. Establishes and enhances external linkages, such as technological partners or other stakeholders.

The management of sustainable innovations consists of the same four basic elements in facilitating innovative activities and resembles the traditional R&D/innovation process of taking a product or service from idea/concept to maturation, production and launch.

However, in managing sustainable innovation particularly three elements differ:

1. The way sustainable innovation is facilitated, managed and measured should ensure higher levels of societal, ethical and environmentally friendly IM approaches (e.g. through diversity in workforce, inclusion/empowerment of stakeholders in the R&D process, recycling/reuse of resources, sustainable KPIs/CSR reporting etc.)
2. The outcome of the innovation management and innovation processes should target sustainable innovations that assist in reducing/improving societal, ethical and/or environmental issues and/or provide more sustainable alternatives to existing, non-sustainable products and services.
3. The innovation management process should incorporate external collaborations and partnerships more effectively and strategically to gain and get access to the types of knowledge, competences, partnerships and networks that enables the development and marketing of new, sustainable (radical) innovations.

During the last decade, the boundaries between a company and its customers, partners, and the environment have been reduced, among others due to social media and other technologies that support sharing of ideas and knowledge. Thus, customers, suppliers, and other relevant stakeholders can and should be included more strategically and actively in the company's innovation management to ensure that their ideas as well as their knowledge, demands, and requirements are used and integrated more effectively in creating new and more sustainable innovations. As such, the basic idea behind open source innovation is that in a knowledge society and in a global world of knowledge, companies cannot afford to solely build their innovation on the company's own research and internal knowledge resources. Companies should therefore put more effort to obtain licenses on processes, product ideas, and patents from other companies and stakeholders. Similarly, internal inventions that are not used in a business should be sold or transferred to other companies, for example, through licensing, joint ventures, or spin-off so that the restrictions that the company framework sets do not stop innovation (Aagaard 2011).

Open source innovation is a perquisite for sustainable innovation, as SI is targeted at society and embraces the needs, challenges and requirements of the company's stakeholders (Aagaard 2016). Chesbrough first coined the term open source innovation in his 2005 book *Open innovation: The new imperative for creating and profiting from technology*, and Chesbrough et al. (2006) later defined open innovation (OI) as the use of purposive inflows and outflows of knowledge to accelerate internal innovation and expand the markets for external use of innovation, respectively. Prevalent reviews on the topic emphasize that OI entails both exploration of external knowledge and ideas, so-called inbound activities, as well as exploitation of internal knowledge outside organizational boundaries, referred to as outbound activities (Dahlander and Gann 2010; Bengtsson et al. 2015). OI is disseminated both through the academic community (Dahlander and Gann 2010) as well as among practitioners (Van de Vrande et al. 2009). From the present OI research, it appears that companies are pursuing radical innovation through different types of open source innovation activities and across different types of partners (Rass et al. 2013; Chesbrough and Brunswicker 2014).

Engaging in partnerships and collaborations is therefore an integrated part of managing sustainable innovation. The collaborations and partnerships of a company are part of the company's public image, and they may quickly put either positive and/or negative imprints on the organization's brand and reputation, if not selected and managed properly. It is therefore necessary to manage, control, and develop the sustainability of the partners that the company cooperates as well. This implies that the company has to establish joint agreements and alignments of the partners' expectations with clarification of their attitudes, activities and values in relations to sustainability. In practice, this is often done through the application of codes of conducts as part of the partnership agreement. However, code of conducts are not guarantees of successful collaborations, and thus strategic management and development of partnerships is needed (Aagaard 2016).

Sustainable collaborations with suppliers throughout the value chain are an integral part of managing sustainable innovation and of sustainable supply chain management (Aagaard 2016). A company's supply chain has economic, environmental, and social consequences for several stakeholders and society, which is also why sustainable supply chain management represents a central pillar of sustainable business and in managing sustainable innovation, which is emphasized by Carbone et al. (2012, p. 477), *"the sustainable supply chain represents a further step towards the institutionalization of a CSR strategy."* Sustainable supply chain management also deals with the dialogue that companies engage with their suppliers in order to prevent violations of fundamental human rights and international environmental standards and in accordance with internationally recognized principles and rights, which is defined by the UN. Berkhout (2014) stresses that the concept of sustainable innovation management is deeply related to the way in which a firm manages its external relationships and that the analysis of innovation management needs to be seen as being embedded in a network of relationships related to regulatory and social governance, value chains and the socio-technical regimes within which it is embedded.

3.1 Sustainable Innovation Through Business-NGO Collaborations

A more recent development in management and collaborations of sustainable innovation relates to sustainable innovations being generated through partnerships between businesses and NGO's (Lodsgård and Aagaard 2016, 2017). The existing research on inter-organizational collaborations between businesses and NGOs is deeply embedded in the field of strategic alliances emphasizing external knowledge acquisition and the development of internal capabilities (Das and Teng 2000; Faems et al. 2005). Interestingly enough, Businesses and NGOs have had a long history of battling against each other across different environmental, social, and ethical issues. Yet, during the last decade more and more businesses and NGOs are collaborating on the basis of philanthropy and volunteerism (Austin 2000). It appears that in the wake of global environmentalism, a shift has taken place from the 1980s, where especially environmental NGOs assumed a far more visible role on the international political scene (Doh and Guay 2006). Throughout the 1980s and 1990s, these activist groups employed a new antagonistic strategy toward sustainable business practices in terms of new environmental and social regulative standards (Argenti 2004; Doh and Guay 2006).

Recent studies reveal new developments in the way NGOs and companies perceive each other and how they get involved in collaborations (Yaziji 2004; Pedersen et al. 2011, 6; Aagaard 2017). Some of the key drivers for businesses to enter into collaborations with NGOs is the opportunity to gain competitive advantage, to meet local and global sustainable pressures and to preserve legitimacy etc. (Yaziji and Doh 2009) and to mimic the way successful companies explore new innovative business models (Holmes and Moir 2007; Dahan et al. 2010). This shift in collaboration patterns and interaction processes provides new opportunities for different types of interaction outcomes. One unique and multi-faceted outcome of business–NGO partnerships is the co-development of sustainable innovation. Practice reveals that a growing number of businesses collaborate with NGOs as part of their CSR development (Kovacs 2006). These NGOs are private/public, voluntary nonprofit organizations that represent specific social movements, ethical ideals and standards (e.g. WWF, Greenpeace, Amnesty International, Care, Oxfam). These NGO's all vary widely in size, mission, and strategy, and are characterized by the specific cases that they are passionate about and support.

Thus, managing these SI collaborations and selecting the proper NGO's that can provide knowledge of and matches the type of sustainable innovation that the company wants to pursue, becomes critical for the success of the collaboration. Another benefit of these SI collaborations, relate to the fact that NGOs are characterized by their extensive networks, which can assist businesses in generating new knowledge, business networks and sustainable innovations (Aagaard 2016). Today, an increasing number of companies collaborate with NGOs in the context of CSR in order to gain value though increased reputation, third part endorsement, code of conducts, sustainable innovation and to gain access to NGO tacit knowledge and networks (Leigh and Waddock 2006; Holmes and Moir 2007; Van Burren and Patterson 2012). Thus, the SI potentials of business-NGO collaborations are therefore vast, if managed effectively.

4 Sustainable Business Models as Platforms for Sustainable Innovations

Effective development and adoption of sustainable innovation often requires new and sustainable business models and platforms as traditional non-sustainable business models seldom support the full business potential of sustainable innovation (Aagaard 2017, 2018). In defining and exploring the theoretical concept of sustainable business model innovation (SBMI) the starting point would be existing definitions of business model innovation. Business models appear in many different forms. They can be applied as a core unit of analysis extending beyond the firm boundaries (e.g. Zott and Amit 2007, 2010). In addition, business models may be viewed as a construct between strategy and implementation (Baden-Fuller and Morgan 2010). Business models can also be a mean for commercializing new technologies (Chesbrough and Rosenbloom 2002; Chesbrough 2007, 2010) and as an intermediary between different innovation actors such as companies, financiers, research institutions, etc., i.e., actors who shape innovation networks (Doganova and Eyquem-Renault 2009).

Business models can therefore be subject to innovation themselves, or a template for implementing managerial initiatives (Zott and Amit 2010). Furthermore, they can be used to depict current realities ("as is") or used for simulations to decide on a preferred future ("to be") (Osterwalder 2004; Chatterjee 2013), i.e. as role exemplars (Baden-Fuller and Morgan 2010). Business models (real) can then be seen as a representation of strategic decisions, which have been implemented through tactical choices (Casadesus-Masanell and Ricart 2010), which may create self-enforcing "virtuous circles" in processes and resources, as stressed by Casadesus-Masanell and Ricart (2011).

Business models can also have a narrative role (Magretta 2002), serving as boundary objects (Doganova and Eyquem-Renault 2009) and as conventions (Verstraete and Jouison-Lafitte 2011) or theories of performative actions (Perkman and Spicer 2010) in which stakeholders become motivated to participate in the joint realization of a venture. As such, the core idea of the business model concept addresses many classic questions of strategic nature such as market relevance (value proposition), what customers to serve, and how to serve them, how to make a profit, technology (Magretta 2002; Morris et al. 2005; Verstraete and Jouison-Lafitte 2011).

Baden-Fuller and Morgan (2010) stress that from a holistic and systemic concept a business model perspective may be expected to contribute to a sustainable innovation agenda by opening up new approaches to overcoming internal and external barriers. Birkin et al. (2009a, b) identified in their study on North European and Chinese companies that societal and cultural demands of sustainable development evolve outside the economic sphere as drivers for organizational change in business enterprises. Their findings reveal that as social and natural needs become institutionalized as concrete societal and cultural demands, these models will change radically and companies are expected to ensure adaptations in order to secure legitimacy, legality as well as business success. Stubbs and Cocklin (2008) define that sustainable business models encompasses both a system's and firm-level perspective, draws on economic, environmental, and social aspects of sustainability in defining an organization's purpose and measuring performance, considers the needs of all stakeholders and treats nature as a stakeholder.

The definitions of sustainable business model innovation originate from different scientific areas. Looking into the literature on sustainable entrepreneurship and corporate sustainability management the concept of sustainable business models is still used in a fuzzy way (Lüdeke-Freund 2009; Schaltegger et al. 2012). In understanding sustainable business models as a way to build linkages between actors that are necessary to successfully market a sustainable product or service (Boons and Mendoza 2010), various elements being open to multiple interpretations may actually be a useful quality in developing sustainable innovations (e.g., Tukker and Tischner 2006; Boons 2009; Hansen et al. 2009). However, these statements are in sharp contrast with the research attempts to map and identify sustainable business models, and further research on how sustainability is constructed by actors involved in value creation is therefore requested (Boons and Lüdeke-Freund 2013).

However, earlier work reveals the first developments in mapping the concept of and movements towards sustainable business model innovation. Lovins et al. (1999) for one propose a four step agenda to align business practice with environmental needs, which they labeled 'Natural Capitalism'. The four steps constitute: increase of natural resources' productivity; imitation of biological production models; change of business models; and reinvestment in natural capital. Important for our review and mapping of the concept is the fact that Lovin's and colleagues see a change towards sustainable business models as crucial to realizing Natural Capitalism.

Another early contribution that emphasizes the same understanding of sustainable business model innovation is Hart and Milstein (1999), who see sustainable development as a force of industrial renewal and progress. They conclude that "simply transplanting business models" from one economy to another will run counter to sustainable development (Hart and Milstein 1999, p. 29). Common for these two classic articles is how they see changing business models as a way to reduce negative social and ecological impacts as well as a way to achieve sustainable development.

Later scientific developments of the concept reveal different definitions and explanations. Yunus et al. (2010) reason that for social businesses a specific business model framework is needed that integrates a social profit equation. According to their concept, social businesses apply business models that above all recover their full costs and pass profits on to customers who shall benefit from low prices, adequate services and better access to maximize the social profit equation: *"It is a no-loss, no-dividend, self-sustaining company that sells goods or services and repays investments to its owners, but whose primary purpose is to serve society and improve the lot of the poor (Yunus et al. 2010, p. 311)"*. Another interesting contribution on defining SBMI is from Boons and Lüdeke-Freund (2013), who define three different types of sustainable business models, which create social value and maximize social profit focusing on (pp. 14–15):

Technological innovation: creating a fit between technology characteristics and (new) commercialization approaches that both can succeed on given and new markets.

Organizational innovation: implementing alternative paradigms that shape the culture, structure and routines of organizations and thus change the way of doing business towards sustainable development.

Social innovation: that helps in creating and further developing markets for innovations with a social purpose.

In managing sustainable innovation with technical, organizational and/or social characteristics the three different sustainable business models typologies provide the proper platforms for management and strategic approach. In this context, business model refers to the logic of the firm, the way it operates and how it creates value for its stakeholders, where strategy refers to the choice of business model through which the firm will compete in the marketplace (Casadesus-Masanell and Ricart 2010).

Berkhout (2014) argues that sustainable innovation is linked to corporate strategy and that sustainable system innovation needs to be seen as a co-evolutionary process involving not only innovative firms, but also a broader context of institutions, infrastructures, and consumer practices. Thus, the choices of business models and strategy have a clear impact on the type of innovation management to emphasize, and the match between these elements is critical to successful management of sustainable innovations (Aagaard 2017). Alkemade et al. (2007) emphasize that the success of sustainable innovations depends to a large part on their environment, and the structure, and dynamics of the innovation system. As such, the actors involved in the innovation system include not only the innovating firms and their shareholders, but also their various stakeholders (Farla et al. 2012). In the words of Schaltegger et al. (2016, p. 6): "(. . .) a business that contributes to sustainable development needs to create value to the whole range of stakeholders and the natural environment, beyond customers and shareholders". By definition, sustainable business model innovation requires change in existing systems. The development and management of the innovation system, the ecosystems and stakeholders of the company are therefore critical to successful sustainable innovation and the integration of sustainable business models.

5   Concluding Remarks

The growing local, national and global pressures for sustainability from society and stakeholders forces companies and organizations to reinvent the ways that they do business and innovate. This has managerial implications, as sustainable innovation requires new ways of measuring due to the fact that traditional performance management and performance evaluations typically do not account for non-monetary types of performances (e.g. branding value, customer- and employee loyalty, license to operate, investor attraction). In addition, sustainable innovation demands an open source innovation approach, as sustainable innovation emphasizes the inclusion of multiple stakeholder needs. Managing multiple stakeholders throughout an innovation process underlines the need for new types of management capable of facilitating and managing open sources of knowledge of stakeholders that are not employed by the company, but with influence on the company. Thus, managers have to apply new management tools and innovation processes to incorporate, engage and motivate the relevant stakeholders in the respective phases of the ideation, innovation, development, adoption/implementation and evaluation of a new sustainable innovation. Furthermore, new networks and partnerships have to be identified, established and managed to provide the necessary types of knowledge and capabilities to develop and mature new and radical ideas for new, sustainable products, -services and/or -business models, which can accommodate the needs and requests of different stakeholders. One example hereof as provided in the chapter are business-NGO collaborations. NGO's are typically advocates for a specific cause and therefore care about influence and not profit. Consequently, these collaborations require a new management approach, as the traditional business logics do not apply with this type of partner. However, NGO's can provide companies with knowledge, capabilities and legitimacy and where the NGOs gain influence and impact, so there are common interests in collaborations. However, there is a trade-off and a 'blending' of logics that has to be made to ensure a success collaboration, and this requires new ways of negotiating and collaborating from management. Sustainable innovation is often treated in literature as just another type of innovation. Yet this chapter challenges this perception, as sustainable innovation ultimately requires a change of business models, innovation processes, partners and potentially also of managers/management capabilities in ensuring successful (and sustainable) performance and impact of the company's sustainable innovations in the short and long run.

References

Aagaard, A. (2011). *Idea and innovation management & leadership*. Copenhagen: Hans Reitzels Forlag.

Aagaard, A. (2016). *Sustainable business – integrating CSR into business and functions*. River Publishers.

Aagaard, A. (2017). *Understanding and evaluating sustainable business models: A cross-industry case study*. Conference paper, The 24th Innovation and Product Development Management Conference (IPDMC), June 11–13, 2017, Reykjavik, Island.

Aagaard, A. (2018). *Sustainable business models – Innovation, implementation and success*. Basingstoke: Palgrave Macmillan.

Adams, R., Bessant, J., & Phelps, R. (2006). Innovation management measurement: A review. *International Journal of Management Reviews, 8*(1), 21–47.

Alkemade, F., Kleinschmidt, C., & Hekkert, M. (2007). Analysing emerging innovation systems: A functions approach to foresight. *International Journal of Foresight and Innovation Policy, 3*(2), 139–168.

Argenti, P. A. (2004). Collaborating with activists: How starbucks works with NGOs. *California Management Review, 47*(1), 91–116.

Austin, J. E. (2000). *The collaboration challenge: How nonprofits and businesses succeed through strategic alliances*. San Francisco: Jossey Bass.

Baden-Fuller, C., & Morgan, M. S. (2010). Business models as models. *Long Range Planning, 43*, 156–171.

Bansal, P. (2005). Evolving sustainability: A longitudinal study of corporate sustainable development. *Strategic Management Journal, 26*, 197–218.

Bengtsson, L., Lakemond, N., Lazzarotti, V., Manzini, R., Pellegrini, L., & Tell, F. (2015). Open to a select few? Matching partners and knowledge content for open innovation performance. *Creativity and Innovation Management, 24*(1), 72–86.

Berkhout, P. H. (2011). Eco innovation: Reflections of an evolving research agenda. *International Journal of Technology, Policy and Management, 11*(3/4), 191–197.

Berkhout, F. (2014). Sustainable innovation management. In M. Dodgson, D. M. Gann, & N. Phillips (Eds.), *The Oxford handbook of innovation management*. Oxford: Oxford University Press.

Birkin, F., Cashman, A., Koh, S. C. L., & Liu, Z. (2009a). New sustainable business models in China. *Business Strategy and the Environment, 18*, 64–77.

Birkin, F., Polesie, T., & Lewis, L. (2009b). A new business model for sustainable development: An exploratory study using the theory of constraints in Nordic organizations. *Business Strategy and the Environment, 18*, 277–290.

Boons, F. (2009). *Creating ecological value. In An evolutionary approach to business strategies and the natural environment*. Cheltenham: Elgar.

Boons, F., & Lüdeke-Freund, F. (2013). Business models for sustainable innovation: State-of-the-art and steps towards a research agenda. *Journal of Cleaner Production, 45*, 9–19.

Boons, F., & Mendoza, A. (2010). Constructing sustainable palm oil: How actors define sustainability. *Journal of Cleaner Production, 18*, 1686–1695.

Boons, F., Montalva, C., Quist, J., & Wagner, M. (2013). Sustainable innovation, business models and economic performance: An overview. *Journal of Cleaner Production, 45*, 1–8.

Carbone, V., Moatti, V., & Vinzi, V. E. (2012). Mapping corporate responsibility and sustainable supply chains: An exploratory perspective. *Business Strategy and the Environment, 21*(7), 475–494.

Carroll, A. B., & Shabana, K. M. (2010). The business case for corporate social responsibility: A review of concepts, research and practice. *International Journal of Management Reviews, 12*(1), 85–105. https://doi.org/10.1111/j.1468-2370.2009.00275.x.

Casadesus-Masanell, R., & Ricart, J. E. (2010). From strategy to business models and onto tactics. *Long Range Planning, 43*, 195–215.

Casadesus-Masanell, R., & Ricart, J. E. (2011). How to design a winning business model. *Harvard Business Review, 89*, 100–107.

Charter, M., & Clark, T. (2007). Sustainable innovation—Key conclusions from sustainable innovation conferences 2003–2006. The Centre for Sustainable Design, University College for the Creative Arts, May 2007. www.cfsd.org.uk

Charter, M., Gray, C., Clark, T., & Woolman, T. (2008). Review: The role of business in realising sustainable consumption and production. In *Perspectives on radical changes to sustainable consumption and production. System innovation for sustainability* (pp. 46–69). Sheffield: Greenleaf Publishing.

Chatterjee, S. (2013). Simple rules for designing business models. *California Management Review, 55*, 97–124.

Chesbrough, H., & Rosenbloom, R. (2002). The role of the business model in capturing value from innovation. *Industrial and Corporate Change, 11*(3), 529–556.

Chesbrough, H. W. (2005). *Open innovation: The new imperative for creating and profiting from technology*. Boston: Harvard Business Review Press.

Chesbrough, H. (2007). Business model innovation: It is not just about technology anymore. *Strategy and Leadership, 35*(6), 12–17.

Chesbrough, H. (2010). Business model innovation: Opportunities and barriers. *Long Range Planning, 43*, 354–363.

Chesbrough, H., & Brunswicker, S. (2014). A fad or a phenomenon? The adoption of open innovation practices in large firms. *Research-Technology Management, 57*(2), 16–25.

Chesbrough, H., Vanhaverbeke, W., & West, J. (2006). *Open innovation: Researching a new paradigm*. Oxford: Oxford University Press.

Choi, J., Nies, L., & Ramani, K. (2008). A framework for the integration of environmental and business aspects toward sustainable product development. *Journal of Engineering Design, 19*(5), 431–446.

Christensen, C. M. (1997). *The innovators dilemma: When new technologies cause great firms to fail*. Boston, MA: Harvard Business School Press.

Christensen, L. T., Morsing, M., & Thyssen, O. (2011). The polyphony of corporate social responsibility. Deconstructing accountability and transparency in the context of identity and hypocrisy. In G. Cheney, S. May, & D. Munshi (Eds.), *Handbook of communication ethics* (pp. 457–474). New York: Lawrence Erlbaum Publishers.

Dahan, N. M., Doh, J. P., Oetzel, J., & Yaziji, M. (2010). Corporate-NGO collaboration: Co-creating new business models for developing markets. *Long Range Planning, 43*(2/3), 326–342.

Dahlander, L., & Gann, D. M. (2010). How open is innovation? *Research Policy, 39*, 699–709.

Das, T. K., & Teng, B. S. (2000). A resource-based theory of strategic alliances. *Journal of Management, 26*(1), 31–61.

De Sousa, M. C. (2006). The sustainable innovation engine. *Journal of Information and Knowledge Systems, 34*(6), 398–405.

Doganova, L., & Eyquem-Renault, M. (2009). What do business models do? Innovation devices in technology entrepreneurship. *Research Policy, 38*, 1559–1570.

Doh, J. P., & Guay, T. R. (2006). Corporate social responsibility, public policy, and NGO activism in Europe and the United States: An institutional stakeholder perspective. *Journal of Management Studies, 43*(1), 47–73.

Dryzek, J. S. (2005). *The politics of the Earth: Environmental discourses. Oxford: Oxford University Press.*

Faber, N., Jorna, R., & Van Engelen, J. (2005). The sustainability of "sustainability" – A study into the conceptual foundations of the notion of "sustainability". *Journal of Environmental Assessment Policy and Management, 7*(1), 1–33.

Faems, D., Van Loovy, B., & Debackere, K. (2005). Interorganizational collaboration and innovation: Toward a portfolio approach. *Journal of Product Innovation Management, 22*, 238–250.

Farla, J., Markard, J., Raven, R., & Coenen, L. (2012). Sustainability transitions in the making: A closet look at actors, strategies and resources. *Technological Forecasting and Social Change, 79* (6), 991–998.

Flynn, M., Dooley, L., O'Sullivan, D., & Cormican, K. (2003). Idea management for organisation innovation. *International Journal of Innovation Management, 7*(4), 417–442.

Hall, J., & Vrendenburg, H. (2003). The challenges of innovating for sustainable development. *MIT Sloan Management Review, 45*, 61–68.

Hansen, E. G., Große-Dunker, F., & Reichwald, R. (2009). Sustainability innovation cube – A framework to evaluate sustainability-oriented innovations. *International Journal of Innovation Management, 13*, 683–713.

Hart, S. L., & Milstein, M. B. (1999). Global sustainability and the creative destruction of industries. *Sloan Management Review, 41*, 23–33.

Hishop, H., & Hill, J. (2011, October). *Reinventing the wheel: A circular economy for resource security.* Report. Green Alliance. ISBN 978-1-905869-46-6.

Hollensen, S. (2008). *Essentials of global marketing.* Harlow: Pearson Education.

Holmes, S., & Moir, L. (2007). Developing a conceptual framework to identify corporate innovations through engagement with non-profit stakeholders. *Corporate Governance, 7*(4), 414–422.

Hull, C. E., & Rothenberg, S. (2008). Firm performance: The interactions of corporate social performance with innovation and industry differentiation. *Strategic Management Journal, 29* (7), 781–789.

Igartua, J. I., Garrigós, J. A., & Hervas-Oliver, J. L. (2010). How innovation management techniques support an open innovation strategy. *Research Technology Management, 53*(3), 41–52.

Kaebernick, H., Kara, S., & Sun, M. (2003). Sustainable product development and manufacturing by considering environmental requirements. *Robotics and Computer-Integrated Manufacturing, 19*(6), 461–468.

Kanter, R. M. (1999). From spare change to real change: The social sector as a beta site for business innovation. *Harvard Business Review, 77*, 123–132.

Kovacs, R. (2006). Interdisciplinary bar for the public interest: What CSR and NGO frameworks contribute to the public relations of British and European activists. *Public Relations Review, 32* (4), 429–431.

Leigh, J., & Waddock, S. (2006). The emergence of total responsibility management systems: J Sainsbury's (plc) voluntary responsibility management systems for global food retail supply chains. *Business and Society Review, 111*(4), 409–426.

Lodsgård, L., & Aagaard, A. (2016). *The four archetypes of business-NGO collaborations in creating sustainable innovation.* Conference paper at 23rd Innovation and Product Development Management IPDMC Conference, University of Strathclyde, Glasgow, June 12–14, 2016.

Lodsgård, L., & Aagaard, A. (2017). Creating value through CSR across company functions and NGO collaborations: A Scandinavian cross-industry study. *Scandinavian Journal of Management, 33*(3), 162–174.

Loorbach, D., van Bakel, J. C., Whiteman, G., & Rotmans, J. (2009). Business strategies for transitions towards sustainable systems. *Business strategy and the environment, 19*(2), 133–146.

Lovins, A. B., Lovins, L. H., & Hawken, P. (1999). A road map for natural capitalism. *Harvard Business Review, 1–14* (HBR paperback reprint 2000).

Lüdeke-Freund, L. (2009). *Business model concepts in corporate sustainability contexts from rhetoric to a generic template for "Business models for sustainability"*. Lüneburg: Lüneburg University, Centre for Sustainability Management.

Magretta, J. (2002). Why business models matter. *Harvard Business Review, 80*, 86–92.

Morris, M., Schindehutte, M., & Allen, J. (2005). The entrepreneur's business model: Toward a unified perspective. *Journal of Business Research, 58*, 726–735.

Nidumolu, R., Prahalad, C. K., & Rangaswami, M. (2009). Why sustainability is now the key driver of innovation. *Harvard Business Review, 87*(9), 56–64.

Osterwalder, A. (2004). The business model ontology. In: *A proposition in a design science approach*. Lausanne: Université de Lausanne.

Pavelin, S., & Porter, L. A. (2008). The corporate social performance content of innovation in the U.K. *Journal of Business Ethics, 80*, 711–725.

Pedersen, G. R. E., Pedersen, T. J., & Jacobsen, Ø. P. (2011). Partnerskaber mellem virksomheder og NGOer. *Ledelse og Erhvervsøkonomi, 76*(4), 33–48.

Perkman, M., & Spicer, A. (2010). What are business models? In N. Phillips, G. Sewell, & D. Griffiths (Eds.), *Research in the sociology of organizations*. Bingley: Emerald Group Publishing.

Rass, M., Dumbach, M., Danzinger, F., Bullinger, A. C., & Moeslien, K. M. (2013). Open innovation and firm performance: The mediating role of social capital. *Creativity and Innovation Management, 22*(2), 177–194.

Schaltegger, S., Hansen, E. G., & Lüdeke-Freund, F. (2016). Business models for sustainability: Origins, present research, and future avenues. *Organization & Environment, 29*(1), 3–10.

Schaltegger, S., Lüdeke-Freund, F., & Hansen, E. G. (2012). Business cases for sustainability – The role of business model innovation for corporate sustainability. *International Journal of Innovation and Sustainable Development, 6*(2), 95–119.

Schaltegger, S., & Wagner, M. (2011). Sustainable entrepreneurship and sustainability innovation. Categories and interactions. *Business Strategy and the Environment, 20*(4), 222–237.

Shove, E. (2003). *Comfort, cleanliness and convenience: The social organization of normality*. Oxford: Berg.

Steketee, D. M. (2010). Disruption or sustenance? An institutional analysis of sustainable business network in west Michigan. In J. Sarkis, D. V. Brust, & J. J. Cordeiro (Eds.), *Facilitating sustainable innovation through collaboration: A multi-stakeholder perspective*. Dordrecht: Springer.

Stubbs, W., & Cocklin, C. (2008). Conceptualizing a sustainability business model. *Organization & Environment, 21*(2), 103–127.

Tidd, J., & Bessant, J. (2013). *Managing innovation – integrating technological, market and organisational change* (5th ed.). London: Wiley.

Tukker, A., Charter, M., Vezzoli, C., Stø, E., & Andersen, M. M. (Eds.). (2008). *Perspectives on radical changes to sustainable consumption and production*. Sheffield: Greenleaf.

Tukker, A., & Tischner, U. (Eds.). (2006). *New business for Old Europe. Product-service development, competitiveness and sustainability*. Sheffield: Greenleaf.

Van Burren, H. J., & Patterson, K. D. W. (2012). Institutional predictors of and complements to industry self-regulation with regard to labor practices. *Business & Society Review, 117*(3), 357–382.

Van de Vrande, V., de Jong, J. P. J., Vanhaverbeke, W., & de Rochemont, M. (2009). Open innovation in SMEs: Trends, motives and management challenges. *Technovation, 29*, 423–437.

Venn, R., & Berg, N. (2013). Building competitive advantage through social entrepreneurship. *South Asian Journal of Business Research, 2*(1), 104–127.

Verstraete, T., & Jouison-Lafitte, E. (2011). A conventionalist theory of the business model in the context of business creation for understanding organizational impetus. *Management International/International Management/Gestión International, 15*, 109–124.

Wagner, M. (2010). Corporate social performance and innovation with high social benefits: A quantitative analysis. *Journal of Business Ethics, 94*, 581–594.

WEF. (2010). *Redesigning business value: A roadmap for sustainable consumption*. Geneva: World Economic Forum (WEF).

Yaziji, M. (2004). Turning gadflies into allies. *Harvard Business Review, 82*(2), 110–115.

Yaziji, M., & Doh, J. (2009). *NGOs and corporations: Conflict and collaboration*. Cambridge: Cambridge University Press.

Yunus, M., Moingeon, B., & Lehmann-Ortega, L. (2010). Building social business models: Lessons from the Grameen experience. *Long Range Planning, 43*, 308–325.

Zink, K. J., Steimle, U., & Fisher, K. (2008). Human factors, business excellence and corporate sustainability: Differing perspectives, joint objectives. In K. J. Zink (Ed.), *Corporate sustainability as a challenge for comprehensive management*. Heidelberg: Physica-Verlag.

Zott, C., & Amit, R. (2007). Business model design and the performance of entrepreneurial firms. *Organization Science, 18*, 181–199.

Zott, C., & Amit, R. (2010). Business model design: An activity system perspective. *Long Range Planning, 43*, 216–226.

Annabeth Aagaard PhD, MSc is an Associate Professor at Aarhus University, Denmark and holds a PhD in Pharmaceutical front-end innovation. She is the director of the research centre, Centre for Business Development (CBD), counting 10 employees and her 4 PhD students at Aarhus University. CBD and Annabeth Aagaard bridge business development, digital technologies and sustainability in research, projects, teaching and international networks. She has a total of 20 years of experience with business modeling, innovation management, and sustainable innovation drawing on experiences in the academic, public and private sectors, where she was formerly a manager and management specialist designing and implementing strategies for top 100 Scandinavian companies. She has published extensively on innovation management, business model innovation and sustainability in highly ranked international scientific journals and has authored ten academic books and management handbooks.

CSR and Innovation: A Holistic Approach From a Business Perspective

Hans-Jürgen August

1 Introduction

1.1 Terminology: CSR and Its Relatives

Even though the actual impact of Corporate Social Responsibility (CSR) activities is sometimes questioned in literature (Hahn et al. 2011), it is undisputed that CSR is gaining more and more attention in academic research as well as in business (Altenburger 2013; Carroll 2015). Porter and Kramer (2008) stated that "myriad organizations rank companies on their performance of their corporate social responsibility [...]. As a result, CSR has emerged as an inescapable priority for business leaders in every country." Yet, the term is still ill-defined. "No unique definition emerged in last few decades in the history of CSR that can be used for all purposes", writes Rahman in his overview paper on CSR definitions (Rahman 2011), and Schneider even captions a book paragraph stating that the "term CSR [is] not stable and clearly defined" (Schneider 2015), even though co-editor Schmidpeter emphasizes the increasing importance of CSR: "The discussion about social responsibility of companies [...] is in full swing. Corporate boards, politicians and academics debate on the responsibility companies assume and on how sustainable management may contribute to solving current social challenges, but also on how it helps to improve competitiveness" (Schmidpeter 2015).

Early definitions of companies' social responsibility usually center on the question asked by Bowen in 1953: "What responsibility to society may businessmen reasonably be expected to assume?" (Bowen 1953). The very formulation of this question—addressing "businessmen"—relates to early-capitalism roots of social responsibility and links nineteenth century approaches to modern concepts. As

H.-J. August (✉)
TTTech Computertechnik AG, Vienna, Austria

© Springer International Publishing AG, part of Springer Nature 2018   29
R. Altenburger (ed.), *Innovation Management and Corporate Social Responsibility*,
CSR, Sustainability, Ethics & Governance,
https://doi.org/10.1007/978-3-319-93629-1_3

early as in 1960, Frederick argues that "businessmen should oversee the operation of an economic system that fulfills the expectations of the public", and he continues by demanding that "production and distribution should enhance total socio-economic welfare" (Frederick 1960).

Schneider (2015) cites Dow Votaw (Votaw and Sethi 1973): "The term is a brilliant one; it means something, but not always the same thing to everybody. To some it conveys the idea of legal responsibility or liability; to others it means socially responsible behavior in an ethical sense; to still others, the meaning transmitted is that of 'responsible for', is a casual mode; many simply equate it with a charitable contribution"—and he declares that this statement is still valid.

No progress in more than 40 years? Among the countless efforts to define the scope of CSR and possibly even some methodologies to manage CSR, at least Elkington's book "Cannibals with forks" shall be mentioned, in which he coins the "triple bottom line" notion, spanning the triangle of social, environmental and economic aspects, or, catchier, promotes the "people, planet, profit" perspective (Elkington 1997, 2004). This very basic concept is still used and discussed also in recent publications (e.g. Carroll 2015; Hansen and Große-Dunker 2013; Schaltegger et al. 2012).

Even though there is still no commonly agreed unique definition of CSR in place, respective attempts have been made by multinational organizations, e.g. by the European Commission (EC) or the International Standards Organization (ISO). According to the EC definition, CSR aims at "maximising the creation of shared value for their owners/shareholders and for their other stakeholders and society at large" and at "identifying, preventing and mitigating their possible adverse impacts". This shall cover at least human rights, labour and employment practices, environmental issues, and combating bribery and corruption. Following the EC statements, community involvement and development, the integration of disabled persons, and consumer interests, including privacy, and are also part of the CSR agenda (European Commission 2011).

The standard ISO 26000 is titled "Guidance on social responsibility", and provides a guideline on the underlying principles of social responsibility, recognizing social responsibility and engaging stakeholders, the core subjects and issues pertaining to social responsibility, and on ways to integrate socially responsible behavior into the organization (ISO 26000: 2010). As usual with ISO standards, the document was developed on a multinational level, involving stakeholders from around 90 countries. The standard proposes to omit the term "corporate" and instead to use the term "social responsibility" to widen the scope of applicability: "The view that social responsibility is applicable to all organizations emerged as different types of organizations, not just those in the business world, recognized that they too had responsibilities for contributing to sustainable development" (ISO 26000: 2010). In the "terms and definitions" part of the standard, "social responsibility" is defined as "responsibility of an organization for the impacts of its decisions and activities on society and the environment, through transparent and ethical behaviour that

- contributes to sustainable development, including health and the welfare of society;
- takes into account the expectations of stakeholders;
- is in compliance with applicable law and consistent with international norms of behaviour; and
- is integrated throughout the organization and practised in its relationships."

So far we have used the letter "R" in "CSR" for "responsibility". Yet, some authors argue that "responsibility" may mislead the discussion. As early as in 1975, Sethi introduced the term "corporate social responsiveness" (Sethi 1975). Corporations—or, following the ISO 26000 perspective, organizations in general—should not just take over responsibility, but respond to the needs of society.

As Archie B. Carroll points out in a recent paper (Carroll 2015), other competing concepts have been promoted in the past decades, including Business Ethics, Stakeholder Management, Sustainability, and Corporate Citizenship. We will discuss these notions in more detail in the next section.

1.2 History: Development of Concepts

Even though discussions on CSR as well as the modeling of respective concepts started in the 1950s, the main roots of businessmen's (and businesswomen's) concern for social topics can be traced back at least to the nineteenth century. As Carroll states, "then, and now, it is sometimes difficult to differentiate what organizations are doing for business reasons, i.e. making the workers more productive, and what the organizations are doing for social reasons, i.e. helping to fulfill their needs and make them better and more contributing members of society" (Carroll 2008)—nonetheless there is quite some evidence for philanthropic activities in nineteenth century capitalism. Management historian Wren (2005) describes examples including the provision of medical care, bathhouses, recreational facilities, and even profit sharing. "In Britain, visionary business leaders in the aftermath of the Industrial Revolution built factory towns, such as Bourneville (founded by George Cadbury in 1879) and Port Sunlight (founded by William Lever in 1888 and named after the brand of soap made there), that were intended to provide workers and their families with housing and other amenities when many parts of the newly industrialized cities were slums", writes Smith (2002). Other examples, including Macy's, the Pullman Car Company or the foundation of the YMCA have been described e.g. by Heald (1970). Supporting Carroll's view on philanthropic efforts in general, also Smith (2002) states that "philanthropic industrialists of the Victorian period were motivated by a desire to do good, but they were also motivated by enlightened self-interest." It is clear that an improved working environment would improve the company's performance also from a business perspective. Nonetheless industrialists such as Cadbury, Lever, or Salt—who built the new city Saltaire close to Bradford, from where he moved the wool textile production to the new location comprising 850 houses for his workers—were pioneers in the dark era of capitalism, and provided their workers a comparatively safe and healthy working environment, decades before governments initiated respective legislation. A brief history of industrialist philanthropy, spanning from the early days of capitalism to the present (e.g. the Bill and Melinda Gates foundation) may be found in the paper of Acs and Phillips (2002).

More comprehensive and systematic Corporate Social Responsibility (CSR) approaches emerged in the 1950s. Carroll (1979) declares Howard R. Bowen's publication "Social Responsibilities of the Businessmen" (1953) as the first milestone in CSR history, "considered by many to be the first definite book on the subject." Since, as stated above, the definition of CSR is still blurry, it was all the more in the "childhood" of this concept. In the early 1960s it was at least clear that CSR has something to do with responsibility taken beyond just the economic aspects. The topic was undoubtedly on the agenda and intensively discussed, which is also reflected by Milton Friedman's reaction to it. In his best-selling book, "Capitalism and Freedom" (1962), the later Nobel-prize winner argues: "The view has been gaining widespread acceptance that corporate officials and labor leaders have a "social responsibility" that goes beyond serving the interest of their stockholders or their members. This view shows a fundamental misconception of the character and nature of a free economy. [...] Few trends could so thoroughly undermine the very foundations of our free society as the acceptance by corporate officials of a social responsibility other than to make as much money for their stockholders as possible. This is a fundamentally subversive doctrine."

Even though Friedman still ranks as one of the most influential economists of the twentieth century, his harsh rejection of the CSR concept did not really affect the further development of the topic, at least not in the mid and long term. In fact, the CSR concept became more and more comprehensive. McGuire declared in 1963 that social responsibilities should go beyond just economic goals and legal compliance. As already mentioned above, the term "responsibility" was criticized as just covering the assumption of an obligation, but not focusing on the outcome of related activities. Ackerman and Bauer (1976) thus propose the use the term "responsiveness" instead. A similar view was promoted by Sethi (1975).

In his seminal 1979 paper, Archie B. Carroll proposes a four-step classification of CSR performance. The basic and most relevant responsibility is still the economic one. Needless to argue—if the company does not survive in the market, any discussion on its social responsibility is obsolete. Secondly, there is the responsibility to comply with the law. Going beyond these basic levels, Carroll distinguishes between ethical and discretionary responsibilities. Even though "ethical responsibilities are ill defined and consequently are among the most difficult for businesses to deal with", they are clearly demanded by society. "Discretionary (or volitional) are those about which society has no clear-cut message for business" (Carroll 1979).

"During the 1990s, CSR practices became commonplace, more formalized, varied, and more deeply integrated into business practices", states Carroll in 2015, and discusses some alternative concepts: "Business ethics is a system of thought that is rooted in moral duty and obligations. It can also be seen as principles or values. Business ethics is concerned with the rightness or fairness of business, manager and employee actions, behaviors and policies taking place in a commercial context" (Carroll 2015). Translated into corporate reality, business ethics are reflected e.g. in anti-bribery rules, fairness rules and codes of conduct. Business ethics thus have a narrower focus, dealing primarily with how to act in a correct way when doing business. Main operative areas concerned may be sales, contracting, accounting, and the relations to stakeholders, which brings us to the next concept discussed:

According to Carroll, the stakeholder approach became popular in the mid-1980s, and is still important today. This is—by the way—also reflected in the new revisions of the ISO 9001 and ISO 14001 standards (released in 2015), in which stakeholder identification, consideration and management gained much importance.

The term "sustainability" was coined in the report of the so-called Brundtland Commission, which published its results in 1987. The commission headed by former Norwegian prime minister Gro Harlem Brundtland put environmental topics on the political agenda, and asked for "sustainable development" (Brundtland 1987). Nowadays—and following John Elkington's notion (1997)—sustainability is usually defined as the triad of economic, environmental, and social issues. Corporate reports covering CSR or related concepts nowadays are quite often published as "sustainability reports". According to Carroll (2015), 95% of the Global Fortune 250 companies issue publications dealing with their performance not just on economic level, but also related to environmental and social aspects.

A fifth concept discussed by Carroll is that of "Corporate Citizenship" (CC), distinguishing between a "broad view", in which CC is more or less identical to CSR, and a "narrow view", in which CC just covers the discretionary or philanthropic level of his four-level concept presented in 1979.

Another important facet of more recent discussions is pointed out by Altenburger (2013), referring to the corporate responsibility initiative of the Harvard Kennedy School (2013): Stakeholders do not just ask how profits are used for social or environmental topics, but how these profits are earned.

1.3   Business: Does CSR Pay Off?

But all discussions about CSR and Business ethics, sustainability and corporate citizenship may be interfered by a very simple question: Does CSR pay off? This reminds us of Milton Friedman's view that the sole responsibility of a businessman is to make profit—the more the better. Although the importance of CSR and similar concepts is in fact increasing, the tension between serving the shareholders by making profit and also considering other stakeholders by assuming additional responsibility has not vanished. In fact, the discussion became more and more prominent around 20 years ago, as Epstein and Roy argued in 2003. Schaltegger and co-workers (2012) trace back the discussion to the mid-1990s, citing publications e.g. by Burke and Logsdon (1996), Hamilton (1995), or Porter and van der Linde (1995a, b). Meanwhile, several authors have examined the relationship between financial performance and sustainability, discussing also the "CSR business case" (e.g. Altenburger 2013; Boons et al. 2013; Hockerts 2007; Schaltegger et al. 2012). As mentioned by Altenburger (2013) Kurucz and co-workers (2008) analyzed CSR studies and identified four major discussion threads: cost and risk reduction, gaining competitive advantages, legitimacy and reputation, and achieving win-win situations for corporations and the society. An overview of concepts, review, and practice can be found in the paper of Carroll and Shabana (2010).

The issue was also addressed by Michael Porter and Mark Kramer in their influential work published in 2006 in Harvard Business Review. The authors declare that "CSR has emerged as an inescapable priority for business leaders in every country" and "debates about CSR have moved all the way into corporate board-rooms", but deplore that "efforts have not been nearly as productive as they could be". For two reasons: first, because business and society are treated separately, although they are interdependent; and second, because companies are forced to treat CSR topics in a generic way instead in that most appropriate to the company's strategy. To resolve this tension, Porter and Kramer see just one option: To escape the zero-sum game and create win-win situations. Business and society have to be integrated in a way that fits the company, its business, its strategy and its possibilities to contribute best: "The essential test that should guide CSR is not whether a cause is worthy but whether it presents an opportunity to create shared value." Porter and Kramer advocate going beyond "responsive CSR" (discussed above) to "strategic CSR", characterized by "strategic philanthropy that leverages capabilities to improve salient areas of competitive context" and by transforming "value chain activities to benefit society while reinforcing strategy". How can this be achieved? "The interdependence of a company and society can be analyzed with the same tools used to analyze competitive position and develop strategy", answer Porter and Kramer and suggest e.g. to evaluate all company processes along the value chain and to integrate CSR aspects whenever appropriate.

2 Models: Understanding and Approach

As shown above, the discussion on CSR is still multifaceted and extensive. The still blurry definition of CSR on the one hand and the increasing importance on the other, has opened a broad field for concepts, approaches and interpretations. It seems to be a natural reaction to complexity to try to systemize phenomena and activities. As statistician and quality management theorist George Pelham Box put it, "all models are wrong, but some are useful"; therefore some models treating different perspectives on CSR shall be discussed.

2.1 Sustainability Innovation Cube

As discussed above, CSR and sustainability are undisputedly important topics on today's business and society agenda. The same applies to innovation: Thousands of papers are published year by year on innovation, a myriad of books try to find the key to successful innovation. Concepts like "disruptive innovations" (Christensen 1997; Christensen and Overdorf 2000; Christensen and Raynor 2003), "open innovation" (Chesbrough 2003, 2006), lead user and customer involvement approaches (von Hippel 2005; Schweitzer 2014), empathic design (Leonard and Rayport 1997; Littman and Peters 2001), the Blue Ocean concept (Kim and Mauborgne 1999, 2005), the Business Model Canvas (Osterwalder and Pigneur 2010), of the Lean Startup approach (Ries 2011)—for an overview on concepts see August et al. (2015)—are not just discussed in the inner circle of innovation management groups, but are almost common parlance.

As CSR, also innovation, innovation management, and, related to innovation, creativity are vast fields of research, and combining CSR with innovation does not really make it easier to keep an overview. Aiming for a systematic view on it, Erik Hansen, Friedrich Grosse-Dunker and Ralf Reichenwald developed the "Sustainability Innovation Cube" (2009) to structure innovations' sustainability effects. The tool is meant to be used by decision-makers in companies to reduce innovation risks related to economic, environmental, and social aspects. The model consists of three dimensions:

- Target dimension: This dimension analyses the effects of innovations of the target area of sustainability. The triple bottom line approach of Elkington (1997, 1998, 2004) is used, i.e. Hansen and co-workers categorize the target dimensions into ecological capital, social capital, and economic capital. The ecological dimension is determined by the consumption of resources and influences on the ecosystem (Fichter 2005), including pollution, changes to the atmosphere, buildings, effects on biodiversity etc. It is clear that the social impact is related to stakeholders both inside and outside the organization (Achterkamp and Vos 2006).
- The life cycle dimension allows for all phases of the life of a product or service, including manufacturing, use/maintenance, and end of life (disposal or recycling).
- As a third dimension, Hansen and co-workers consider the kind of innovation and differentiate between technological, product/service and business model innovations.

Combining these three dimensions with three aspects each, Hansen and co-workers arrive at the Sustainability Innovation Cube (SIC), consisting of 27 single cubes or intersection areas. To assess the impacts in each of these areas, the authors offer a portfolio of 76 methods that may be applied, either as "support methods" to assess effects ex-ante, or as "methods for analysis" to assess existing products and services ex post.

2.2 Business Model for Sustainability Concept

Does CSR pay off? This question was already raised above, and beyond pure philanthropy the answer to this question is crucial for the long-term success of CSR.

Stefan Schaltegger et al. (2012) picked up this question and discuss business cases for sustainability, aiming at increasing corporate economic value *through* environmental and social activities. As the authors state, "a business case for sustainability has to be created and managed—it does not just happen". This leads the authors to business model innovation and a concept that supports creating business models which lead to corporate financial profits, while serving the environment and society. Schaltegger and co-workers first explore the drivers of business cases, referring also to the work of Hansen (2010), Collins et al. (2010), and Revell et al. (2010):

- Cost and cost reduction
- Risk and risk reduction
- Sales and profit margin
- Reputation and brand value
- Attractiveness as employer
- Innovative capabilities

The next step is to "map out interrelations between business case drivers and the business model." For this, "a general business model has to be introduced". Schaltegger and co-workers identify four key elements:

- Value proposition (offerings representing a value to the customer)
- Customer relationship
- Infrastructure (network of partners that are needed to create the value proposition and manage the customer relationship)
- Financial aspects (e.g. cost and revenue structures)

Relating the six business case drivers with the four elements of a business model results in a 24 field matrix, for which the authors describe the value enhancing impact. Combining e.g. "attractiveness as employer" with "customer relationship" results in "Better customer service as a result of higher employee motivation"; "Risk and risk reduction" combined with "financial aspects" leads to "improved risk and credit rating resulting from lowered sustainability risks".

What are the main obstacles to implement sustainability business cases? Schaltegger and co-workers recall that in spite of business model innovation being hyped as the next big thing in innovation management, a variety of hurdles has to be overcome as also reviewed by Chesbrough (2010). Schaltegger and co-workers conclude that "sustainability-oriented innovations are obviously predisposed to not fit with the dominant logic of an established business model". Yet they see support by using accommodative and proactive sustainability strategies.

The model presented by Schaltegger and co-workers surely helps to build business models that lever the synergies between economic and operational advantages on the one hand and environmental and societal benefits on the other hand. Thus, it may also be used to advocate sustainable business models in an organization. An interesting extension of the model presented is to use Osterwalder and Pigneur's (2010) nine sector "Business Model Canvas" instead of the four element model used by Schaltegger, Lüdeke-Freund and Hansen. Of course this adds complexity to the model. On the other hand, it links the concept to a model that is quite popular in strategic and innovation management, thus lowering board members' and managers' inhibition thresholds to adopt the concept of business cases for sustainability. Two related approaches will be discussed below.

2.3   Classification of CSR Relevant Innovations

While Hansen et al. (2009) construct a 27 sector cube based on the three dimensions target, life cycle and innovation type and Schaltegger et al. (2012) combine business case drivers and business case elements, Ulrike Gelbmann et al. (2013) ask which areas of an organization are influenced by innovation and which effects may be found related to sustainability aspects. Partly based on the work of Grieshuber (2012) and Teece (2010) they differentiate between six impact areas:

- Organization, Management (e.g. introduction of respective strategies and management systems)
- Social (e.g. internal: working time model, work-life-balance; external: sponsoring, volunteering)
- Processes (e.g. improvement of process performance in terms of sustainability)
- Portfolio offered, products and services (e.g. improvement of products and services)
- Business model (e.g. changes of business models or introduction of new ones)
- Innovation system (e.g. stakeholder management, partnering)

In addition to identifying what is changed and which effect this may have, the authors also shed a light on possible reasons or goals that drive these innovations.

2.4   Impact of Innovation on CSR: Assessment Based on Business Model Canvas

In 2010, Alexander Osterwalder and Yves Pigneur presented their book "Business Model Generation" (Osterwalder and Pigneur 2010), which became a huge success and added new methodologies to the (innovation) manager toolbox. Even though Peter Drucker described as early as in 1994 a business model as "theory of a business" (Drucker 1994), it was Osterwalder's work (Osterwalder 2004; Osterwalder et al. 2005, 2014; Osterwalder and Pigneur 2010) which fueled also scholar discussion on business models. According to Upward and Jones (2016), more than one million copies of the book "Business Model Generation" were sold—a number that may even impress quite successful novelists—and as Upward and Jones state, "the widely known business model canvas (BMC) [. . .] has become a de facto reference standard and is taught in management and entrepreneurship education worldwide". Compared to decades ago, innovations of business models have gained much more importance (see also the review of Bocken et al. 2014), and major game-changing innovations are not so much based on new technologies, but on new ways to offer products and services and to create revenue. Considering the widespread use of the business model canvas—which may be interpreted as covering the "profit" aspect of the triple bottom line—it offers itself to use this model as a basis to consider and cover also the dimensions "planet" and "people". In recent years, a couple of related approaches have been suggested in literature, as by Nancy M. P. Bocken and co-workers, who develop a set of so-called "sustainable business model archetypes" (Bocken et al. 2013). In her very comprehensive master thesis Lara Obst (2015) states that Osterwalder and his co-workers (e.g. Osterwalder and Pigneur 2011) mention sustainability aspects in connection with their business model canvas, but also cites Nancy M. P. Bocken and co-workers (Bocken et al. 2014), who state that the business model canvas is "poorly suited for assessing a firm in generating wider sustainability". Responding to this issue, a couple of business model canvas alternatives have been developed recently. In the following I will present two comprehensive approaches and suggest the use of an approach that covers "planet" and "people" aspects when using the original Osterwalder and Pigneur business model canvas as a tool for innovation.

The "Strongly Sustainable Business Model Canvas" was presented by Antony Upward in 2014 (Upward 2013), mapping the four pillars "process", "values", "people" and "outcome" of their own business model canvas approach to the CSR dimensions of "economy", "society", and "environment". In subsequently published articles, Antony Upward and Peter Jones (Jones and Upward 2014; Upward and Jones 2016) investigate ontologies for sustainability-oriented business model designs, supporting the "Flourishing Business Model Canvas".

An approach that refers more directly to the model of Osterwalder and Pigneur was presented as "Triple Layered Business Model Canvas" ("TLBMC") by Alexandre Joyce and Raymond Paquin (2016), who claim to present "a practical tool for coherently integrating economic, environmental, and social concerns into a holistic view of an organization's business model". The authors start from the business model canvas of Osterwalder and Pigneur as representing the economic layer, and add layers with similar structures to cover environmental and social aspects. As a proof of concept, they analyze the Nespresso business model in terms of environmental as well as social aspects using the TLBMC approach. Based on experiences gained in field testing, the authors conclude that "the TLBMC seems well suited to support creatively developing more sustainable business models through a two-step approach", namely first, the analysis and communication of the current situation, and second, the possibility to explore possible innovations. Joyce and Paquin refrain from trying to integrate "multiple types of value into a single canvas", arguing that the separation into three canvasses, linked vertically by the 9-segment structure, allows to explicitly investigate the economic, environmental, and social dimensions of a business model. The author of the current article supports the approach of Joyce and Paquin, acknowledging especially the closeness to the original business model canvas of Osterwalder and Pigneur, since this may facilitate the adoption of the methodology in the many organizations which already use the business model canvas to analyze and innovate their business models. Yet, from a practitioner's point of view, there may be a need for a simpler and even more integrative methodology, which thus can be applied easily by practitioner teams that are already using the business model canvas of Osterwalder and Pigneur.

2.5 Integrated Balanced Sustainability Business Model Canvas

In the following the principles of the proposed "Integrated Balanced Sustainability Business Model Canvas" ("IBSBMC") shall be discussed. The methodology suggests to start from the original model of Osterwalder and Pigneur and to complement the current situation as well as considered changes of the business model with environmental and social perspectives.

2.5.1 Basic Ideas and Intended Benefits

Before going into detail, let us discuss some basic questions:

- Why should other CSR aspects be integrated into one business model canvas along with the "profit" perspective?

  The use of business model canvasses as tools for making current business mechanisms transparent and for further developing these mechanisms is widely used in practice. Since the focus so far lies on the "profit" perspective, the direct integration of the "people" and "planet" aspects into one canvas suggests to ensure as much as possible that these additional aspects are treated as integral part of the entire business case, reducing the risk of simply neglecting or even suppressing the "planet" and "people" dimensions.
- Why base the new approach on that of Osterwalder and Pigneur?

  As mentioned above—and supported by the impressive figures reported by Upward and Jones (2016)—the model of Osterwalder and Pigneur has not just gained much attention in academia, but also wide acceptance as a management tool in business.

  Additionally, the approach of Osterwalder and Pigneur is explicitly designed and used as a methodology for business model *innovation*, thus laying the foundation for a comprehensive analysis and consideration of the effects of business model *changes*. Integrating "people, planet and profit" into one single canvas and using the model of Osterwalder and Pigneur as starting point ensures that the analysis and innovation teams work jointly using a methodology that is already well established in many companies.

- What is meant with "balanced"?

Every change to a system will impact "profit, planet, and people"—otherwise one could not speak of "change". From a CSR perspective, the objective of changes to the current system may be described based on the European Commission view of "maximising the creation of shared value for their owners/shareholders and for their other stakeholders and society at large" and of "identifying, preventing and mitigating their possible adverse impacts". Using the term "balanced" which may remind us of the Balanced Scorecard of Kaplan and Norton (1992, 1996) is on purpose. The aim of Kaplan and Norton was to design a "set of measures that gives top managers a fast but comprehensive view of the business", complementing "the financial measures with operational measures on customer satisfaction, internal processes, and the organization's innovation and improvement activities". "By combining the financial, customer, internal process and innovation, and organizational learning perspectives, the balanced scorecard helps managers understand, at least implicitly, many interrelationships." (Kaplan and Norton 1992).

Analogously, the Integrated Balanced Sustainability Business Model Canvas aims at helping managers to understand the impact of business model changes not just on profit, but also on people and planet, and to find balanced solutions.

2.5.2 Integration: Considering All CSR Aspects

As mentioned in the example above, to cover the environmental and social impacts of changes in the economic level of the business model canvas would imply that all sectors of the environmental and social layer be mapped into each single sector of the economic layer. This view diverges to some extend from the "vertical coherence argument" of Joyce and Paquin. The impact of changes to the single sectors of Osterwalder and Pigneur's canvas can be identified using a questionnaire:

Key Partners:
- Does the new supplier offer a better overall CSR performance? (And in more detail some examples:)
- Does the new supplier (or partner in general) subscribe to our own value charter?
- Under which conditions are the products manufactured?
- Which materials are used to produce these products?
- What social impact does a change of supplier have at the current supplier?
- What impact would purchasing products from a new supplier have on transportation and logistics?

Key Activities
- How does a change of our key activities change our CSR performance? (And in more detail some examples:)
- What effect does a change of key activities have on our own staff (e.g. need for higher qualification and other competences; restructuring)?
- If we expand / reduce our key activities, which effect does this have on our partners?
- If we reduce our key activities (by outsourcing of activities), is there a risk of losing governance and steering possibilities related to CSR topics?

Key Resources
- How does a change of our key resources change our CSR performance? (And in more detail some examples:)
- If we change the financing system, which effect would this have on CSR (e.g. dependence from possibly less transparent sources)?
- Does the exchange of certain materials needed in production offer the possibility to substitute minerals extracted under problematic circumstances in terms of ecology or labor environment?
- What effect does a change of key activities have on our own staff (e.g. need for higher qualification and other competences; restructuring)?

Value Proposition
- How does a change of our value proposition change our CSR performance? (And in more detail some examples:)
- Does our altered value proposition help our customers to live up to their ethical convictions?
- How does a modified value proposition change our customer relationships and our addressed customer segments?
- Which environmental and social risks are related to our new value proposition?

Customer Relationship
- How does a change of our customer relationship change our CSR performance? (And in more detail some examples:)
- Does a change of our customer relationship improve the possibilities of our customers to give us feedback on CSR topics?
- Do we reduce the autonomy of our customers (e.g. free choice of offers)?
- Does a change in customer relationship increase customer dependency to increase profitability?

Channels
- How does a change of our channels change our CSR performance? (And in more detail some examples:)
- What effect does a change in our marketing strategy have on complying with our values?
- Are new redistributors complying with our value codex?
- What social effects may result from re-arranging our sales channels?

Customer Segment
- How does a change of our customer segment change our CSR performance? (And in more detail some examples:)
- Does addressing new customer segments make our products unaffordable for our current customers?
- May there be ethical concerns when addressing new customer segments?
- Can we expect new customer groups to handle our products in an environmentally friendly way, e.g. when it comes to correct disposal?

Cost Structure
- How does a change of our cost structure change our CSR performance? (And in more detail some examples:)
- What social impact does cost saving by restructuring have?
- Do financial gains in procurement thwart our CSR programs and goals?
- How do increases of efficiency or organizational changes impact e.g. employees' health?

Revenue Stream
- How does a change of our revenue stream change our CSR performance? (And in more detail some examples:)
- Is a secured revenue stream based on customers depending on us?
- Will a change in the pricing mechanism be accepted by our customers as "fair"?
- If revenues are coming from new sources, are these compliant with our CSR values and regulations?

CSR related questions that may arise from changes to the Osterwalder and Pigneur business model canvas sectors are very diverse, in many cases also addressing CSR issues that are related to other sectors of the canvas. The example questions shown above give just a very rough impression of which consequences may have to be considered when developing a new or altered business model. It is very clear that the list of questions presented here is a by far not exhausting, and a questionnaire or checklist still needs to be developed. From a practitioner's point of view, sources to compile related catalogues include of course academic literature,

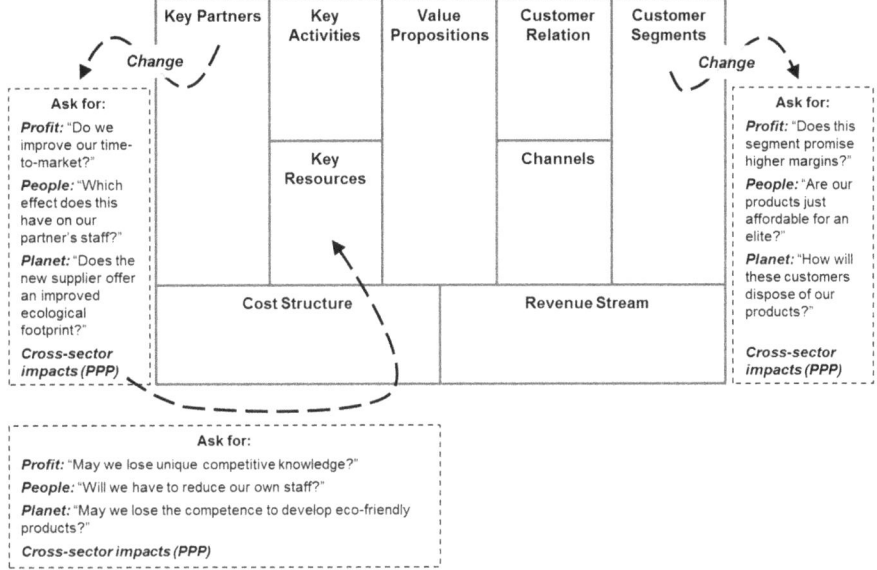

Fig. 1 Simplified illustration of an IBSBMC application

but also questionnaires on CSR already used in practice when assessing suppliers for qualification purposes and in the course of regular evaluation.

Figure 1 shows a schematic and for illustration purposes very much simplified example of how to use the IBSBMC: A substitution of a key partner in the business model may lead to a variety of effects. A set of questions should be answered related to profit, people and planet. Of course it is important to keep in mind that such changes may have direct and indirect effects, the latter including

- effects that materialize somewhere up or down the value chain (e.g. having an impact on the supplier of our supplier), and
- effects in other sectors of the business model canvas (e.g. change of key partners may have an impact on own key resources).

2.5.3   Balance: Supporting Sustainable Decision-Making

Looking at CSR and innovation the Integrated Balanced Sustainability Business Model Canvas (IBSBMC) can be used to:

- Consider the impact of economically driven changes of the business model on CSR aspects and performance (asking, in a generic way: "If we change this part of the business model, what are the consequences in terms of CSR performance?"—implicit approach)

- Consider how scenarios could look like that improve the CSR performance while also offering economic gains (asking, again generically: "Which positive move in terms of CSR would also improve our business performance?"—explicit approach).

Using this scheme also strongly and inherently supports the claims and requirements stated by Michael Porter and Mark Kramer (2006). The authors identified that treating business and society separately is one of the reasons for low effectiveness of CSR efforts and argue that "the interdependence of a company and society can be analyzed with the same tools used to analyze competitive position and develop strategy"—in our case using the business model canvas integrating "people, planet, and profit".

Decision-making of course may and shall be supported by an if possible quantitative assessment of alternatives (which reflects, by the way, also the philosophy of ISO 9001). The model under discussion therefore proposes to use an "Innovation Sustainability Balance Sheet" (ISBS) to provide as much clarity as possible. It is consciously avoided to suggest that such an analysis will result in undisputable and final quantitative data, because many elements of this balance sheet cannot be precisely quantified. Furthermore, the "values" of different elements may not be easily comparable, since the quantification of these effects is done in so-to-speak "different languages": While immediate business effects will be quantified in monetary units, ecological footprint impacts may be reported in CO_2 equivalents—an approach that already includes some "translation" uncertainties (not every ecological effect may be properly quantified as CO_2 equivalent). Of course CO_2 equivalents may be converted into monetary units by simply using the current market price for CO_2 certificates, but again the appropriateness of this conversion may be disputed. Yet, we are still moving on comparatively safe ground. But what about quantifying the effect of providing new jobs and related education offerings in some developing country? Is this correctly quantified by the "product value" generated by these jobs? Concluding it is clear that the quantification of innovation effects in terms of sustainability entails uncertainties and needs decisions based on management judgement. Nonetheless, the use of an innovation sustainability balance sheet is suggested to—if not scientifically quantify—at least make transparent CSR impacts of innovations and deliver a semi-quantitative basis for discussion and decision making.

For illustration purposes, let us briefly discuss the effects of a change to the business model of a—purely fictitious—company. Let us think about a European company that produces high-end electronic systems which have to comply with very demanding requirements. Even though these systems (physically, electronic boards mounted in racks) are not for military use, they are critical to the customers' operations. A failure of these systems may lead to large business losses and in the end to liabilities and loss of reputation. Let us assume that these boards are currently produced in Europe, employing 60 people.

In our fictitious example, the company is looking for cost savings in the production of these boards. The idea is to outsource the production to a manufacturer in a developing country in the Far East—a country that does not yet possess a large-scale end experienced industry in this sector (meaning: not Japan, Korea, China or the like).

Let us think about some questions that may be relevant in this scenario from a CSR perspective. Talking about production costs: How much cheaper would just the production of a single board be? Considering that most electronic components on a board are manufactured by some specialized supplier, there will be almost no gain in terms of materials savings. As in many cases, cost reductions will mainly result from lower labor costs. Since we talk about small batches of high-end electronic systems that have to comply with demanding functionality and quality requirements, the share of personnel costs to the total production costs will be considerably higher than in mass production. Thus, it seems reasonable to suggest that in terms of "profit" the idea to outsource the production is a good one—so far. What has not yet been considered are a couple of other costs that may not be as obvious at first sight. Of course one-time investments will be necessary to develop the new provider to meet the standards required. Even if all monetary investments e.g. in production lines and manufacturing machines are done by the local company, the outsourcing company will have to invest at least time and effort—and related expenses—to build up the competences of the subcontractor. Supposing that the co-operation works fine, most of these expenses will be one-time costs. Yet, even in the age of global communication, the continuous alignment between customer and contractor will require more efforts, flexibility and cultural understanding than if this is just done between teams working in possibly even the same building.

In the case of outsourcing, inevitably the question about the fate of the own employees arises. In fact, many different scenarios are conceivable, depending on the business situation of the company, the employability of the staff, the possibility to further develop the employees' competencies, the regional environment etc. For illustration purposes, let us depict two extreme scenarios, starting with the unfavorable one: In this case, the company is one of the few employers in an already disadvantaged region. The business performance of the company is critical, that is why cost saving potentials are desperately sought. The life cycle analysis of the company's portfolio reveals that the product in question is more or less the only relevant one; currently there are no products in early stages of the life cycle. The staff's education is focused on, not to say limited to performing the specific tasks needed (operation of production equipment, inspection of boards, cabling of racks, inspection of incoming material etc.). Working for years on the same tasks employees may also not be willing to further develop, learn and accept new challenges. It is clear that in this environment the outsourcing of production parts would result in negative impacts on the "people" perspective, even taking into account the quite good social and educational system established in most European countries.

What would a favorable scenario look like? The production of the boards and racks forms just a small part of the company's business, and the order situation is so good that one of the main organizational challenges is to manage work overload and deliver in time. Even if the education of the staff working on the racks in question is limited, training and further education would allow the transfer to other areas of operations with possibly even higher added value. This would also support employ-ability in case that the company fails despite the currently good performance. Of course, a corporate and personal culture of continuous learning and flexibility would be the foundation to open new professional perspectives. Since some of the employees currently working in the possibly outsourced operational area would be needed to manage the co-operation with the supplier, also intercultural experience would complement the person's competence basket.

From an ecological and health & safety viewpoint, it may be challenging to develop a production system that meets European standards as defined e.g. by the European Union. For instance, complying with "REACH" (Registration, Evaluation, Authorisation and Restriction of Chemicals), the EU regulation on chemicals, is not required throughout the world (simply because European Union laws of course are not binding in other countries). Similarly, the content of the ISO 45001 standard, defining minimum requirements to an occupational health and safety management system, is much disputed, which for some years impeded the alignment and common understanding on global level. Since outsourcing may not be understood as simple transfer of responsibilities, but includes taking responsibility to guide and develop suppliers, the outsourcing customer usually will have to invest—be it time for consulting and guidance, be it travel costs, audit costs etc.

Obviously, several other issues have to be considered when assessing the ficti-tious example discussed above. And of course, also the time perspective would have to be defined: Which period should be considered when evaluating these aspects in terms of a business case (for profit, people and planet)? Since some consequences are one-time, others are developing (or disappearing), and again others are sustainable effects, the overall balance will depend on the time period taken as basis.

Summarizing, it is evident that many questions have to be answered and pre-requisites for the balancing to be defined by the organization itself, based on its objectives, culture, values etc. Based on these definitions the "Innovation Sustain-ability Balance Sheet" helps to make effects of changes and innovations transparent. At least within the organization's system of definitions, specifications and values it becomes possible to compare different scenarios, be it to contrast the current situation with an intended future one, or to evaluate the impacts of different scenarios and thus to decide on which to pursue.

Figure 2 shows a schematic illustration of two scenarios reflecting possible impacts of the outsourcing plan discussed above. The somewhat trivial and ordinary example of the outsourcing of electronic board production is chosen on purpose to show what variety of effects not just on profit (the driver for the outsourcing activities), but also on planet and people may materialize even in simple and widespread business model changes. As discussed above, the current situation and environment of the company will influence the triple bottom line balance. This forms the easier part of the evaluation, which can rely on existing information and should result in quite accurate forecasts. The less quantifiable and less foreseeable balance items usually will emerge in the course of the implementation of the plan, due to possibly sloppy analysis and planning and due to incidents that could not have been reasonably anticipated.

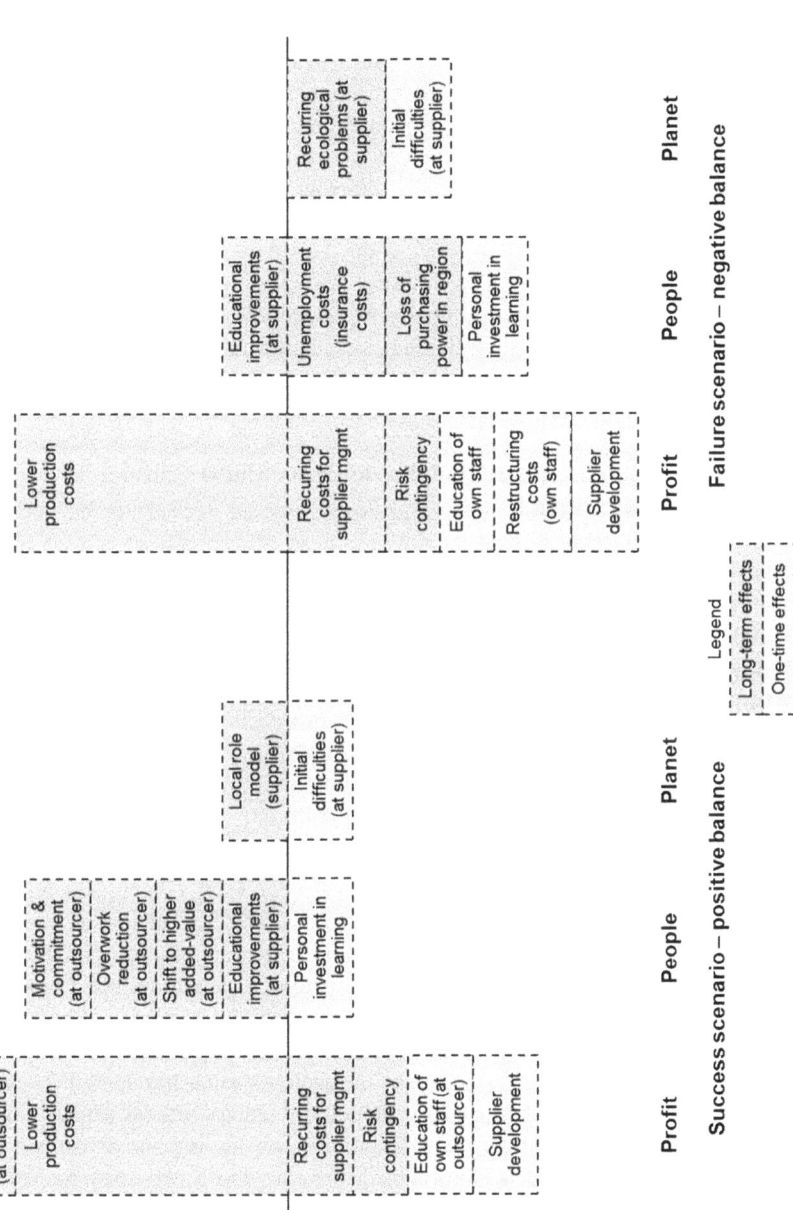

Fig. 2 Simplified illustration of an Innovation Sustainability Balance Sheet (ISBS) analyzing possible impacts of an intended business model change (outsourcing)

The contrasting juxtaposition of two scenarios in Fig. 2 reflects all kinds of effects. Some balance items are strongly influenced by the initial environment and boundary conditions such as the availability of working alternatives at the outsourcer and the willingness of employees to further develop their competences. Some items may not have been foreseen in the initial business case design, such as recurring environmental management problems at the supplier's side. The success scenario also shows some kind of positive chain reaction across the triple bottom line: The outsourcing leads to a reduction of chronic work overload, which makes it possible to further educate the employees and to motivate them to keep learning. This increases the employability as well as the value of the work done by the highly skilled personnel. Commitment to stay at the company increases, which again improves the return on investment of training programs and reduces costs arising from educating new employees. And finally, higher motivation leads to better work results in terms of effectiveness and efficiency, again adding to the company's financial performance.

2.5.4 Integrated Balanced Sustainability Business Model Canvas: Summary and Outlook

In conclusion, the proposed Integrated Balanced Sustainability Business Model Canvas (IBSBMC) offers not just a methodology to analyze existing business models, but particularly to evaluate the impact of possible changes to the business model, and thus the impact of innovations on the triple bottom line. The complementing Innovation Sustainability Balance Sheet is a tool designed to illustrate the consequences of different scenarios providing an at least semi-quantitative overview. By this, it offers the possibility to compare these scenarios in a very transparent and clear way, supporting business model innovation decisions that aim at considering not just profit, but also people and planet aspects.

The model intends to provide a tool at hand to create a win-win-win situation as demanded by authors discussing the business case *for* CSR (Porter and Kramer 2006; Altenburger 2013; Boons et al. 2013; Hockerts 2007; Schaltegger et al. 2012) or even the business case *of* CSR (Schaltegger and Hasenmüller 2006; Schaltegger 2015). Due to its nature of being based on the model of Osterwalder and Pigneur it does not primarily aim to provide a tool for dedicated and explicit sustainability innovations. This is surely better covered by the methodology of Joyce and Paquin (2016), since this allows to start from the respective "environmental life cycle business model canvas" and the "social stakeholder business model canvas". Yet, if the generation of a new economic business model considering and possibly even building on sustainability aspects is in focus of the company's activities, the use of the Integrated Balanced Sustainability Business Model Canvas (IBSBMC) may offer the advantage of a leaner methodology and thus faster adoption in a company (which may already be used to employ the business model canvas of Osterwalder and Pigneur) and of a more integrated approach, understanding the triple bottom line as inherent to every business model. This is also in line with the claim of Porter and

Kramer (2006) that "the interdependence of a company and society can be analyzed with the same tools used to analyze competitive position and develop strategy", in our case the business model canvas as proposed by Osterwalder and Pigneur.

While the Integrated Balanced Sustainability Business Model Canvas (IBSBMC) offers a straightforward approach to consider multi-dimensional effects of business models and possible changes, the likewise proposed Innovation Sustainability Balance Sheet (ISBS) is to be understood as first draft of how CSR impacts can be quantified and depicted. Further development of this model is suggested, and it is clear that no matter how refined this model will be, in the end especially the "currency exchange" policies (e.g. "how much is enhancement of education worth for our company?") will result and be dependent from the organization's values, commitment and—based on this—its decisions.

3 Management System Standards and Innovation

Standards issued by e.g. the International Standards Organization (ISO) or national bodies increasingly gain importance in daily business operations, as the author of the current article can testify from everyday experience and practice exchange with managers in other corporations. At least large companies in European countries make their choice on selecting suppliers more and more also based on extensive assessments of the standard compliant CSR performance or on the question whether a company is certified or not. In some cases, the results of a CSR evaluation may even be decisive, proven compliance to international standards thus becoming an important competitive differentiator.

From a practitioner's point of view it is a bit surprising that the role of international standards and certifications is not much discussed in academic literature. While ISO 26000, focused on "social responsibility" of organizations, and ISO 14001, treating environmental management, is at least mentioned in several publications (e.g. Perera 2008; Schwartz and Tilling 2009; Hahn 2012; Altenburger 2013; Schmiedeknecht and Wieland 2015; Dal-Bianco 2015; Lorentschitsch and Walker 2015; Ebner and Goiser 2015), BS OHSAS 18001 and its successor ISO 45001 (occupational health and safety management) are discussed in only few papers related to CSR (e.g. Jørgensen et al. 2006). Interestingly, information security issues—covered by the respective standard ISO 27001—which may be assigned also to the social dimension of Elkington's triple bottom line, is virtually non-existent in CSR literature. In the next chapters some light shall be shed on the role of standards in CSR strategies and activities.

3.1 ISO 26000

Of course ISO 26000, the "guideline on social responsibility", issued in 2010 by ISO, is a comparatively often reviewed standard in the context of CSR in general (e.g. Altenburger 2013; Schmiedeknecht and Wieland 2015). The advantage of this standard is that it addresses the topic itself, the drawback is that it is not possible to be certified according to ISO 26000. For companies this means that the standard may be used internally as a guidance to set up, maintain and continually improve (and innovate) a CSR management system, but since no certification is intended by ISO, a company may not get an external and independent confirmation that a respective management system is in place and working well. As a means of a formalized, CSR oriented supplier evaluation and, based on this, supplier selection and management, ISO 26000 is thus of limited value. Some national certification bodies have derived certifiable standards, but in a global business environment respective certifications may not be considered and accepted by customers, so that only few companies go for certification. This may be illustrated by the amount of currently issued certificates: The respective Austrian standard is ONR 192500. According to the database of Quality Austria—the leading Austrian certification body and Austrian representative of IQNet, the network of 36 international certification bodies—at the end of 2016 just two companies were certified according to ONR 192500—compared to 9360 companies certified according to ISO 9001 (standard revisions of 2008 and of 2015).

Analyzing the connection between ISO 26000, strategic management, and CSR, Rüdiger Hahn (2012) first of all confirms that only a "few peer-reviewed scholarly articles have been published" on ISO 26000. Looking at the strategic dimension, Hahn concludes that "the process of strategic management is only partly covered by the standard". Overall, Hahn acknowledges that "ISO 26000 is an important step in broadly improving sustainability performance since the standard provides a basis especially for companies that have not dealt with CSSR in depth before", but suggests that the standard may not offer much help for companies that already have achieved an advanced level of CSR management.

In the comprehensive book, "Corporate Social Responsibility" (edited by Schneider and Schmidpeter 2015), Maud H. Schmiedeknecht and Josef Wieland (2015) contribute an entire chapter on ISO 26000. An overview on the development process of this standard, engaging hundreds of contributing experts and observers from 99 countries, leads the authors to the conclusion that the legitimacy of this guideline is based on the "participation of relevant stakeholders, on the balance between the interests of these stakeholders, on the balance of the interests of developing, emerging, and industrial countries, and on the consensus-oriented and democratic process." The authors emphasize that "the core aim of the guideline is to define globally accepted standards of good organizational behavior". Considering the broad approval by quite different stakeholders and most of the ISO member countries, the standard may serve as a basis for a common understanding of CSR, shared all over the world by not just companies, but organizations in general: "SR is not a task exclusive to companies, but the result of the efforts of all actors in society and their interconnections."

3.2 ISO 14001

The roots of ISO 14001, the ISO standard on environmental management systems, lie in the early 1990s. It was the British Standard 7750, published in 1994, that served as basis for the launch of the first ISO 14001 version in 1996 (Campos et al. 2015). In parallel, the European Eco-Management and Audit Scheme System (EMAS) was developed and adopted by the European Union Council (EC) in 1993. Companies in the European Union are invited to participate in this program, which is also open for non-EU organizations. Compliance with both management systems may be approved by certifications from accredited bodies.

As of 2015, by far most ISO 14001 certifications are assigned to organizations in Europe (119,754 or 37.5%) and in East Asia and Pacific (165,616 or 51.9%) (International Standards Organization 2017), while e.g. in North America ISO 14001 certificates are seldom found (8712 or 2.7%). The influence of an ISO 14001 compliant management system on environmental performance is much disputed in literature (e.g. Dahlström et al. 2003; Potoski and Prakash 2005; Arimura et al. 2016). Research results vary; some studies show no dependence between a certified environmental management system and environmental performance, while others do. In fact, at least in countries that are members of the European Union (EU), common as well as country specific legislation covers environmental issues quite extensively. This may also explain why the implementation of an ISO 14001 complying management system does not improve environmental performance dramatically. A partly similar approach to explain divergent findings on the dependency between ISO 14001 certifications and environmental performance is also found in the paper of Arimura et al. (2016) on "resolving equivocal findings".

The revision of the ISO 14001 standard, resulting in the current ISO 14001: 2015 version, requires a more comprehensive management of environmental issues: Particularly the strongly emphasized lifecycle perspective on the environmental impact of products and services extends the area of corporate responsibility. The "cradle to cradle" principle (Braungart and McDonough 2009) is also reflected by the "Manifesto for a resource-efficient Europe" released by the European Commission in 2012 (European Commission 2012).

3.3 BS OHSAS 18001/ISO 45001

The British Standard OHSAS (Occupational Health and Safety Assessment Series) 18001 was developed on the basis of the British Standard BS 8800:1996, following the structures of the ISO 9001 and the ISO 14001. Even though it is not an international (ISO), but a British standard, it has gained de facto validity on a global scale. This also reflects the importance of globally harmonized standards on the CSR aspect of occupational health and safety: Since no official standard has been released by a supranational standardization body such as the International Standards Institute (ISO), companies have adopted the British standard as a blueprint for the design and implementation of an appropriate management system—and ask their suppliers for evidence that they act accordingly down the supply chain.

It is remarkable that literature on CSR hardly deals with occupational health and safety management systems, in spite of the importance that related certifications play in business. The relevance of this standard is also reflected by the challenges encountered in defining the successor to OHSAS 18001, the ISO standard 45001. Since mid-2013, the working group ISO/PC 283 is defining the content of this standard. The first "draft international standard (DIS)" version was presented in February 2016, but rejected. "71% of members voted in favor of the DIS, with 28% against and 1% abstaining. In order for the DIS to be passed, two-thirds had to be in favor with less than a quarter against, taking into account abstentions." (SGS, press release, 3.6.2016). More than 3000 comments were submitted. In past years, planned publication dates were postponed several times. Finally, the standard ISO 45001 was published in March 2018.

3.4 ISO 27001

In times of global availability of data, big data analysis and a still increasing business value of data, information security in general and data privacy in particular are under discussion on different levels. For companies, it is clear that the protection of business critical data is a key concern. Security systems aim at ensuring appropriate confidentiality, integrity and availability of data ("CIA" principle). Cyberattacks which may be initiated by individual criminals, but also by companies or authorities and intelligence services may result in considerable damages. On the other hand, the availability of data e.g. on consumer behavior forms the basis of business models of many, mainly US based, companies. As an example, search engine services are offered free of (financial) charge and "paid" for by provisioning of personal data. The growing business of big data analytics (e.g. Mayer-Schönberger and Cukier 2013) that allows, among others, to predict consumer behavior or characteristics of complex systems in general depends of course on the availability of huge amounts of data. While the artificial intelligence system "Deep Blue" defeated the world chess champion Gary Kasparow in 1999, the growing "competence" of artificial intelligence systems such as IBM's "Watson" led to its victory against human experts on the quiz show "Jeopardy!" in 2011. The rise of machine learning (e.g. Flach 2012; Goodfellow et al. 2016) starts to change not just industries, but societies in general (Brynjolfsson and McAffee 2016; Davenport and Kirby 2016).

The ethical dimension of data usage, especially of personal data is very intensively disputed worldwide, with considerable differences seen on global scale. While

European Union regulations are comparatively strict—with a tendency to become even stricter, as indicated by the General Data Protection Regulation which applies from May 25, 2018 without having to be explicitly translated into local legislation— the handling of data seems to be less restricted and regulated e.g. in the US. This certainly roots in differing cultural traditions, but economic considerations on country levels may also have a certain influence.

Looking at CSR related aspects of data handling, it is clear that treating the data of customers as agreed in contracts or as (possibly just implicitly) expected by clients is a facet of acting in a responsible way. This can of course be generalized from customers to any stakeholders, including own employees, suppliers, research partners etc. Since the economic dimension forms part of the triple bottom line, treating the own company's data in a way that prevents damages is a prerequisite for responsible behavior. Against this background it is no surprise that the importance of the "information security management standard" ISO 27001 is rapidly growing, as is also reflected by the increasing number of certifications worldwide (more than doubled between 2009 and 2015, according to ISO). As of 2015, by far the most certifications are assigned to organizations in Europe (10,446 or 37.9%) and in East Asia and Pacific (11,994 or 43.6%) (ISO 2018), while North American organizations hold 1445 or 5.2% of the globally issued ISO 27001 certificates.

This being said, it is quite surprising that the ISO 27001 standard and data protection and privacy in general are very rarely treated in connection with CSR on academic level. This may lead to the assumption that the relevance of data handling in the context of social responsibility is not yet fully recognized in academia. Considering the rapidly growing importance of data as a key resource for many businesses and its impact on society (just to mention industry 4.0, artificial intelligence systems, substantial changes to the world of work) it is to be expected that CSR research will investigate the realm of data in more detail soon.

3.5   ISO 9001

The standard on quality management systems is so-to-speak the mother of all system management standards. The first version (subdivided into three standards) was published in 1987 and offered three models for quality management systems, depending on the organization's area of activity. Four revisions later, ISO 9001 has come a long way and extended considerably in its scope. Since "quality" is defined as the "degree to which a set of inherent characteristics of an object fulfils requirements" (ISO 9000: 2015), the requirements of ISO 9001: 2015 to a "quality management system" in fact include most of the aspects of managing an organization in general (apart from business administration topics), covering leadership as well as competence management, understanding the organization's context and stakeholder expectations as well as risk and opportunity management, communication topics as well as continual improvement, process management as well as R&D, etc.

Inherently, also ISO 9001 addresses an important aspect of CSR: Mapping the three standards discussed to Elkington's "triple bottom line", OHSAS 18001 covers the "people" aspect, ISO 14001 deals with safeguarding our "planet" and ISO 9001 intends to support organizations in achieving their goals, which in company environments usually include financial goals and "profit".

3.6   Do CSR Related Management System Standards Support Innovation?

On a very generic level, it can be stated that the requirement to continually improve the performance of the company's management system or—more specifically—e.g. the environmental performance of course implies innovation activities. Yet, these activities may result in just incremental changes, and since the requirement addresses the performance of the organization, innovation will mostly be about processes improvements.

However, some more specific requirements aim at the innovation of products and services offered to customers. The lifecycle perspective of ISO 14001: 2015 requires that the environmental impact of a product is analyzed and managed from design to the end of life. Following the spirit of the circular economy approach, end of usage of a product should not result in disposal, but in recycling of as much components as possible. The facilitation of an easy, resource saving recycling of products has already to be considered in product design. Management system standards very much address the question of how things are done, how activities are defined, steered and performed. In other words: The process approach is a key element of management system standards. By requiring that ecological aspects be considered in processes (e.g. via process definitions or checklists), these standards at least highlight challenges in product design and development, which again may spark innovative creativity.

This brings us to the development and production requirements defined in ISO 9001. The subchapters to Chapter 8 of this standard define e.g. that product and service requirements shall be determined in close cooperation with the customer. The standard gives guidelines to ensure that these requirements are designed, developed, produced and delivered according to the customer's needs and expectations (ISO 9001: 2015, Chap. 8). If products and services are ordered by a customer based on detailed technical descriptions, sometimes even defining not just what shall be produced, but even how it shall be produced, there will not be much room for own creativity and innovative approaches. If, on the other hand, customer requirements are identified, analyzed and translated into features and offerings by the company, ISO 9001 may serve as a guideline to develop product and service innovations in a structured, reproducible and sustainable way.

Another important prerequisite for innovation has been introduced in the 2015 revisions of ISO 9001 and ISO 14001: Unlike the previous versions of these standards the new ones explicitly ask for understanding "the organization and its context" (ISO 9001: 2015 and ISO 14001: 2015, Sect. 4.1) and the "needs and expectations of interested parties" (ISO 9001: 2015 and ISO 14001: 2015, Sect. 4.2). Even though strategic management is not addressed as explicit topic, it is clear that the standards' requirements suggest the use of strategy development tools such as a PESTEL analysis, stakeholder analysis, possibly a core competence and a SWOT analysis etc. This again may serve as basis for the definition of a strategic innovation framework. The consideration of stakeholders is crucial to innovation as well as to CSR. The combination of these topics even increases the importance of thorough stakeholder analysis and management (Altenburger 2013)—which again is required and supported by the new revision ISO 9001: 2015.

In summary, we may state that the growing demand of customers towards their suppliers to let their management systems be certified according to the "people, planet, profit" standards OHSAS 18001 (and ISO 45001 as successor), ISO 14001, and ISO 9001 forces companies to innovate at least incrementally—which may be called "improvement". Requirements related to occupational health and safety as well as environmental product and service features are usually in line with e.g. European laws. The standards do not explicitly require substantial or break-through innovation. Whenever substantial innovation is required by other than market and competition forces, this will most probably be triggered simply by legal requirements.

Yet, the new standards' requirements to analyze the organization's context and stakeholder needs and expectations may support a strategic framework for innovation. Considering that successful innovations have to meet market needs, the strong emphasis on identifying and meeting customer expectations as defined in ISO 9001 is of course a prerequisite for successful product and service innovations.

In the era of digitalization and big data it shall again be emphasized that ISO 27001, defining information security requirements, should be considered as important standard addressing CSR topics (such as data privacy). Looking at the battlefield of cyberattacks, defense and counterattacks it is clear that innovation is at the same time exposing more and more system vulnerabilities and remediation.

4 Sustainability and Innovation at Siemens

Siemens is a technology company with core business in the fields of electrification, automation and digitalization, and activities in nearly all countries of the world. The company is incorporated in Germany, with the corporate headquarters situated in Munich. Siemens consists of the divisions Power and Gas, Wind Power and Renewables, Energy Management, Building Technologies, Mobility, Digital Factory, and Process Industries and Drives as well as the separately managed business Healthineers (formerly called Healthcare), which together form the Industrial Business (Siemens Annual Report 2016). With more than 351,000 employees in more than 200 countries worldwide and a revenue of 79,644 million € in fiscal year 2016, Siemens positions itself as "global powerhouse positioned along the electrification value chain—from power generation, transmission and distribution to smart grid solutions and the efficient application of electrical energy—as well as in the areas of medical imaging and laboratory diagnostics". On June 17th, 2016, Siemens and Spanish company Gamesa announced to merge their wind power businesses to create a leading wind power player with the legal domicile and global headquarters in Spain and Siemens holding a 59% stake.

Siemens is acting as a truly global company; just 13% of the revenue is earned in Germany, 39% in European countries other than Germany, CIS, Africa and Middle East, 29% in the Americas, and 19% in Asia and Australia. Accordingly, Siemens assumes responsibility for CSR on a global scale.

An analysis of Siemens' business activities shows that the company has contributed—directly and indirectly—around 250 billion € to the global gross domestic product (GDP), laying the foundation for around 4.3 million jobs worldwide (Siemens 2017).

4.1   Corporate Level

Since its foundation in 1847, Siemens always has followed the conviction of Werner von Siemens that the company shall apply science and engineering for the common good. The first invention already supported bringing the people closer together: It was a substantial improvement of the telegraph that laid the foundation of a more than 170-year success story. A couple of years later, Siemens introduced the first dynamo without permanent magnets and subsequently broadened the portfolio, offering a variety of products in the electric domain, from light bulbs to tramways and trains, power plants and electricity networks. Siemens pioneered in electronics and telecommunication networks and entered many other business areas that bring value to people. Healthcare innovations include the heart pacemaker in 1958, the computer tomograph in 1974, full body MR tomography in 2003, and a new generation of MR-PET systems in 2010.

Siemens' R&D activities aim at developing innovative, sustainable solutions for its customers and simultaneously safeguarding competitiveness. Current focus topics include:

- Economically sustainable energy supplies and innovative solutions solutions for smart grids and for the storage of energy from renewable sources.
- Supporting energy efficiency especially in building technology, industry and transportation, e.g. through highly efficient drives for production facilities or for local and long-distance trains.
- Highly flexible, connected factories using advanced automation and digitalization technologies (context of "Industry 4.0").
- Use of intelligent analytical systems to turn unstructured data into valuable information e.g. for predictive maintenance (context of "big data analysis").

- Making medical imaging technology, in vitro diagnostics and IT for medical engineering an integral part of results-oriented treatment plans.

To promote promising ideas in Siemens' growth business areas of electrification, automation, and digitalization, the company has set up the innovation ecosystem "next 47", which shall identify new trends, invest in promising initiatives and develop future-oriented business together with innovative partners (Siemens Press Release 2016). "next 47" receives a 1 billion € funding over 5 years to foster new ideas (Siemens Sustainability Report 2016). For the fiscal year 2016 (ending on 30.9.2016), Siemens reported R&D expenses on 4.7 billion €, as compared to 4.5 billion € one year before. In relation to the revenue, this results in an R&D intensity of 5.9%. Around 33,000 employees were active in R&D worldwide (Siemens Annual Report 2016). Innovation and R&D activities in all of Siemens' businesses consider sustainability aspects, be it by directly providing products that support protecting the environment (e.g. wind turbines), be it by improving the environmental performance of the products (e.g. improving energy efficiency).

Siemens' reference for societal value creation is the United Nations' 2030 Agenda for Sustainable Development, defining 17 Sustainable Development Goals (SDGs). The "Business to Society" approach identifies issues that are relevant to the development of a country and describes the company's contribution. The 17 SDGs are mapped to Siemens' "Business to Society" impact areas:

- "Strengthening the economy", addressing the SDGs "affordable and clean energy", "decent work and economic growth", and "industry, innovation and infrastructure".
- "Developing local jobs and skills", addressing the SDGs "quality education" and "decent work and economic growth".
- "Driving innovations", addressing the SDG "industry, innovation and infrastructure".
- "Sustaining the environment", addressing the SDGs "clean water and sanitation", "affordable and clean energy", "responsible production and consumption", "climate action", "life below water", "life on land".
- "Improving quality of life", addressing the SDGs "no poverty", "zero hunger", "good health and well-being", "industry, innovation and infrastructure", "sustainable cities and communities".
- "Shaping societal transformation", addressing the SDGs "gender equality", "reduced inequalities", peace, justice and strong institutions".
- The SDG "partnership for the goals" is seen as an overarching activity.

"Business to Society" results include:

- Siemens' global operations contribute to about 250 billion € in GDP creation and more than 4.3 million jobs (12 times more than own employees).
- 40% of the purchasing volume in Germany is attributable to small and medium enterprises.
- 1270 million patients worldwide have access to Siemens imaging systems.
- More than 400,000 UK students were reached by education projects in 2015.

Siemens' "Business to Society" program complements the wide range of the company's sustainability activities. Naturally, environmental protection, including the reduction of CO_2 emissions is addressed by Siemens' sustainability activities. Siemens is the first major industrial company to commit to cutting its CO_2 emissions by half by 2020 and to being carbon neutral by 2030. In this context, Siemens will invest a total of 100 million €. Considering climate change as a major trend, Siemens not just sets goals to reduce CO_2 emissions from own operations, but also supports customers in doing so. In fiscal year 2016, Siemens helped to save more than 521 Mt. CO_2 with products installed in previous years and still in use (Siemens Sustainability Report 2016). Since fiscal year 2015, Siemens discloses sustainability information with reference to the guidelines (G4) of the Global Reporting Initiative (GRI).

The company's aim to combine engineering excellence, innovation, and corporate responsibility is also reflected by the Siemens brand claim "Ingenuity for Life", which was presented in late 2015. It describes Siemens' "unrelenting drive and promise to create value for customers, employees and societies", as explained in the 2016 sustainability report: "'For life' relates to our role in society: to make real what matters. We deliver on this promise by combining our innovation with our knowhow—in the areas of electrification and automation, enhanced by digitaliza-tion—aiming at improving the lives of people today and creating lasting value for future generations."

Siemens efforts and successes in the area of sustainability are clearly acknowl-edged: For the 17th consecutive year, Siemens is a member of the DJSI World Index of RobecoSAM/Dow Jones Sustainability Indices, receiving top scores in seven of nineteen categories. The "Carbon Disclosure Project" (CDP) rates Siemens with "A-", the Financial Times Stock Exchange (FTSE) included Siemens in its FTSE4Good series. EcoVadis rated Siemens with the "Silver recognition level", underlining the performance as a sustainable supplier (Siemens Sustainability Report 2016).

4.2   Siemens Convergence Creators Level

Siemens Convergence Creators is a Siemens subsidiary headquartered in Vienna, Austria. The first predecessor organization was founded in 1961 as part of Siemens AG Austria, providing mainly internal services to other Siemens units. Thus, the company's expertise is based on more than half a century of experience and expertise. Answering the emerging challenges of globalization quite early, the predecessor organization Siemens Program and Systems Engineering (PSE) established R&D centers in Eastern Europe. Subsidiaries in Budapest and Bratislava were founded as early as in 1991, Prague followed 1 year later. In 1995, the first Croatian location was opened in Zagreb, a Romanian site in Braşov followed in 2001. Thus, Siemens PSE contributed to the development of highly-skilled person-nel and a sustainable information technology industry. In 2010, the organization's business mandate and consequently its strategy were substantially redefined, now focusing on own products and solutions, including related services. The success of this transformation program strongly roots in innovation, resulting in products and solutions that meet the most demanding requirements. The company provides its customers with turnkey solutions and services in the fields of communication networks, service and customer management, public safety and security, multimedia infotainment, as well as space technology. At the end of fiscal year 2016, Siemens Convergence Creators had about 850 employees at 16 locations in 11 countries: Austria, China, Croatia, Czech Republic, Germany, India, Romania, Saudi Arabia, Slovakia, United Arab Emirates and USA. The company supplies more than 300 customers in 70 countries with communication and media products and solutions. Among the most important customers are the top players in their respective industrial sectors, i.e. telecommunications, media (TV, publishing houses), transportation (railways, aircraft manufacturers, airlines and airports), space, public safety (action forces) and energy (wind power, oil and gas).

4.2.1 Framework

Being a 100% subsidiary of Siemens AG, all regulations of Siemens (except for very few) also apply to Siemens Convergence Creators, thus setting the frame and ensuring company-wide standards also in terms of social responsibility. Related regulations are defined in so-called "circulars" which also contain clearly defined requirements for checks to be performed on different levels of governance and detail. Within this framework, Siemens' organizational units shall implement appropriate systems to comply with the global rules. The concrete realization may depend on the size of the organizational unit or legal entity, on the business area etc.

4.2.2 Innovation

Innovation is at the core of Siemens Convergence Creators—the further development and new creation of products and services that quickly proof successful on the market was and is the key for the company's transformation from an organization primarily providing internal R&D services to other Siemens entities to a stand-alone company positioning itself as an important market player providing innovative solutions in clearly defined market and technological areas. The successful transformation is based on the combination of innovative strength and a supporting corporate culture, fostering and organizing collaboration on a global scale (August 2017).

The success of the innovation efforts is reflected by the business transformation achieved in the past 6 years, but also by the patent portfolio: The company doubled the number of patent families (filed or granted) as compared to that of the predecessor organization (which was founded in 1961) in just 5 years. Currently, Siemens Convergence Creators holds around 40 patent families.

Siemens Convergence Creators' innovative solutions include SIECAMS ILS, the first working single-satellite system to geo-locate sources of interference in satellite communication, the Smart Video Engine including artificial intelligence to enhance the consumers' experience watching content over the web, communication systems in offshore windfarms to manage operations and assets with the highest level of security, reliability, and effectiveness.

4.2.3   Integrated Management System and CSR

The substantial revision of the organization's business mandate and strategy in 2010 was also accompanied by a profound re-organization and new structural setup. Until then all local companies acted according to locally defined and maintained management systems. To improve effectiveness and to ensure that customers all over the world obtain the same high level quality deliveries, the opportunity of the organization's re-organization was seized also to establish a global Integrated Management System (IMS).

Starting in fall 2010 the complete system was newly set up as "in vivo" operation, since the operational business of course had to be maintained. It soon turned out that designing and implementing an Integrated Management System that balances the different stakeholders' interests and the organization's requirements is like walking the tightrope. "All entities move and nothing remains still", said Heraclitus about two and a half millennia ago: Maintaining the balance on the tightrope was (and still is) only possible by continuously identifying and making the right moves. In a period of just around 8 months the new, globally valid Integrated Management System was set up—at least covering all main processes and those required by the respective standards—and certified in summer 2011 as compliant with ISO 9001 and ISO 14001. At this point in time, all European countries were incorporated into the global system, while the Indian organization was integrated about 1 year later. The system consists of a centrally governed "Handbook"—in fact a web-based, mainly graphical description of processes and procedures—also allowing for local amendments that may be required e.g. due to local laws (August 2014, 2015).

The main enablers for the success of the transformation were the close alignment with strategy and operations and the co-operation with all stakeholders, including the consideration of cultural topics, the balancing of global and local concerns. Undoubtedly, the team spirit within the global quality and process managers' community and the clearly communicated support by the top management were crucial for fast and effective implementation of this project (August 2014, 2015).

Following Siemens regulations, the Integrated Management System of Siemens Convergence Creators fully complies with the standards OHSAS 18001, ISO 14001, and ISO 9001. As discussed above, this results in covering the "people, planet, profit" aspects of Elkington's triple bottom line. Additionally, the requirements of ISO 27001 are fully implemented to meet the needs of data privacy and information security, which Siemens Convergence Creators judges to be an integral part of the CSR management system.

Siemens Convergence Creators decides on certifications depending on the current business need. Nonetheless, all local companies—adhering to the globally valid Integrated Management System regulations—operate in full compliance with the standards mentioned above. The global harmonization of the management system is of course a main lever for efficient and effective coverage of these topics. Thus, the term "integrated" in "Integrated Management System" can be understood in at least two respects: Siemens Convergence Creators' management system integrates at least four management system standards, and it integrates regulations for all country entities of the company.

In line with the company's strategy, five key processes have been defined: The Sales Process, the Innovation Process, the Product Management Process, the Competence Management Process and the main value generating process, the so-called Deliver Process, which defines how product development, solutions, service and maintenance projects shall be managed. The detailed consideration of CSR topics in these processes is discussed below.

4.2.4 Organization

Siemens Convergence Creators integrates quality, CSR and innovation management perspectives to achieve a comprehensive holistic framework and approach. This is also reflected by the organizational setup: The central function unit "Innovation Management & Quality Management" in fact covers not only the named areas of responsibility, but also environmental protection as well as health and safety topics on global level. If we assume that the standards ISO 9001, OHSAS 18001 and ISO 14001 reflect the triple bottom line, the responsibility for CSR on management level in fact is assumed by this central unit. This includes ensuring compliance with standard and legal requirements, guidance, and of course continual improvement. Close collaboration within the network of operational business, other central functions and country representatives is a key to success and to effective as well as efficient operation of the management system (August 2017).

The global head of "Innovation Management & Quality Management" reports directly to the CEO of the global company, the same principle is translated on country level, where the local innovation and quality managers report directly to their respective country CEOs (and to the global head of innovation and quality, of course). The integration of all standards, including all CSR aspects, is also reflected in a joint Management Review as required by all standards.

This organizational setup and the assignment of both innovation and CSR management responsibilities on personal level weld together these topics regarding management system definition as well as in daily operations.

4.2.5 Organizational Integration of CSR and Innovation

A perspective that may provide additional insight into possible dimensions of CSR in general and of CSR related innovations in particular is to approach the topic from outside, i.e. not to analyze current materializations of CSR and to derive a system or model to classify existing phenomena (inside-out perspective), but to take a

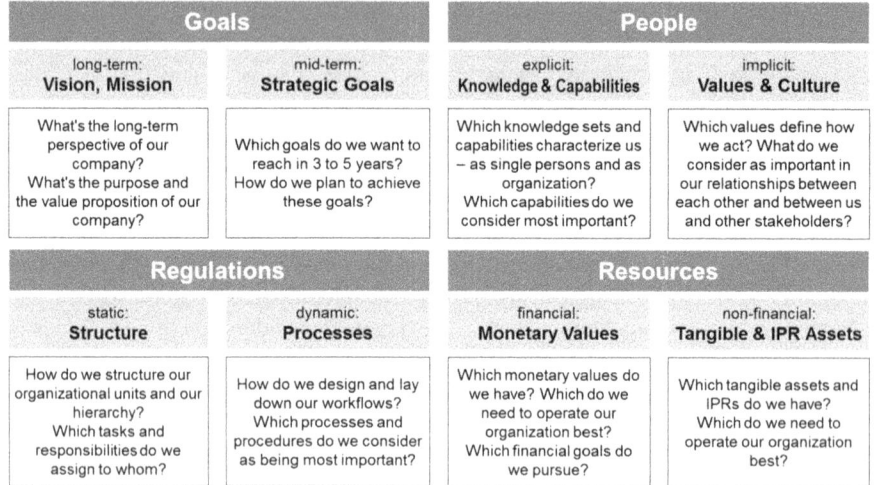

Fig. 3 4 × 2 organizational model (originally for strategy management, Kohlöffel and August 2012)

management model established outside the CSR world and to analyze how this fits CSR aspects (outside-in perspective). The advantage of this approach is that it may help expand the view and perhaps even identify some blind spots. This approach is also (at least partly) used in the models mentioned above, e.g. by taking over classic models on life cycle, innovation type, value drivers etc.

A quite holistic approach to any organizational topic is to use so-called organizational models as basis. These models try to identify the various dimensions of an organization. Well-known concepts include those of Kotter (1978), Peters and Waterman (1982), French and Bell (1994), and Glasl and Lievegoed (2004). Based on a discussion of these models and combining all aspects covered by the single models, Kohlöffel and August (2012) derived a new model. The 4 × 2 sectors cover "goals", "people", "regulations", and "resources". The model can be applied to various management fields in every organization, and is independent of size, industrial sector, profit orientation etc. (Fig. 3).

Even though this model was developed for strategy management, it can be used for any aspect or sub-system of an organization, including innovation as well as CSR management. The use of this concept to perform a comprehensive analysis of an innovation ecosystem is discussed elsewhere (August and Buljubasic 2012; August 2015).

It should be emphasized that this model just depicts internal aspects. It is clear that all of these aspects are interrelated with the external world (Fig. 4):

How are these aspects considered and translated into operations at Siemens Convergence Creators?

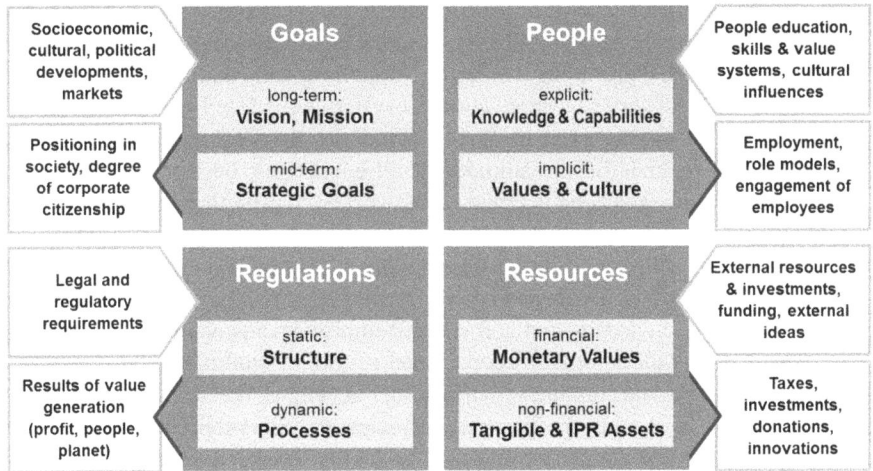

Fig. 4 Schematic illustration of external interrelations in the 4 × 2 model

Goals
- As mentioned above, innovation is seen as decisive for Siemens Convergence Creators' business success. Accordingly, innovation is defined as one of the company's six core values and of course is found in mission, vision and strategy statements. The other core values refer to acting in a responsible and sustainable way: "We care", "ONE" (reflecting the spirit of close co-operation across all boundaries), "trust", "speed", and, of course, "customer orientation". In summary, all six core values relate to innovation and CSR aspects.
- On an operational level, the impact of innovation on business is of course tracked by analyzing respective figures. Innovation as well as ecological, health and safety and information security key performance indicators are monitored and part of the company's Balanced Scorecard. Following the spirit of continual improvement and aiming for excellence, usually on a yearly basis the company's top management defines new CSR goals and programs to be implemented. Since consumption of electricity and travelling have been identified as the two main "environmental aspects" following the ISO 14001 terminology, improvement programs mainly aim at the reduction of electricity consumption and the CO_2 footprint caused by travelling. Health and safety activities reflect the "zero-harm" culture and program pursued by Siemens in general.

People
- It is well acknowledged that employees' satisfaction, motivation and commitment as well as CSR activities influence the performance of organizations (Harter et al. 2002; Korschun et al. 2014). This holds true all the more for Siemens Convergence Creators, being active in an area for which Peter Drucker more than half a

century ago coined the term of "knowledge work" (Drucker 1959, 1999). Innovation would simply be impossible without the creativity of employees and other stakeholders. Thus, providing a creative, supporting and conducive environment is a key to success for every innovation ecosystem (Amabile 1988, 1997; Shalley and Gilson 2004; August et al. 2014). Similarly, the CSR performance of an organization depends on the attitudes of all employees, be it top management defining company goals, be it a blue-collar worker contributing an idea to reduce waste.

- In a high-tech company, offering training and education is an important prerequisite for the ability to create and deliver cutting-edge products and solutions and to contribute to environmental and societal improvements. Continuous learning on personal as well as organizational level results in multiple advantages: The employee's skills and competences are further developed (ensuring employability also in higher ages and offering new challenges that fit competence levels), and the organization maintains and improves its ability to take up new developments and opportunities quickly.

- Correspondingly, the trainings offered at Siemens Convergence Creators cover a vast field: Technological knowledge and engineering capabilities are of course continuously developed and form the basis for the creation and implementation of innovative products and solutions. Additionally, aspects of innovation management are trained by internal instructors who also teach at universities. Thus, comprehensive and up-to-date knowledge on innovation management is spread throughout the organization. This includes basics on creativity and innovation, customer value orientation, strategic innovation methodologies such as Disruptive Innovations, Business Model Canvas, Lean Startup, etc., sources of innovation, propagation of innovation and cultural aspects. The importance of environmental protection is frequently communicated; regular health and safety training is performed regularly to support the "zero-harm" culture.

- Speaking of culture, of course the tone from the top is decisive. It is only logical that top management supports this culture also by showing its commitment to innovation and CSR. This is also reflected by the top management's decision to let Siemens Convergence Creators be certified according to ISO 9001, ISO 14001, OHSAS 18001, and ISO 27001—of which the CSR dimensions have been discussed above.

Regulations
- As mentioned above, innovation managers are in place to support creative employees also on local basis. These managers, best described as "innovation coaches" also cover other tasks and responsibilities, mainly in quality and EHS management, thus ensuring a close coupling of these topics in each Siemens Convergence Creators entity.
- Even though the role of processes is sometimes disputed in literature on creativity and innovation, Siemens Convergence Creators applies processes to manage

ideas, to implement products and solutions and to harvest from innovative business. Since R&D, focusing on development, is our daily business, the innovation management process at Siemens Convergence Creators just covers the period from the submission of an idea to the final decision of the management board to invest in an innovative project. The process is structured following a milestone logic, here called "decision gates". While decision gate 0 reflects the submitter's decision to file a proposal, decision gate 4 results in the budgeting decision. It is a major concern to keep processing times as short as possible, addressing two quite different challenges: In terms of business, quick, yet well-founded and sound decisions are paramount to stay ahead of competition; in terms of innovation culture prompt, transparent and understandable feedback to an idea proves the organization's and management's seriousness and commitment, which again is a prerequisite to keep a high motivational level.

- The idea handling process also lays the foundation for the Siemens Convergence Creators award system for innovations, inventions, and improvements. While many companies have such programs in place, three specifics shall be mentioned:

  First, not only successfully marketed ideas are acknowledged. In the course of the processing of ideas, passing one milestone after another, award points are assigned to the idea for each gate passed. The amount of points increases as more demanding gates are passed, yet an idea (and its submitter) may collect a considerable amount of points even if the idea is rejected in a final decision. This practice shall convey the message that submitting ideas is acknowledged even if the proposal is not chosen to be implemented.

  Secondly, a team approach in developing ideas is fostered by assigning more points to an idea if this is submitted by more than one person. The amount of points assigned is not simply doubled (which could lead to simply inviting colleagues to benefit for free), but depends on the number of team members, following a clearly defined and transparent mathematical formula.

  Thirdly, the assignment of points is fully transparent and does not depend on an explicit management judgement. Naturally, the top management decision whether to invest in an idea or not implicitly influences the amount of point collected. The calculation mode for award points is freely accessible on the innovation pages of the company, each employee can also see the current amount of points as well as the current ranking compared to other colleagues.

- Innovation and CSR aspects are also interwoven on operational level, i.e. in R&D activities as well as in delivering services and solutions to customers. Operational processes follow internal regulations, which are continually monitored, improved and adapted to business needs. The core value generating process at Siemens Convergence Creators is the already mentioned "Deliver" process, steering the development and delivery of products, solutions and services. This process is a good example to illustrate how innovation and international management standards interact in daily operations. The vast majority of Siemens Convergence Creators' business is project-based, delivering customized solutions meeting the specific needs of each single customer request. Thus, innovation is inherent in virtually all activities of the company, ranging from small improvements and the design of project-specific architectures to the development of cutting-edge innovations—always following the Deliver process. This process fully complies with the requirements of ISO 9001: 2015, ISO 14001: 2015, OHSAS 18001: 2007, and ISO 27001: 2013, which again means that all "triple bottom line" perspectives are considered. In everyday life, this is reflected e.g. in project milestone checklists that address technological and project management topics as well as environmental, occupational health and safety, and information security aspects. In fact, these issues are already taken into account in the preceding Sales process, since environmental, health and safety, and information security aspects shall be evaluated in the course of preparing an offer and cleared prior to closing of a contract.

Resources
- As in each and every company, Siemens Convergence Creators investments in innovation follow a business rationale. Budgeting of a product's further development according to a feature roadmap is done in the course of the yearly planning. "Unforeseen" innovation ideas and those not yet assigned to a product of the company's portfolio may be funded based on the innovation process discussed above.
- As already mentioned, Siemens Convergence Creators holds and continuously develops a patent portfolio. Employees are encouraged to submit invention proposals, which will all be processed by a dedicated member of the headquarters' innovation and quality management team. Again, the "coaching" aspect is seen as important, since writing an invention disclosure and even defining the right scope of the invention may be quite a challenge for employees not familiar with patents.
- Funding of CSR relevant activities forms part of the yearly budgeting process. Major improvement steps are planned and steered as projects with dedicated human as well as financial resources. Recent examples include the further development and implementation of a very comprehensive occupational health and safety system in projects and the full consideration of ISO 27001 information security requirements in all activities of the organization.

5   Conclusion and Outlook

In past decades, innovation as well as Corporate Social Responsibility (CSR) have gained more and more importance in business life—meeting the competition driven need to continually offer enhanced or new products and solutions and reflecting the increasing societal and political demand to not just consider the "profit" perspective in Elkington's triple bottom line, but also "people" and "planet". This development is fully acknowledged and supported by Siemens Convergence Creators, fully conforming with the company's values and commitments. As Michael Porter and

Mark Kramer postulated, sustainability of CSR efforts will only be ensured in an integrative win-win-win approach for the three bottom line elements.

Therefore, innovation and CSR are treated from a real life business perspective, based on methodologies that are already well-known in companies. This results in proposing the "Integrated Balanced Sustainability Business Model Canvas" ("IBSBMC") along with the "Innovation Sustainability Balance Sheet" (ISBS) for the evaluation of innovations' impacts on CSR performance. The first methodology is based on the widely-used Business Model Canvas of Osterwalder and Pigneur, thus keeping adoption barriers low. The second tool simply refers to usual balancing methodologies, depicting innovation impacts on CSR in a straightforward way and offering a three-dimensional view on positive and negative effects. It thus helps to design innovations and changes in general in a way that optimizes balanced win-win-win outcomes.

Particularly in Europe and Asia compliance with ISO management standards is a prerequisite for business or at least a competitive advantage gaining more and more importance. Since there is an international management standard in place dealing explicitly with social responsibility (ISO 26000), it is understandable that academic research and publications focus on this standard. From a business perspective, there is one major drawback: Being designed as a guidance rather than a requirements document, ISO 26000 cannot be used for certification purposes. This means that for companies or organizations in general it is not possible to obtain a proof of compliance from an independent, accredited certification body. The good news from a business perspective is that there are other widely acknowledged, certifiable management standards to cover Elkington's triple bottom line: OHSAS 18001 (occupational health and safety, substituted by ISO 45001) for "people", ISO 14001 (environmental management) for "planet", and ISO 9001 (quality management) roughly for "profit". Complementing, ISO 27001 (information security) addresses CSR topics such as proper and responsible handling of data. Any company that complies with these standards implicitly has to also consider CSR topics in all strategic and operational activities—including innovation and product development. The commitment to "continual improvement" as required by all standards also supports at least enhancements of products and solutions.

At Siemens Convergence Creators, the interrelation of innovation and CSR is well established and anchored in the company's values and Integrated Management System (IMS). Innovation is at the core of the company, being active in high-technology business areas, in which continuous innovation is a key to sustainable success. Complying with the management standards named above, the IMS integrates CSR aspects in all relevant processes and regulations, including those describing R&D activities, product and solution development and delivery, sales and innovation. The close coupling is also illustrated on organizational level: In the headquarters as well as on country level the responsibility for innovation management and the overall guidance of the Integrated Management System lies within the same organizational unit.

Closing this article with a brief look at possible future developments, the author again shares the conviction that innovation and CSR will continue gaining importance, also as interrelated forces. From a business perspective, a tighter, more integrated connection with existing strategy management activities and daily operations is desirable. Two approaches supporting this aim have been discussed in more detail in this article: Firstly, the importance of certifiable international management standards such as ISO 9001, ISO 14001, OHSAS 18001 / ISO 45001, and ISO 27001 as not just guidance, but requirement systems related to CSR topics. Extended research looking at the dependencies and correlations between compliance with these standards and actual CSR and innovation performance could help to further integrate these topics also in business environments. Secondly, the use of a well-known strategic methodology such as the Business Model Canvas of Osterwalder and Pigneur as basis for a system integrating CSR aspects into business model innovations leads to the "Integrated Balanced Sustainability Business Model Canvas" ("IBSBMC"), which is so-to-speak ready to use. The complementing methodology, the proposed "Innovation Sustainability Balance Sheet" (ISBS) for the evaluation of innovations' impacts on CSR performance, is surely just at an early stage and needs further development. In particular, the question of how to convert different "CSR currencies" such as financial units, CO_2 equivalents, working conditions, higher education levels etc. opens a field for further research.

References

Achterkamp, M., & Vos, J. (2006). A framework for making sense of sustainable innovation through stakeholder involvement. *Journal of Environmental Technology and Management, 6* (6), 525–538.

Ackermann, R. W., & Bauer, R. A. (1976). *Corporate social responsiveness: The modern dilemna*. Reston: Reston Publishing.

Acs, Z. J., & Phillips, R. J. (2002). Entrepreneurship and philanthropy in American capitalism. *Small Business Economics, 19*(3), 189–204.

Altenburger, R. (2013). Gesellschaftliche Verantwortung als Innovationsquelle. In R. Altenburger (Ed.), *CSR und Innovationsmanagement: Gesellschaftliche Verantwortung als Innovationstreiber und Wettbewerbsvorteil*. Berlin: Springer.

Amabile, T. M. (1988). A model of creativity and innovation in organizations. In B. M. Staw & L. L. Cummings (Eds.), *Research in organizational behavior* (Vol. 10, pp. 123–167). Greenwich: JAI Press.

Amabile, T. M. (1997). Motivating creativity in organizations: On doing what you love and loving what you do. *California Management Review, 40*(1), 39–58.

Arimura, T. H., Darnall, N., Ganguli, R., & Katayama, H. (2016). The effect of ISO 14001 on environmental performance: Resolving equivocal findings. *Journal of Environmental Management, 166*, 556–566.

August, H.-J. (2014). '. . .but rather the one most adaptable to change.' How to set up and maintain an effective quality management system in stormy times of change. In *Proceedings of the 58th EOQ Congress*, Gothenburg, June 11–12, 2014.

August, H.-J. (2015, April). Transformation meistern. *Q1—Magazin für Qualitätsmanagement und Integrierte Managementsysteme, 2*, 30–35.

August, H.-J. (2017). Case study "global knowledge exchange and collaboration as key to business transformation". In K. Jensen (Ed.), *Leading global innovation: Facilitating multicultural collaboration and international market success*. London: Palgrave Macmillan.

August, H.-J., & Buljubasic, T. (2012). Setting up an innovation ecosystem supporting a new organization's strategy. In *Proceedings of the XXIII ISPIM Conference*, Barcelona, June 17–20, 2012.

August, H.-J, Hohl, E. K., & Platzek, B. (2014). Creativity traits and innovation beliefs: Implications for companies and universities. In *Proceedings of the XXV ISPIM Conference,* Dublin, June 8–11, 2014.

August, H.-J., Dinc, O., & Hohl, E. K. (2015). A toolbox to effectively improve an organization's innovation ecosystem. In *Proceedings of the XXVI ISPIM Conference*, Budapest, June 14–17, 2015.

Bocken, N., Short, S., Rana, P., & Evans, S. (2013). A value mapping tool for sustainable business modelling. *Corporate Governance, 13*(5), 482–497.

Bocken, N. M. P., Short, S. W., Rana, P., & Evans, S. (2014). A literature and practice review to develop sustainable business model archetypes. *Journal of Cleaner Production, 65*, 42–56.

Boons, F., Montalvo, C., Quist, J., & Wagner, M. (2013). Sustainable innovation, business models and economic performance: An overview. *Journal of Cleaner Production, 45*, 1–8.

Bowen, H. R. (1953). *Social responsibilities of the businessman.* New York: Harper & Row.

Braungart, M., & McDonough, W. (2009). *Cradle to cradle.* London: Vintage Books.

Brundtland, G. (1987). *Our common future.* UN Brundtland Commission Report.

Brynjolfsson, E., & McAffee, A. (2016). *The second machine age: Work, progress, and prosperity in a time of brilliant technologies.* New York: Norton and Company.

Burke, L., & Logsdon, J. M. (1996). How corporate social responsibility pays off. *Long Range Planning, 29*(4), 495–502.

Campos, L. M., de Melo Heizen, D. A., Verdinelli, M. A., & Miguel, P. A. C. (2015). Environmental performance indicators: A study on ISO 14001 certified companies. *Journal of Cleaner Production, 99*, 286–296.

Carroll, A. B. (1979). A three-dimensional conceptual model of corporate performance. *Academy of Management Review, 4*(4), 497–505.

Carroll, A. B. (2008). A history of corporate social responsibility: Concepts and practices. In A. Crane, A. McWilliams, D. Matten, J. Moon, & D. Siegel (Eds.), *The Oxford handbook of corporate social responsibility* (pp. 19–46). Oxford: Oxford University Press.

Carroll, A. B. (2015). Corporate social responsibility. *Organizational Dynamics, 44*(2), 87–96.

Carroll, A. B., & Shabana, K. M. (2010). The business case for corporate social responsibility: A review of concepts, research and practice. *International Journal of Management Reviews, 12*(1), 85–105.

Chesbrough, H. W. (2003). The logic of open innovation: Managing intellectual property. *California Management Review, 45*(3), 33–58.

Chesbrough, H. W. (2006). *Open innovation: The new imperative for creating and profiting from technology.* Boston: Harvard Business Press.

Chesbrough, H. W. (2010). Business model innovation: Opportunities and barriers. *Long Range Planning, 43*(2-3), 354–363.

Christensen, C. M. (1997). *The innovator's dilemma.* Boston: Harvard Business School Press.

Christensen, C. M., & Overdorf, M. (2000). Meeting the challenge of disruptive change. *Harvard Business Review, 78*(2), 66–77.

Christensen, C. M., & Raynor, M. E. (2003). *The innovator's solution: Using good theory to solve the dilemmas of growth.* Boston: Harvard Business School Press.

Collins, E., Roper, J., & Lawrence, S. (2010). Sustainability practices: Trends in New Zealand businesses. *Business Strategy and the Environment, 19*(8), 479–494.

Dahlström, K., Howes, C., Leinster, P., & Skea, J. (2003). Environmental management systems and company performance: Assessing the case for extending risk-based regulation. *European Environment, 13*(4), 187–203.

Dal-Bianco, E. (2015). Nachhaltigkeit messbar machen—Integration von ISO 26000 in die sustainability balanced scorecard. In A. Schneider & R. Schmidpeter (Eds.), *Corporate social responsibility* (pp. 311–324). Berlin: Springer.

Davenport, T. H., & Kirby, J. (2016). *Only humans need apply: Winners and losers in the age of smart machines*. New York: HarperCollins.

Drucker, P. F. (1959). *The landmarks of tomorrow*. New York: Harper and Row.

Drucker, P. F. (1994). The theory of the business. *Harvard Business Review, 72*(5), 95–104.

Drucker, P. F. (1999). Knowledge worker productivity: The biggest challenge. *California Management Review, 41*(2), 78–94.

Ebner, G., & Goiser, T. (2015). CSR und Risikomanagement. In A. Schneider & R. Schmidpeter (Eds.), *Corporate social responsibility* (pp. 571–580). Berlin: Springer.

Elkington, J. (1997). *Cannibals with forks. The triple bottom line of 21st century*. Oxford: Capstone.

Elkington, J. (1998). Partnerships from cannibals with forks: The triple bottom line of 21st-century business. *Environmental Quality Management, 8*(1), 37–51.

Elkington, J. (2004). Enter the triple bottom line. In H. Henriques & J. Richardson (Eds.), *The triple bottom line: Does it all add up?* (pp. 1–16). Abingdon: Earthscan.

Epstein, M. J., & Roy, M. J. (2003). Making the business case for sustainability. *Journal of Corporate Citizenship, 9*(1), 79–96.

European Commission. (2011). *A renewed EU strategy 2011–14 for Corporate Social Responsibility*.

European Commission. (2012). *Manifesto for a resource-efficient Europe*.

Fichter, K. (2005). *Interpreneurship. Nachhaltigkeitsinnovationen in interaktiven Perspektiven eines vernetzenden Unternehmertums*. Marburg: Metropolis.

Flach, P. (2012). *Machine learning: The art and science of algorithms that make sense of data*. Cambridge: Cambridge University Press.

Frederick, W. C. (1960). The growing concern over business responsibility. *California Management Review, 2*, 54–61.

French, W. L., & Bell, C. H. (1994). *Organisationsentwicklung: sozialwissenschaftliche Strategien zur Organisationsveränderung*. Bern: Haupt.

Friedman, M. (1962). *Capitalism and freedom*. Chicago: The University of Chicago Press.

Gelbmann, U., Rauter, R., Engert, S., & Baumgartner, R. J. (2013). CSR-Innovationen in kleinen und mittleren Unternehmen. In R. Altenburger (Ed.), *CSR und Innovationsmanagement* (pp. 31–54). Berlin: Springer.

Glasl, F., & Lievegoed, B. (2004). *Dynamische Unternehmensentwicklung*. Bern: Haupt.

Goodfellow, I., Bengio, Y., & Courville, A. (2016). *Deep learning (Adaptive computation and machine learning series)*. Boston: The MIT Press.

Grieshuber, E. (2012). CSR als Hebel für ganzheitliche Innovation. In A. Schneider & R. Schmidpeter (Eds.), *Corporate social responsibility* (pp. 371–384). Berlin: Springer.

Hahn, R. (2012). Standardizing social responsibility? New perspectives on guidance documents and management system standards for sustainable development. *IEEE Transactions on Engineering Management, 59*(4), 717–727.

Hahn, T., Figge, F., Pinske, J., & Preuss, L. (2011). Trade-offs in corporate sustainability: You can't have your cake and eat it. *Business Strategy and the Environment, 19*(4), 217–229.

Hamilton, J. T. (1995). Pollution as news media and stock market reactions to the toxic release inventory data. *Journal of Environmental Economics and Management, 28*(1), 98–113.

Hansen, E. G. (2010). *Responsible leadership systems: An empirical analysis of integrating corporate responsibility into leadership systems*. Wiesbaden: Gabler.

Hansen, E. G., & Grosse-Dunker, F. (2013). Sustainability-oriented innovation. In S. O. Idowu, N. Capaldi, L. Zu, & A. Das Gupta (Eds.), *Encyclopedia of corporate social responsibility* (pp. 2407–2417). Berlin: Springer.

Hansen, E. G., Grosse-Dunker, F., & Reichwald, R. (2009). Sustainability innovation cube—A framework to evaluate sustainability-oriented innovations. *International Journal of Innovation Management, 13*(04), 683–713.

Harter, J. K., Schmidt, F. L., & Hayes, T. L. (2002). Business-unit-level relationship between employee satisfaction, employee engagement, and business outcomes: A meta-analysis. *Journal of Applied Psychology, 87*, 268–279.

Heald, M. (1970). *The social responsibilities of business: Company and community 1900–1960.* New Brunswick: Transaction Publishers.

Hockerts, K. (2007). *Managerial perceptions of the business case for corporate social responsibility.* Frederiksberg: CBS Center for Corporate Social Responsibility.

International Standards Organization. (2017). *The ISO survey.* https://www.iso.org/the-iso-survey.html

ISO 27001. (2013). *Information technology ⬜ Security techniques ⬜ Information security management systems ⬜ Requirements.* Geneva: ISO.

ISO 9000. (2015). *Quality management systems—Fundamentals and vocabulary.* Geneva: ISO.

ISO 9001. (2015). *Quality management systems—Requirements.* Geneva: ISO.

ISO 14001. (2015). *Environmental management systems.* Geneva: ISO.

ISO 26000. (2010). *Guidance on social responsibility.* Geneva: ISO.

ISO 45001. (2018). *Occupational health and safety management systems.* Geneva: ISO.

Jones, P., & Upward, A. (2014). Caring for the future: The systemic design of flourishing enterprises. In *The Third Symposium of Relating Systems Thinking and Design (2014, RSD3)* (pp. 1–8).

Jørgensen, T. H., Remmen, A., & Mellado, M. D. (2006). Integrated management systems—Three different levels of integration. *Journal of Cleaner Production, 14*(8), 713–722.

Joyce, A., & Paquin, R. L. (2016). The triple layered business model canvas: A tool to design more sustainable business models. *Journal of Cleaner Production, 135,* 1474–1486.

Kaplan, R. S., & Norton, D. P. (1992). The balanced scorecard: Measures that drive performance. *Harvard Business Review, 1992,* 71–79.

Kaplan, R. S., & Norton, D. P. (1996). *The balanced scorecard: Translating strategy into action.* Boston: Harvard Business Press.

Kim, W. C., & Mauborgne, R. (1999). Creating new market space. *Harvard Business Review, 77* (1), 83–93.

Kim, W. C., & Mauborgne, R. (2005). *Blue Ocean strategy: How to create uncontested market space and make the competition irrelevant.* Boston: Harvard Business School Press.

Kohlöffel, K. M., & August, H.-J. (2012). *Veränderungskonzepte und strategische Transformation: Trends, Krisen, Innovationen als Chancen nutzen.* Erlangen: Wiley.

Korschun, D., Bhattacharya, C. B., & Swain, S. D. (2014). Corporate social responsibility, customer orientation, and the job performance of frontline employees. *Journal of Marketing, 78,* 20–37.

Kotter, J. P. (1978). *Organizational dynamics—Diagnosis and intervention.* London: Prentice Hall.

Kurucz, E. C., Colbert, B. A., & Wheeler, D. (2008). The business case for corporate social responsibility. In A. Crane, A. McWilliams, D. Matten, J. Moon, & D. Siegel (Eds.), *The Oxford handbook of corporate social responsibility* (pp. 83–112). Oxford: Oxford University Press.

Leonard, D., & Rayport, J. F. (1997). Spark innovation through empathic design. *Harvard Business Review, 75,* 102–115.

Littman, J., & Peters, T. (2001). *The art of innovation: Lessons in creativity from IDEO, America's leading design firm.* New York: Currency.

Lorentschitsch, B., & Walker, T. (2015). Vom integrierten zum integrativen CSR-Managementansatz. In A. Schneider & R. Schmidpeter (Eds.), *Corporate social responsibility* (pp. 299–310). Berlin: Springer.

Mayer-Schönberger, V., & Cukier, K. (2013). *Big data: A revolution that will transform how we live, work, and think.* Boston: Houghton Mifflin Harcourt.

McGuire, J. W. (1963). *Business and society.* New York: McGraw-Hill.

Obst, L. (2015). *Utilizing the business model canvas to enable sustainability measurement on the business model level: An indicator framework supplementing the business model canvas* (Master thesis). University of Twente, Twente.

OHSAS 18001. (2007). *Occupational health and safety assessment series.* London: British Standards Institute.

Osterwalder, A. (2004). *The business model ontology—A proposition in a design science approach* (Thesis). Université de Lausanne, Lausanne.

Osterwalder, A., & Pigneur, Y. (2010). *Business model generation: A handbook for visionaries, game changers, and challengers*. New York: Wiley.

Osterwalder, A., & Pigneur, Y. (2011). Aligning profit and purpose through business model innovation. In G. Palazzo & M. Wentland (Eds.), *Responsible management practices for the 21st century* (pp. 61–76). Paris: Pearson International.

Osterwalder, A., Pigneur, Y., & Tucci, C. L. (2005). Clarifying business models: Origins, present, and future of the concept. *Communications of the Association for Information Systems, 16*(1), 1.

Osterwalder, A., Pigneur, Y., Bernarda, G., & Smith, A. (2014). *Value proposition design: How to create products and services customers want*. New York: Wiley.

Perera, O. (2008). *How material is ISO 26000 social responsibility to small and medium-sized enterprises (SMEs)?* Winnipeg: International Institute for Sustainable Development.

Peters, T. J., & Waterman, R. H. (1982). *In search of excellence*. New York: Harper and Row.

Porter, M. E., & Kramer, M. R. (2006). Strategy and society. *Harvard Business Review, 84*(12), 78–92.

Porter, M. E., & Kramer, M. R. (2008). Strategy & society—The link between competitive advantage and corporate social responsibility. In M. E. Porter (Ed.), *On competition* (p. 479). Boston: Harvard Business Press.

Porter, M. E., & van der Linde, C. (1995a). Toward a new conception of the environment-competitiveness relationship. *Journal of Economic Perspectives, 9*(4), 97–118.

Porter, M. E., & van der Linde, C. (1995b). Green and competitive: Ending the stalemate. *Harvard Business Review, 73*(5), 120–133.

Potoski, M., & Prakash, A. (2005). Green clubs and voluntary governance: ISO 14001 and firms' regulatory compliance. *American Journal of Political Science, 49*(2), 235–248.

Rahman, S. (2011). Evaluation of definitions: Ten dimensions of corporate social responsibility. *World Review of Business Research, 1*(1), 166–176.

Revell, A., Stokes, D., & Chen, H. (2010). Small businesses and the environment: Turning over a new leaf? *Business Strategy and the Environment, 19*(5), 273–288.

Ries, E. (2011). *The lean startup: How today's entrepreneurs use continuous innovation to create radically successful businesses*. New York: Crown Business.

Schaltegger, S. (2015). Die Beziehung zwischen CSR und Corporate Sustainability. In A. Schneider & R. Schmidpeter (Eds.), *Corporate social responsibility* (pp. 199–209). Heidelberg: Springer.

Schaltegger, S., & Hasenmüller, P. (2006). Nachhaltiges Wirtschaften aus Sicht des "Business Case of Sustainability". In E. Tiemeyer & K. Wilbers (Eds.), *Berufliche Bildung für nachhaltiges Wirtschaften. Konzepte, Curricula, Methoden, Beispiele* (pp. 71–86). Bielefeld.

Schaltegger, S., Lüdeke-Freund, F., & Hansen, E. G. (2012). Business cases for sustainability: The role of business model innovation for corporate sustainability. *International Journal of Innovation and Sustainable Development, 6*(2), 95–119.

Schmidpeter, R. (2015). Unternehmerische Verantwortung—Hinführung und Überblick. In A. Schneider & R. Schmidpeter (Eds.), *Corporate social responsibility* (pp. 1–21). Berlin: Springer.

Schmiedeknecht, M. H., & Wieland, J. (2015). ISO 26000, 7 Grundsätze, 6 Kernthemen. In A. Schneider & R. Schmidpeter (Eds.), *Corporate social responsibility* (pp. 299–310). Heidelberg: Springer.

Schneider, A. (2015). Reifegradmodell CSR—eine Begriffsklärung und–abgrenzung. In A. Schneider & R. Schmidpeter (Eds.), *Corporate social responsibility* (pp. 22–42). Berlin: Springer.

Schwartz, B., & Tilling, K. (2009). 'ISO-lating'corporate social responsibility in the organizational context: A dissenting interpretation of ISO 26000. *Corporate Social Responsibility and Environmental Management, 16*(5), 289–299.

Schweitzer, J. (2014). Leadership and innovation capability development in strategic alliances. *Leadership and Organization Development Journal, 35*(5), 442–469.

Sethi, S. P. (1975). Dimensions of corporate social performance: An analytic framework. *California Management Review, 17*, 58–64.

Shalley, C. E., & Gilson, L. L. (2004). What leaders need to know: A review of social and contextual factors that can foster or hinder creativity. *The Leadership Quarterly, 15*, 33–53.

Siemens. (2017). *hi!tech, 1*.

Siemens Annual Report. (2016). Accessible via the Siemens website.

Siemens Sustainability Report. (2016). Accessible via the Siemens website.

Siemens Press Release. (2016, June 28). Siemens founds next47 as separate unit for startups

Smith, N. C. (2002). *Corporate social responsibility: Not whether, but how? Centre for marketing working paper*. London: London Business School.

Teece, D. (2010). Business models, business strategy and innovation. *Long Range Planning, 43* (2/3), 172–194.

Upward, A. (2013). *Towards an ontology and canvas for strongly sustainable business models: A systemic design science exploration*. Toronto: York University.

Upward, A., & Jones, P. (2016). An ontology for strongly sustainable business models: Defining an enterprise framework compatible with natural and social science. *Organization and Environment, 29* (1), 97–123.

Von Hippel, E. (2005). *Democratizing innovation*. Cambridge: The MIT Press.

Votaw, D., & Sethi, S. P. (1973). *The corporate dilemma: Traditional values versus contemporary problems*. New Jersey: Prentice-Hall.

Wren, D. A. (2005). *The history of management thought*. Hoboken: Wiley.

Hans-Jürgen August was Vice President Innovation and Quality Management at Siemens Convergence Creators. Leading the headquarters central unit his responsibilities included the further development of the company's globally valid Integrated Management System (IMS) as well as innovation, intellectual property, quality management, environmental, health and safety (EHS) systems and activities throughout the company's 16 locations in 11 countries. Prior to that, he led numerous projects and central units in the areas of strategic management, innovation, change management and business and organizational transformation.

He shares insights and experiences as lecturer at universities as well as at conferences and in publications, bringing together research and business practice. Together with Prof. Klaus Kohlöffel, he authored the book "Veränderungsmanagement und Strategische Transformation".

In May 2018 Hans-Jürgen August moved to TTTech Computertechnik AG, headquartered in Vienna, where he assumed the position as Senior Director Quality, being responsible for the global company's Quality and Integrated Management System.

Integrating CSR in Innovation Value Networks

Karen L. Janssen, Vera Blazevic, and Kristina Lauche

1 Introduction

Innovating in today's complex, globalized, interconnected markets requires collaborating in stakeholder value networks. It is increasingly unlikely that a single organization possesses all the required resources, power and competences to effectively conduct the full innovation process on its own (Sarkis et al. 2010). Organizations have to find new forms of collaboration to create more sustainable outcomes of the innovation process and to establish long-term relations with their partners and other stakeholders. This development is enhanced by pressure from all kinds of stakeholders to integrate corporate social responsibility (CSR) into organizations' innovation efforts. The growing interest in CSR among business firms is evident from the high number of organizations that now participate in evaluative firm rankings that benchmark their CSR performance (Chabowski et al. 2011). A BCG study among executives of globally operating firms found that 70% of the participating firms have placed CSR permanently on their management agenda (BCG 2012). For example, environmentally legitimate firms have been found to incur less unsystematic stock market risks (Bansal and Clelland 2004), investors have also realized the added value of organizational CSR efforts. Likewise, consumers are becoming increasingly aware of the effects of their own choices (Peloza et al. 2013) and stricter legislation on environmental impact and labor rights are other external triggers for increasing CSR efforts.

K. L. Janssen (✉)
Avans University of Applied Science, Breda, The Netherlands
e-mail: kl.janssen@avans.nl

V. Blazevic · K. Lauche
Nijmegen School of Management, Radboud University Nijmegen, Nijmegen, The Netherlands
e-mail: v.blazevic@fm.ru.nl; k.lauche@fm.ru.nl

© Springer International Publishing AG, part of Springer Nature 2018   75
R. Altenburger (ed.), *Innovation Management and Corporate Social Responsibility*,
CSR, Sustainability, Ethics & Governance,
https://doi.org/10.1007/978-3-319-93629-1_4

However, integrating CSR in innovation value networks is a non-trivial task that requires a complex and multidisciplinary development process. Organizing this integration calls for network orchestration. Orchestration is defined by Dhanaraj and Parkhe (2006) as a coherent set of deliberate actions to create value with and extract value from a network, and consists of directing, influencing and coordinating activities. Previous literature on interorganizational networks, e.g. Provan and Kenis (2008) discussed mainly two ways to orchestrate networks: (1) central orchestration, where the network is managed by a hub or lead firm, and (2) decentralized orchestration, where network members commonly manage the activities and tasks of the innovation network. Furthermore, orchestration can be done by participants of the network or by external partners.

This chapter stresses the importance of *orchestrating practices* to manage the interplay between innovation processes, CSR and stakeholder integration. In order to gain insight into the orchestrating practices of integrating CSR in innovation value networks, we first combine insights from different literature streams, namely network theory, stakeholder theory, CSR and innovation management. Based on our experiences from several, longitudinal case studies, we developed a normative framework of the key practices of orchestrating an innovation value network for CSR.

2 The Importance of CSR in Innovation Processes

2.1 *Strategic Ambition Towards Sustainability*

Sustainability gained prominence through the Brundtland commission (WCED 1987), which defined sustainable development as development that meets the needs of the present without compromising the ability of future generations to meet their own needs. With respect to innovation processes this rather general definition of sustainability has been translated as the so-called 'Triple Bottom Line' (Pujari 2006). This concept addresses the challenge to balance economic, social and environmental performance in value creation (Elkington et al. 2006). In order to create sustainable innovation outcomes, e.g. least harmful products (Pujari 2006), companies create CSR policies that explicitly address how the organization addresses economic, social and environmental aspects in the organization.

Organizations usually set strategic goals at corporate level, which are translated to operations. These strategic visions include often an abstract and a long-term perspective on economic, but also environmental and social values. These points on the horizon may conflict with the financially concrete and short-term targets of operational processes. For a sustainable outcome, it is important that the targets are balanced and feasible. For instance, Slawinski and Bansal (2012) explain firms' reactions to sustainability issues along a continuum from reactive to proactive. Reactive responses imply reacting to for instance changes in environmental legislation or pressures from stakeholder groups (Valente 2012). Proactive actions anticipate future regulations and social trends and try to modify the existing products, processes, and operations in advance. For example, consumer groups and Greenpeace have been addressing the environmental and social issues connected with the non-sustainable production of palm oil. In response, firms in the food sector have founded the Roundtable on Sustainable Palm Oil (RSPO) with the objective of promoting the growth and use of sustainable palm oil products through credible global standards and engagement of stakeholders (http://www.rspo.org/). Valente (2012) proposes a further extension of this scale from reactive to proactive by proposing the 'sustaincentric organization' that adopts a more paradigmatic shift in its sustainability orientation, moving beyond a proactive approach. For the alignment of activities in a value network, Valente (2012: 586) explains that, "*what is sustained is a result of a complex interactive and idiosyncratic process where firms and their stakeholders build cognitive complexity within a network system in a way that creates synergistic value creation*". He suggests that, for a sustaincentric ambition, organizations in value networks have to collaborate in such way that it includes all relevant organizations and embraces all related systems (inclusion) and understands all causes and effects of these systems in interrelationships (interconnectedness). Strategic ambitions of a firm to CSR can hence be classified according to their intensity from a mere reactive 'doing no harm' to proactive 'doing good' or even an systematic 'sustaincentric' ambition.

Organizations use their innovation processes to integrate CSR, because during innovation efforts many decisions are made that determine the environmental and societal impact of future products and production processes. It thus creates more leverage for change than later processes within the organization. Hence, innovation can be a well-suited means to trigger CSR activities. Innovation efforts are an ongoing, core process of organizational activities, which ensures longer-term survival by generating future sales (Henard and Szymanski 2001). Hence, companies can use their ongoing innovation efforts to implement CSR initiatives. Second, innovation at its core is concerned with change and creative and new ideas, and hence it is a suitable route to initiate new CSR activities. Third, to successfully manage the innovation process employees of different functions, such as R&D, marketing, production and supply chain management, need to collaborate and coordinate themselves (Song and Song 2010). Hence, innovation is an organizational activity where information and practices disseminate well into different areas of the organization.

2.2 The Innovation Value Network

When integrating the three aspects of CSR—economic, ecological and social—into an organization's innovation process, organizations need to consider a broader set of requirements, which often affects the number of organizations that need to be involved. Previous efforts towards more integrated innovation processes have led to cooperation and interaction between the traditional R&D department with other departments inside the organization, like marketing, purchasing and manufacturing, but also with external actors, like suppliers, competitors and distributors (Nobelius 2004). However, integrating more CSR also requires more attention towards the entire value network and namely to organizations beyond the classical supply chain, such as NGOs and local communities (Bouman et al. 1997). In order to address this growing complexity in innovation processes, organizations have to join forces. For the development of sustainable products and services, where outside resources and external knowledge play an important role, organizations are moving to a more network-centric setting (Nambisan and Sawhney 2011), which we refer to as 'value networks'. Value networks are "a set of relatively autonomous units that can be managed independently but operate together in a framework of common principles and agreements" (Peppard and Rylander 2006: 132). Value networks are often viewed as loosely coupled systems of autonomous firms (e.g. Dhanaraj and Parkhe 2006). Value is determined as the perceived benefits received by the organizations minus the perceived sacrifices that are made during the production of those benefits (Järvensivu and Möller 2009).

If multiple organizations aim to integrate CSR into a joint innovation process, value creation requires truly joining forces in which value is created simultaneously. All stakeholders play a role in influencing the value captured from the development process (Chesbrough and Rosenbloom 2002). According to Vanhaverbeke and Cloodt (2006), firms that have to invest in new assets should be supported and compensated by the other stakeholders in the network, as these also gain future benefits from this investment. To establish a successful collaboration, organizations in a value network have to pursue a fair distribution of investments and return on investments in terms of economic, ecological and social impact for all organizations (Kriron et al. 2013). If organizations succeed in establishing a fair allocation of resources, opportunities, basic needs, and property rights (Valente 2012), it is more likely that the value network will be viable on the long term.

For example, the Life Cycle Inventory (LCI) project of the European industry organization for paint, powder coatings and artist inks (CEPE) succeeded in fairly allocating required resources and promised benefits. This project was a multi-partner collaboration within the realms of CSR. CEPE is the European industry organization focusing on the paints, powder coatings and inks business. They provide a legal platform for their members to deal with industry topics like legislation, sustainability, standardization and lobbying. Together the members of CEPE represent over 85% of their industry. Members are both small and medium-sized enterprises as well as multinationals like AkzoNobel, BASF, PPG and DOW amongst others. Together they form a real contribution and voice towards the EU commissions, EU industry associations and the UN (CEPE Annual Report 2013). The LCI project aimed to create an industrial standard for a raw material life cycle inventory in order to disclose more data about companies' products. In this innovation value network, they managed to orchestrate their efforts through several means. Firstly, sharing of knowledge and (confidential) information between different partners appeared to be transparent and open. Secondly, goals were more unified which means that all the participants saw the benefits and decided to work together on it. Less conflict and opportunistic behavior was perceived than in 'normal' collaboration. Thirdly, in order to address the big challenges that CSR entails, collaboration along the value chain or network was necessary to achieve significant results. CEPE used orchestration practices to successfully manage all stakeholders of the innovation value network.

2.3 Multiple Stakeholders in an Innovation Value Network

For successful integration of CSR in innovation processes, organizations in a value network need to interact in a more synchronized fashion. We use Murnighan and Conlon's (1991) research on internal dynamics and coordination processes in professional string quartets to illustrate this interplay between different stakeholders in a value network. Members of a string quartet are mutually dependent because they use the output of the other members as an input for their own play and vice versa. The string quartet can only perform as a unit; they cannot perform a composition without all members working together simultaneously. Murnighan and Conlon (1991) identified three main paradoxes that we believe also apply to different stakeholders pursuing CSR in innovation value networks.

The first paradox is the "leadership versus democracy" paradox. Members of a string quartet have a one-fourth contribution in the play (and business decisions), but are often bound by the leader's decision. In value networks, legitimate authority may clarify formal power differences, but at the same time, companies are entities on their own. Cooperation and coordination are essential for good performance; the members should concentrate on their own performance, but at the same time on each other's activities.

The second paradox is called "the paradox of the second fiddle". The second violin is important for the group success, as it always has to support the first violin even if (s)he seems wrong, which provides one of the most salient bases for evaluating the quartet as a whole: "they're only as good as their weakest link—but they are rarely recognized" (p. 169). This paradox could be transferred to organizations in the value network that are not compensated for their investments and therefore are underappreciated. Not paying attention to their concerns might jeopardize the viability of a value network.

The third paradox is that of "confrontation versus compromise". While Murnighan and Conlon (1991) expected to find that a collaborative approach would be most successful that focuses on musical rather than interpersonal conflicts, they in fact found that successful string quartets had a preference for avoidance and compromise. Their findings indicate that members of successful string quartets had learned to understand each other's points of view, to accept each other's differences, and to focus on implicitly managing their interdependencies. To recognize but not openly discuss these kinds of dilemmas may be the essential element for group success in such a context.

We expect to find similar paradoxes in attempts to integrate CSR in innovation processes when conflicts arise from diversity or disagreeing on specific matters, such as strategic goals, organizational cultures, innovation processes, etc. Clarifying conflicts may help, but achieving a stage in which there is sufficient mutual understanding that an implicit compromise will suffice may enable smoother collaboration.

2.4 Stakeholder Engagement

As different organizations produce the essential knowledge on which the innovation is developed (Adner and Kapoor 2010; Dougherty and Dunne 2011), value creation results from the efforts of multiple actors in the network. These interactions between multiple (individual) organizations, sub-tasks and resources give rise to dependencies among the organizations involved (Espinosa et al. 2004).

Adner and Kapoor (2010) explored the external challenges that a focal firm needs to consider when being confronted by partners in its value network. They found that the impact of challenges depends on an organization's position in the value network: "depending on their location, challenges in the external ecosystem can either enhance or erode a firm's competitive advantage from technology leadership" (Adner and Kapoor 2010: 326). Network position is referred to as the location of the firm within this complex set of interacting relations in which it is embedded. What activities an organization in a network can enable and constrain are partly determined by this position through reciprocal dependencies. Often, the position of the organization is characterized in terms of its power, i.e., its ability to control and access key resources in the network, and its role and value as a network partner (Anderson et al. 1994).

Organizations in a value network have to determine which CSR challenges need to be addressed by which organizations, e.g. by the focal firm, its upstream suppliers, and its downstream customers. These stakeholder relationships have often been framed in literature in terms of a power-dependence of some sort (Mitchell et al. 1997). For the integration of CSR into innovation value networks, these power dependency relations between stakeholders become mutual: to fulfill the goals, organizations depend on each other. Due to different forms of collaboration and transaction models, firms have different relations with their stakeholders, depending on the kind of network.

Arevalo et al. (2011) analyzed how organizations develop dynamic capabilities for integrating and managing sustainability into their business models. In their GOLDEN concept, they suggest that managers should realize that "stakeholders could both have a potentially inhibiting role as well as a potentially enabling role" and that effective management will increase profits (Arevalo et al. 2011: 944). A value network in which organizations are aware of their position and that of others and assess others' willingness to invest in the network will result in a network with long-term viability.

3   Conceptual Literature Framework

3.1   Stakeholder Engagement to Integrate CSR in Innovation Processes

How much organizations engage with stakeholders also influence the level of CSR integration in innovation processes. Organizations need other organizations to complete their individual and joint tasks (Dougherty and Dunne 2011). They have to be aware of multiple unknown and unpredictable interactions related to their strategic ambition toward CSR, representing the organizations' commitment to economic, ecologic and social aspects. Organizations in a value network will have to make trade-offs between these aspects in view of the future and social relevance (Birkinshaw 2006; Dietz et al. 2003). According to Byggeth and Hochschorner (2006) trade-offs in CSR occur when alternative solutions emphasize different aspects that have to be balanced against each other, such as a conflict between different environmental targets, or between economic and social criteria.

The general notion is that organizations that aim to integrate CSR in their innovation processes do not see the role of their stakeholders as primary (potential) facilitators of this change process. Rather, they see stakeholders as counterparts that make sustainable development possible (Arevalo et al. 2011). Effective coordination is needed to complete both joint and individual tasks (Dougherty and Dunne 2011; Gulati et al. 2012, b). Garud and Gehman (2012: 983) explore the relational perspective on sustainability journeys, and stress that stakeholders are "neither insiders nor outsiders, but instead are part of on-going entanglements". They point out that organizations are entangled in multiple ever-changing networks, which shape their actions. Such enlarged value networks require orchestration efforts between the different organizations and interests to achieve sustainable innovation and create a fair deal for all parties involved. Figure 1 shows an overview of the literature streams we discuss and how orchestration fits in.

3.2   Orchestration

Orchestration is defined by Dhanaraj and Parkhe (2006) as a coherent set of deliberate actions to create value with and extract value from a network. Orchestrations consists of directing, influencing and coordinating activities and helps to maximize the jointly created value. Making agreements can help to share this jointly created value. Orchestrating is seen as a dynamic process between organizations that are entangled in a multiple ever-changing value network that is shaped by their actions (Garud and Gehman 2012; Jarzabkowski et al. 2012). Nambisan and Sawhney (2011) propose that orchestration is about directing the roles, relationships and boundaries of the actors as much as possible in order to increase value creation.

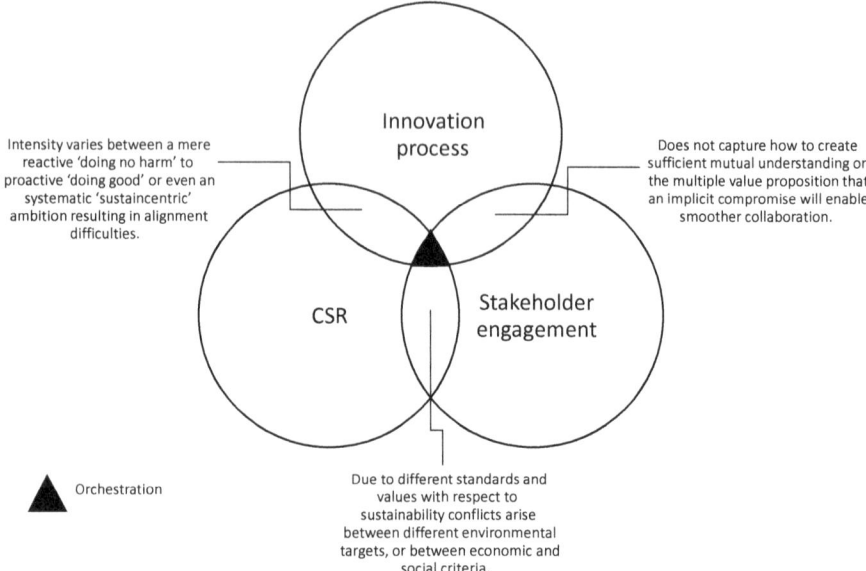

Innovation process

Intensity varies between a mere reactive 'doing no harm' to proactive 'doing good' or even an systematic 'sustaincentric' ambition resulting in alignment difficulties.

Does not capture how to create sufficient mutual understanding on the multiple value proposition that an implicit compromise will enable smoother collaboration.

CSR

Stakeholder engagement

▲ Orchestration

Due to different standards and values with respect to sustainability conflicts arise between different environmental targets, or between economic and social criteria.

Fig. 1 Different approaches to stakeholder engagement to integrate CSR in innovation processes

Networks often do not have a hierarchical authority to ensure this creation and extraction of value. Networks may have hub entities that can play an essential role through individual action to maximize the innovation efforts of their networks. Orchestration can be performed by such a hub entity or even by an organization that is not part of the original network, such as a specialized independent innovation broker. In decentralized networks without a clear hub firm all participants might contribute to orchestration activities. The need for aligning multiple stakeholders' efforts with other partners might change the way organizations manage their innovation journey (Van de Ven et al. 1999).

Based on these considerations, we addressed the following question: *Which orchestrating practices can be identified that allow integrating CSR in innovation value networks?* Based on the findings, we propose a set of best practices that help to integrate CSR when orchestrating such networks.

4 Research Method and Approach for Identifying Best Practices

Our data comes from a longitudinal, qualitative study with several organizations.[1] We used multiple cases to compare how different organizations in different contexts integrated CSR in their innovation value networks. Our case study involved four core companies from different industries. Company A is a leading multinational in the chemical industry. Company B is a multinational food company, operating in more than 200 countries with a well-known major brand as well as many local brands. Company C is a utility company responsible for energy distribution, grid maintenance and installations. Company D is a medium-sized, independent contract research company mainly operating in food research. All four companies engaged heavily in innovation efforts. However, they operated in different sustainability settings. While Companies A and C operated in industrial contexts in which CSR is very important, Companies B and D were in a more ambiguous situation where external forces are not as strict with respect to CSR concerns. The data collection lasted more than 1.5 years and included interviews, company documentation, company website information and regular feedback sessions with company representatives. We performed a thematic content analysis of the collected data through an inductive process (Holsti 1969). We followed the process suggested by Spiggle (1994), moving from categorization to abstraction, comparison, dimensionalization, integration, iteration and refutation.

5 Framework of Best Practices Based on Case Study

5.1 Orchestrating Practices

Organizations that aim to integrate CSR in innovation value networks are confronted with a set of dilemmas arising from the need to demonstrate value for various network-members. This kind of value networks asks for orchestrating efforts between several organizations and stakeholders in order to meet the requirements of sustainability in negotiating a fair deal among all parties involved. The need for aligning external stakeholders' efforts with other partners might change the way organizations manage their innovation journey towards sustainability (Van de Ven et al. 1999). The orchestrating process consists of assembling and developing an inter-organizational network, which should be understood as a set of evolving actions, not as a static structural position (Paquin and Howard-Grenville 2013).

---

[1]The longitudinal case study was a core part of the project "IOP–TOV", sponsored by Rijksdienst voor Ondernemend Nederland (RVO). More information on the project can be found at: http://www.ru.nl/nsm/imr/our-research/hot-spots/innovation-and-entrepreneurship-in-business/selected-projects/projects-0/tools-orchestrating-value-chains-sustainability/

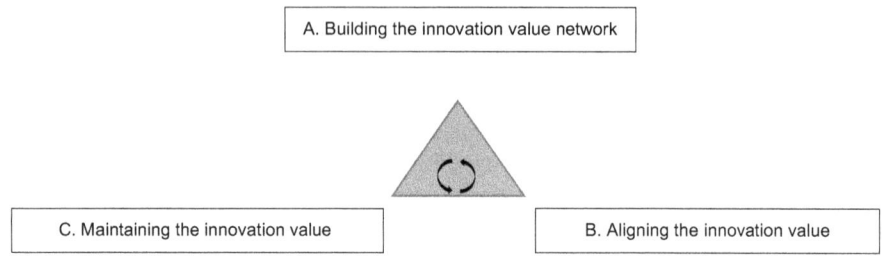

Fig. 2 Framework of core orchestrating practices as identified in case study

Organizations often integrate CSR in a rather evolutionary fashion, rather than through a radical change program. Therefore, the orchestrating practices do not represent a chronologic order. Figure 2 shows our framework of the three main orchestration practices that we identified as important in integrating CSR in innovation value networks.

A. Building the Innovation Value Network

From a more traditional supply chain perspective, organizations often aim to have a large range of suppliers to prevent supply scarcity. For integrating CSR into innovation processes, we believe that the value network depends on the scope and diversity of the search and on the organizations' capacity to move beyond this traditional way of thinking in the industry. Similar to partner selection for discontinuous innovation projects (Birkinshaw 2006), partners required for a value network aiming at CSR are most likely to be found outside the former 'tried and tested' relationships of the organization. For integrating CSR into innovation processes, a broader approach to selecting actors is needed. The integration of CSR into innovation processes implies that such decisions on the selection of organizations will also be shaped by their ecological and social impact. This implies that decisions on new partners will also be shaped by their ecological and social impact. We believe that this will result in a stronger mutual dependency between potential partners. Setting up a value network and selecting partners will be an activity to be handled with caution. Previously unconnected organizations need to be brought together. Other interests that were not on the organizations' priority list challenge the organization to consider a diverse set of stakeholders. The following steps help to build an innovation value network.

Questions

A1. Who are the actors that are needed to achieve the objective?
A2. What types of actors need to be involved?
A3. Who are the actors that should be avoided?

A1 Who Are the Actors That Are Needed to Achieve the Objective? The viability of the value network depends on the scope and diversity of the exploration and on the organizations' capacity to move beyond its traditional way of thinking. Besides the direct suppliers and customers, the stakeholders for the sustainable initiative are most likely to be found outside the former 'tried and tested' relationships of the organization. Other interests that were not on the organizations' priority list challenge the organization to consider a diverse set of stakeholders. This interplay causes a stronger mutual dependency between potential partners.

Tips and Tricks

- Invest time to search for the actors you need
- Think diverse
- Think of:

  - Customers or launching customers
  - Suppliers
  - Critical suppliers
  - Investors
  - Technology providers
  - Waste stream partners
  - NGO's, reducing unexpected reputational risks
  - National associations
  - Communities

A2 What Types of Actors Need to Be Involved? Organizations should be aware that decisions on the selection of organizations should also be shaped by the ecological and social impact of the suppliers. Previously unconnected organizations need to be brought together to get the specific sustainable expertise on board. Moreover, in the period of establishing legitimacy of the network's activities to a wide audience, also potential members who can create ties should be attracted.

Tips and Tricks

- Invest time to search for the actors you need
- Think diverse
- Think of:

  - Innovators
  - Best of class
  - Experts

(continued)

- Consider experience or shared history
- Organizations that contribute to reputational benefits
- Networking by making new contacts yourself or making use of existing personal relations
- Using the networks of others to seek new contacts

A3 Who Are the Actors That Should Be Avoided? To reach a composition of relevant and valuable organizations, some organizations may be invited to the network, while for others access may be denied. For example, companies are confronted with discussions on sustainability in the popular press and at governmental institutions, e.g. from the EU or the UN. Organizations might, in addition to legal standards, feel pressured to abide by voluntary codes, such as the Greenhouse Gas Protocol and the Global Reporting Initiative (GRI) and choose to avoid to once that do not comply with these standards to make sure they reach their sustainable performance goals.

Tips and Tricks

- Consider the reputational risks
- Consider lack of knowledge
- Determine their willingness or possibilities to comply to standards
- Identify who wants Intellectual Property Rights
- Consider the longer term strategy with respect to competition

B. Aligning the Innovation Value Network

Value networks consist of several different organizations that share uniting intentions that bring them together but may have conflicting goals and different approaches of value creation and extraction of the value network. The efforts of the organizations have to be aligned to deliver the desired outcomes (Gulati et al. 2012, b; O'Mahony and Ferraro 2007). Sometimes, organizations fail to focus on the common ground of integrating CSR in innovation processes, as they tend to focus on their own priorities instead of the overall network value. Innovation processes are usually exploratory in nature and outcomes are difficult to determine in advance. Value networks that aim to integrate CSR in innovation processes will have to repeatedly reorient themselves regarding their goals during the course of the collaboration. Reorienting regarding their goals might include adding or removing actors, resources or activities and changing the way the network operates in its environment (Busquets 2010; Järvensivu and Möller 2009). Activities include framing (establishing and influencing the operating rules of the network and altering the perceptions of the network participants), activating (identifying participants for and structuring the network), mobilizing (building commitment among actors) and synthesizing (creating conditions for productive interaction while preventing, minimizing, and removing obstacles to cooperation) (Järvensivu and Möller 2009).

While traditional alliances may be formed between two complementary partners, we argue that for the integration of CSR into innovation processes, different interests of various stakeholders should be taken into account. Because of this broader range of goals, more reorientation and reformulations of goals will be involved than in other types of alliances. The following steps help to align an innovation value network.

Questions

B1. What is the common objective?

B2. What are the stakes per partner and how do they affect the value creation (per partner to whom)?

B3. What are the tasks that need to be allocated to achieve the objective and its deliverables?

B1 What Is the Common Objective?   Value networks consist of several different organizations that share uniting intentions that bring them together but may have conflicting goals and different approaches of value creation and extraction of the value network. Sometimes, organizations fail to focus on the common ground of integrating sustainability in innovation processes, as they tend to focus on their own priorities instead of the overall network value (Doz et al. 2000; Human and Provan 2000). These efforts have to be aligned to deliver the desired outcomes. This even can create synergies: collaboration implies an opportunity for knowledge exchange in which the organizations themselves benefit, while at the same time other goals of the network are realized (Faraj and Xiao 2006; Koschmann et al. 2012). Communication strives at aligning the actions of two or more organizations towards a common goal by continually explaining and discussing. This can be done by "encouraging meaningful participation, managing centripetal and centrifugal forces, and creating distinct and stable identities." (Koschmann et al. 2012: 347).

Tensions can occur between an organization's intention to fulfill the network goals and the need to create value for its own organization. An organization might try to influence other organizations outcomes or promote certain issues to reach their own goals. In the most favorable case, synergies might be created in which the collaboration implies an opportunity for knowledge exchange in which more organizations themselves benefit, while the overall goal is leading.

Tips and Tricks

- Identify a clear common goal at start
- Do not hide information, be open and transparent
- Exhibiting a diversity of goals
- Explicate business of each other (in direct contact)
- Puzzling and balancing to meet a diverse set of goals
- Continually explaining and discussing

B2 What Are the Stakes Per Partner and How Do They Affect the Value Creation (Per Partner to Whom)? Collaboration start with an exploratory nature (March, 1991) and outcomes are difficult to determine in advance. Organizations have to repeatedly reorient themselves regarding the individual and common goals during the course of the collaboration, in relation to CSR. Reorienting regarding their goals might include adding or removing actors, resources or activities and changing the way the network operates in its environment (Busquets 2010; Järvensivu and Möller 2009). Activities include framing (establishing and influencing the operating rules of the network and altering the perceptions of the network participants), activating (identifying participants for and structuring the network), mobilizing (building commitment among actors) and synthesizing (creating conditions for productive interaction while preventing, minimizing, and removing obstacles to cooperation) (Järvensivu and Möller 2009). Paying attention to the expectations and deliverables of each organization, in an open and transparent setting contribute to the process of value creation.

Because of the mutual dependencies and non-equal contributions towards the sustainable outcome of the innovation process, we argue that for the integration of CSR into the innovation process, organizations have to know each other's responsibilities and share in value creation. Sharing knowledge contributes to people's understanding of certain tasks and helps them to anticipate on what is going to happen and who will undertake what activities (Espinosa et al. 2004). Coordination may be defined as the process of managing dependencies between activities (Malone and Crowston 1994). Based on task dependencies and expectations of the value network, organizations can judge their relevant resources, for instance efforts, time, information, strengths or talents (Wittenbaum et al. 1996) and estimate how and what they can contribute.

Tips and Tricks

- Aim to create synergies
- Sharing the ambitions reduces complexity

(continued)

- Balance between different interests, consider trade-off's, create an overall optimum of all stakeholders involved
- Think diverse: in actions, in kind, interest, monetary, etc.
- Create a mechanisms for balancing allocation of efforts and benefits
- Create common ground and clear cut agreement between actors involved in the value network
- Define scope until gate or grave
- Envision future
- Redefine scope of the project in order to include sustainability
- Let go of individualism

B3 Which Tasks Need to Be Allocated to Achieve the Objective and Its Deliverables? For the integration of CSR into innovation processes, it is important that all organizations involved contribute and deliver on their tasks as agreed. Relationships, agreements and processes have to be shaped (Järvensivu and Möller 2009). The value of the network and the individual organizations has to be made transparent for all involved (Capaldo 2007). When networks are established for novel activities, the value for potential partners is uncertain. In such circumstances, the value should be framed for a diverse audience and over time, and involved (potential) partners have to learn from each other (Paquin and Howard-Grenville 2013). Making expectations clear and establishing a basis for trust is essential (Birkinshaw 2006; Järvensivu and Möller 2009). Because of the mutual dependencies and non-equal contributions towards the sustainable outcome of the value network, organizations have to know each other's responsibilities and share in value creation.

Organizations in a value network may vary in terms of structural characteristics, such as centrality or inequality of power (Adner and Kapoor 2010). Power is seen as a means to maneuver between flexibility and loose alignment on the one hand and effectiveness, assurance and balance on the other hand (Busquets 2010). There can also be different dependencies in a network resulting from task decomposition or allocation of resources. Dependencies might constrain the efficiency of task performance, and may be inherent in the structure of the network (Malone and Crowston 1994).

Tips and Tricks

- Create an overview per partner
- Consider if the tasks can be integrated in the mainstream processes of the network partner? If not, check if the work still can be done, and how.
- Consider the approach of events, proactive versus pragmatic

(continued)

- Balance between flexibility and dependencies
- Go beyond the use of formal legitimacy and establishing legitimacy also for issues of future generations

C. Maintaining the Innovation Process

The increased dependency in value networks focused on CSR gives rise to new challenges, like commitment, coordinating with complementary innovators and managing the focal project. Coordination is defined as the effective alignment and adaptation to the actions of organizations (Gulati et al. 2012, b). Network management could contribute to the fulfillment of more complex tasks with substantial dependencies (Espinosa et al. 2004). Coordination and network management are not only performed when assembling a network; they remain relevant through the life of a network (Doz et al. 2000). Orchestrating is an activity that needs to be undertaken repeatedly and that will generate new challenges and dilemmas (Paquin and Howard-Grenville 2013). Time, energy and other resources have to be invested for the development of fruitful relationships. While the network develops, organizations need to be attracted and managed (Birkinshaw 2006; Capaldo 2007; Human and Provan 2000). Moorman et al. (1992: 316) define commitment as "an enduring desire to maintain a valued relationship." Morgan and Hunt (1994: 23) explore the commitment-trust theory of relationship marketing and define a common theme for commitment in relationships: "parties identify commitment among exchange partners as key to achieving valuable outcomes for themselves, and they endeavor to develop and maintain this precious attribute in their relationships". They emphasize that the presence of relationship commitment and trust is central to successful partnerships and that commitment only exists when the relationship is considered important.

Questions

C1. Who will achieve leadership position within the proposed time frame (market leadership) and can this be spread over actors and time?

C2. Who can be made responsible (owner) to guarantee future progress (deliverables) and take care of (new) obstacles?

C3. What are possible contractual agreements that allow for dynamic adjustments of membership and roles?

C1 Who Will Best Placed to Adopt a Leadership Position Within the Proposed Time Frame (Market Leadership) and Can This Be Spread Over Actors and Time? While the network develops, organizations need to be attracted and managed. The presence of relationship commitment and trust is central to successful partnerships and that commitment only exists when the relationship is considered important. Organizations have to highlight the importance of (personal) commitment and invest time, energy and other resources for the development of fruitful relationships. The essence of the collaboration can be spread over the different actors and time to ensure commitment. In this context, it appears important to understand how stakeholders can build on past experiences and future aspirations as they face important challenges in coordination (Bartel and Garud 2009). According to Birkinshaw (2006: 8), the development a value network is "enabled by your past experiences with relationship-building, the strength of your position within your industry, and an open attitude towards knowledge sharing". Therefore, commitment in a value network is a dynamic agreement about the direction, behavior and the way how conflicts are dealt with during a period of change (Busquets 2010). These agreements can be stimulated by determining sanctioning or reward systems; organizations can try to influence the behavior of other organizations, either by forcing them or by stimulating them to pursue goals that they would not have aspired to themselves. Choosing sanctions or rewards should be done carefully (Avadikyan et al. 2001; Tenbrunsel and Messick 1999). Possible forms are contractual agreements that specify incentive and reward schemes such as shared profit distribution agreements or intellectual property rights (Grandori and Soda 1995), or sanctioning schemes (Tenbrunsel and Messick 1999). Despite the fact that many organizations use some kind of sanctioning system, the impact is subject of debate (Tenbrunsel and Messick 1999). On the one hand, the ethics of a value network are disturbed; the likely result is that organizations will comply only to avoid sanctions rather than acting out of their own interests. On the other hand, establishing sanctions can stimulate organizations to refrain from inappropriate behavior. A more fruitful approach may be to use reward systems that can stimulate organizations to get the most out of the collaboration (Dietz et al. 2003).

Tips and Tricks

- Make them investors
- Consider revolving funds
- Consider rewarding systems, for instance shared profit distribution agreements or intellectual property rights
- Consider sanctioning systems
- Plan and give regular updates on the status
- Make a small piece part of a bigger picture

C2 Which Partner Is Best Placed to Take on the Responsibility to Guarantee Future Progress (Deliverables) and to Deal With Unexpected Events and Obstacles? Because collaboration in a value network for sustainability is a dynamic process, the ability to pass on the role of leadership should be considered. Organizations in value network have the possibilities to appoint one of the involved stakeholders as leading party or make use of a third independent agency that takes the lead. Taking the lead for a stakeholder might be difficult to combine with the responsibilities that the organization have to bear, as well as for the own organizational stakes of the individual deliverable. For a successful outcome, the right stakeholder should be responsible at the right time. The value network should consider the possibility to shift ownership with time and objective during the innovation process.

Tips and Tricks

- Define the leading partner, based on stakes involved and recognize the shifting potential or specific situation for network actors
- Incentive a new leading party at the right moment
- Appoint neutral authority or representative
- Consider leadership style, like fairly allocating, persistently persuading
- Consider personal characteristics, like friendly, openly sharing

C3 What Are Possible Contractual Agreements That Allow for Dynamic Adjustments of Membership and Roles? Inter organizational collaboration are characterized by the absence of formal authority. Formal relationship management refers to specifying rules and agreements in a contract. According to Gulati et al. (2012, b) the absence of such formal authority in a value network is because organizations often do not work on the basis of a "employee-employer relationship". They suggest two substitutes for this lack of formal authority: (1) informal authority based on expertise, reputation, status, gate keeping privileges and (2) bargaining power arising from asymmetric dependence. Differences in power and value may lead to conflicts between organizations in a value network. Conflicts may have different effects on value creation; they can lead to disruption but also to lessons and changes (Dietz et al. 2003).

Integrating sustainability into innovation processes is a relative new challenge. Organizations are often confronted with new regulations or are restricted by current regulations, and might have to explore what is legally possible or required in the future. For monitoring the compliance regarding sustainability, organizations in the value network have to determine parameters of control. Once parameters for control have been determined it becomes possible to reprimand organizations that do not adhere to or even violate these rules. Compliance can be monitored through social or formal control. Social control is based on group norms and values, reputation and peer reviews (Jones et al. 1997; Ouchi 1979), while formal control systems often include a planning and control moments of (fixed) release agreements (Grandori and Soda 1995).

Tips and Tricks

- Determine how decisions are made, consensus wise versus based on authority
- Determine parameters of control
- Consider if actors can step in or step out
- Consider if there is a creative way for budgeting or value capturing

6 Conclusion

Pursuing CSR in an innovation value network presents new challenges for organizations and gives rise to a set of different practices. This chapter elaborated on the interplay between CSR, innovation management, and stakeholder relationships, which we argued are important for integrating CSR in innovation value networks. Negotiating a 'fair deal' is important to manage the integration of CSR in innovation value networks. Orchestrating practices contribute to building, aligning and maintaining these networks. While the dependencies within the network present important conditions for such negotiations, orchestrating efforts can dynamically affect in what way these conditions become relevant as joint practices evolve between partners. The enactment of these orchestrating practices in turn helps to create a fair deal in the network and ensures the long-term viability of the network.

Our framework helps organizations in their efforts towards developing sustainable innovations and negotiating the involvement of other organizations. Different innovation value networks use different orchestration practices to organize their joint innovation efforts.

Our findings show that building, aligning and maintaining these networks comprise the central practices that participants need to fulfill. For example, in specific situations, in which dependencies are the source of potential commitments, the need to closely work together is a necessity and organizations then make less use of formal contracts. Based on these dependencies, organizations are connected to each other, and they have to deal with one another in order to integrate sustainability in innovation processes. Our findings show the importance of how actors draw on their personal experience and implicit understanding of responsible project management in their attempts to renegotiate the scope of a project and gain commitment from other organizations. Furthermore, we found that connectedness is important; respondents indicated that frequent meetings resulted in closer relationships and collaboration with organizations, and lead to a more trustful and familiar value network. If the scope of the project is adjusted, it will affect all organizations involved in the network. For a fair distribution of investments and return on investments, it is important to know how to approach other organizations. How to best approach organizations in the value network is influenced by former or other partnerships between the same organizations, but also by the reputation and opinions of other organizations in the value network.

Different dependencies exist between organizations that aim to integrate CSR into their innovation processes in a value network. These dependencies result in a specific position of the firm towards other stakeholders. Furthermore, organization's strategic ambition towards sustainability is determined by what economic, ecological, and social impacts are taken into account. The enactment of appropriate orchestrating practices in turn helped to create a fair deal in the network and ensured the long-term viability of the network. Hence, a fair deal is important to manage the integration of CSR in innovation processes.

References

Adner, R., & Kapoor, R. (2010). Value creation in innovation ecosystems: How the structure of technological interdependence affects firm performance in new technology generations. *Strategic Management Journal, 31*(3), 306–333.

Anderson, J., Håkansson, H., & Johanson, J. (1994). Dyadic business relationships within a business network context. *Journal of Marketing, 58*(4), 1–15.

Arevalo, J. A., Castelló, I., de Colle, S., Lenssen, G., Neumann, K., & Zollo, M. (2011). Introduction to the special issue: Integrating sustainability in business models. *Journal of Management Development, 30*(10), 941–954.

Avadikyan, A., Llerena, P., Matt, M., Rozan, A., & Wolff, S. (2001). Organisational rules, codification and knowledge creation in inter-organisation cooperative agreements. *Research Policy, 30*(9), 1443–1458.

Bansal, P., & Clelland, I. (2004). Talking trash: Legitimacy, impression management, and unsystematic risk in the context of the natural environment. *Academy of Management Journal, 47*(1), 93–103.

Bartel, C. A., & Garud, R. (2009). The role of narratives in sustaining organizational innovation. *Organization Science, 20*(1), 107–117.

Birkinshaw, J. (2006). *Finding, forming, and performing: Creating networks for discontinuous innovation* (Doctoral dissertation). Institute of Management Research Tanaka Business School, Imperial College London.

Boston Consulting Group (BCG). (2012). Sustainability nears a tipping point. *MIT Sloan Management Review Research Report*.

Bouman, M., James, P., & Wolters, T. (1997). Stepping-stones for integrated chain management in the firm. *Business Strategy and the Environment, 6*(3), 121–132.

Busquets, J. (2010). Orchestrating smart business network dynamics for innovation. *European Journal of Information Systems, 19*(4), 481–493.

Byggeth, S., & Hochschorner, E. (2006). Handling trade-offs in Ecodesign tools for sustainable product development and procurement. *Journal of Cleaner Production, 14*(15–16), 1420–1430.

Capaldo, A. (2007). Network structure and innovation: The leveraging of a dual network as a distinctive relational capability. *Strategic Management Journal, 28*(6), 585–608.

CEPE Annual Report. (2013). Retrieved June 2018, from http://www.cepe.org/wpcontent/uploads/2018/01/CEPE_annual_report_2013.pdf

Chabowski, B., Mean, J., & Gonzalez-Padron, T. (2011). The structure of sustainability research in marketing, 1958-2008: A basis for future research opportunities. *Journal of the Academy of Marketing Science, 39*(1), 55–70.

Chesbrough, H., & Rosenbloom, R. S. (2002). The role of the business model in capturing value from innovation: Evidence from Xerox Corporation's technology spin-off companies. *Industrial and Corporate Change, 11*(3), 529–555.

Dhanaraj, C., & Parkhe, A. (2006). Orchestrating innovation networks. *Academy of Management Review, 31*(3), 659–669.

Dietz, T., Ostrom, E., & Stern, P. C. (2003). The struggle to govern the commons. *Science, 302* (5652), 1907–1912.

Dougherty, D., & Dunne, D. D. (2011). Organizing ecologies of complex innovation. *Organization Science, 22*(5), 1214–1223.

Doz, Y. L., Olk, P. M., & Ring, P. S. (2000). Formation processes of R&D consortia: Which path to take? Where does it lead? *Strategic Management Journal, 21*(3), 239–266.

Elkington, J., Emerson, J., & Beloe, S. (2006). The value palette: A tool for full spectrum strategy. *California Management Review, 48*(2), 6–30.

Espinosa, A., Lerch, J., & Krout, R. (2004). Explicit vs. implicit coordination mechanism and task dependencies: One size does not fit all. In E. Salas, S. M. Fiore, & J. A. Cannon-Bowers (Eds.), *Team cognition: Process and performance at the inter- and intra- individual level* (pp. 107–129). Washington: American Psychological Association.

Faraj, S., & Xiao, Y. (2006). Coordination in fast-response organizations. *Management Science, 52* (8), 1155–1169.

Garud, R., & Gehman, J. (2012). Metatheoretical perspectives on sustainability journeys: Evolutionary, relational and duration. *Research Policy, 41*(6), 980–995.

Grandori, A., & Soda, G. (1995). Inter-firm networks: Antecedents, mechanism and forms. *Organization Studies, 16*(2), 183–214.

Gulati, R., Puranam, P., & Tushman, M. (2012a). Meta-organization design: Rethinking design in interorganizational and community contexts. *Strategic Management Journal, 33*(6), 571–586.

Gulati, R., Wohlgezogen, F., & Zhelyazkov, P. (2012b). The two facets of collaboration. *The Academy of Management Annals, 6*(1), 531–583.

Henard, D. H., & Szymanski, D. M. (2001). Why some new products are more successful than others. *Journal of Marketing Research, 38*(3), 362–375.

Holsti, O. R. (1969). *Content analysis for the social sciences and humanities*. Reading: Addison-Wesley.

Human, S. E., & Provan, K. G. (2000). Legitimacy building in the evolution of small-firm multilateral networks: A comparative study of success and demise. *Administrative Science Quarterly, 45*(2), 327–365.

Järvensivu, T., & Möller, K. (2009). Metatheory of network management: A contingency perspective. *Industrial Marketing Management, 38*(6), 654–661.

Jarzabkowski, P. A., Lê, J. K., & Feldman, M. S. (2012). Toward a theory of coordinating: Creating coordinating mechanisms in practice. *Organization Science, 23*(4), 907–927.

Jones, C., Hesterly, W., & Borgatti, S. (1997). A general theory of network governance: Exchange conditions and social mechanism. *Academy of Management Review, 22*(4), 911–945.

Koschmann, M. A., Kuhn, T. R., & Pfarrer, M. D. (2012). A communicative framework of value in cross-sector partnerships. *Academy of Management Review, 37*(3), 332–354.

Kriron, D., Kruschwitz, N., Reeves, M., & Goh, E. (2013). The benefits of sustainability-driven innovation. *MIT Sloan Management Review, 54*(2), 69–73.

Malone, T. W., & Crowston, K. (1994). The interdisciplinary study of coordination. *ACM Computing Surveys (CSUR), 26*(1), 87–119.

Mitchell, R., Agle, B., & Wood, D. (1997). Toward a theory of stakeholder identification and salience: Defining the principle of who and what really counts. *The Academy of Management Review, 22*(4), 853–886.

Moorman, C., Zaltman, G., & Deshpande, R. (1992). Relationships between providers and users of market research: The dynamics of trust within and between organizations. *Journal of Marketing Research, 29*(3), 314–328.

Morgan, R., & Hunt, S. (1994). The commitment-trust theory of relationship marketing. *Journal of Marketing, 58*(3), 20–38. https://doi.org/10.2307/1252308.

Murnighan, J., & Conlon, D. (1991). The dynamics of intense work groups: A study of British string quartets. *Administrative Science Quarterly, 36*(2), 165–186.

Nambisan, S., & Sawhney, M. (2011). Orchestration processes in network-centric innovation. *The Academy of Management perspectives, 25*(3), 40–57.

Nobelius, D. (2004). Towards the sixth generation of R&D management. *International Journal of Product Development, 22*(5), 369–375.

O'Mahony, S., & Ferraro, F. (2007). The emergence of governance in an open source community. *Academy of Management Journal, 50*(5), 1079–1106.

Ouchi, W. G. (1979). A conceptual framework for the design of organizational control mechanisms. *Management Science, 25*(9), 833–848.

Paquin, R. L., & Howard-Grenville, J. (2013). Blind dates and arranged marriages: Longitudinal processes of network orchestration. *Organization Studies, 34*(11).

Peloza, J., White, K., & Shang, J. (2013). Good and guilt-free: The role of self-accountability in influencing preferences for products with ethical attributes. *Journal of Marketing, 77*(1), 104–119.

Peppard, J., & Rylander, A. (2006). From value chain to value network: Insights for mobile operators. *European Management Journal, 24*(2–3), 128–141.

Provan, K., & Kenis, P. (2008). Modes of network governance: Structure, management, and effectiveness. *Journal of Public Administration Research and Theory, 18*(2), 229–252.

Pujari, D. (2006). Eco-innovation and new product development: Understanding the influences on market performance. *Technovation, 26*(1), 76–82.

Sarkis, J., Cordeiro, J. J., & Brust, D. A. V. (2010). Facilitating sustainability innovation through collaboration. In J. Sarkis (Ed.), *Facilitating sustainability innovation through collaboration.* Heidelberg: Springer.

Slawinski, N., & Bansal, P. (2012). A matter of time: The temporal perspectives of organizational responses to climate change. *Organization Studies, 33*(11), 1537–1563.

Song, L. Z., & Song, M. (2010). The role of information technologies in enhancing R&D–marketing integration: An empirical investigation. *Journal of Product Innovation Management, 27*(3), 382–401.

Spiggle, S. (1994). Analysis and interpretation of qualitative data in consumer research. *Journal of Consumer Research, 21*(3), 491–503.

Tenbrunsel, A. E., & Messick, D. M. (1999). Sanctioning systems, decision frames, and cooperation. *Administrative Science Quarterly, 44*(4), 684–707.

Valente, M. (2012). Theorizing firm adoption of sustaincentrism. *Organization Studies, 33*(4), 563–591.

Van de Ven, A. H., Polley, D. E., Garud, R., & Venkataraman, S. (1999). *The innovation journey.* New York: Oxford University Press.

Vanhaverbeke, W., & Cloodt, M. (2006). Open innovation in value networks. In H. Chesbrough, W. Vanhaverbeke, & J. West (Eds.), *Open innovation: Researching a new paradigm.* Oxford: Oxford University Press.

WCED. (1987). *Our common future.* New York: Oxford University Press.

Wittenbaum, G. M., Stasser, G., & Merry, C. J. (1996). Tacit coordination in anticipation of small group task completion. *Journal of Experimental Social Psychology, 32*(2), 129–152.

Karen Janssen works as project manager/researcher at the Centre of Expertise for Sustainable Business, Avans University of Applied Science. She received her Ph.D. from Radboud University in Nijmegen. In her Ph.D. project, she gained insights into effective methods and organizational arrangements for increasing the role of consumers in radical product innovation. She then continued at Radboud University as postdoctoral researcher, where she was responsible for the TOV (Tool for Orchestrating Value Networks) project. The project developed guidelines to help organizations achieve sustainable innovation outcomes and stable relations with their stakeholders. Her current research at Avans focuses on building applied knowledge and skills regarding developing, implementing and upscaling Circular Business Models. Methods and tools are developed and tested aiming at (1) identifying and managing circular value networks, (2) the effectiveness of partnerships and (3) creating value from waste.

Vera Blazevic is Assistant Professor of Marketing at Nijmegen School of Management, Radboud University Nijmegen, and Visiting Professor at RWTH Aachen University in the Technology and Innovation Management Group. Her research interests include co-creation and social processes in innovation management and the management of sustainability in organizations' innovation efforts. Her prior work has been published in various leading journals, such as *Journal of Marketing, Journal of Product Innovation Management, Journal of Service Research, Journal of the Academy of Marketing Science* and *Journal of Interactive Marketing*, among others. She has worked on sustainability-oriented projects sponsored by RVO Netherlands and EU KIC Raw Materials. For these projects, she has cooperated with organizations from different sectors (e.g. chemicals, food, electricity services, etc.). She has developed tools for managing sustainability-oriented innovation within firms and across stakeholder consortia.

Kristina Lauche is the chair of Organizational Development and Design at Nijmegen School of Management, Radboud University. Previously, she held research and teaching positions at the University of Munich, ETH Zurich, University of Aberdeen, and Delft University of Technology. Her research addresses planned and unplanned forms of organizational change in the context of product innovation and technology implementation, and collaboration across organizational boundaries. Her work has been published in journals such as *Organization Science, MIS Quarterly* and *Journal of Product Innovation Management*.

The Art of Responsible Change

Sustainable Entrepreneurship, Tacit Knowing and Improvisation

Wolfgang Stark

Today, we have enough scientific knowledge to opt for change towards sustainability and responsible business. But we fail in changing individual and entrepreneurial mindsets. The majority is still based on rational thinking, growth and effectiveness.

Rational thinking bears the overall assumption that all technical and societal challenges can be solved by an objective step-by-step rational approach. Yet, many entrepreneurial settings are governed by unknown situations, subjective personal creativity and implicit knowing and intuition. The more complex a situation and setting gets, the more planning and rationality is loosing ground in the process of organizing. Complexity then may lead to the use of emergent and creative processes based on the tacit knowing of the arts.

In order to enhance an entrepreneurial mindset for sustainability and social responsibility, we need to enact economical, political, sociological and psychological drivers. Change and transformation processes should not be restricted to rational choice and planning only. In complex systems like entrepreneurial dynamics multidisciplinary approaches always are needed. Therefore, in addition, we should use art-based sources of transformation for entrepreneurial communities for sustainability.

Coping with unpredictable processes is an everyday challenge in organisations and entrepreneurial communities. In addition to codified rational procedures, members of social systems usually will develop a set of tacit procedures which proved to be viable. Similar to improvisation in jazz music, where musicians interact on the

---

W. Stark (✉)
Strascheg Center for Entrepreneurship, Munich, Germany

Organizational Development Lab, University of Duisburg-Essen, Essen, Germany

Steinbeis Center 'Innovation and Sustainable Leadership', Paehl, Germany
e-mail: Wolfgang.Stark@uni-due.de; http://www.sce.de/en/entrepreneurship.html; http://www.
orglab.org/orglab/; http://www.stw.de

© Springer International Publishing AG, part of Springer Nature 2018   99
R. Altenburger (ed.), *Innovation Management and Corporate Social Responsibility*,
CSR, Sustainability, Ethics & Governance,
https://doi.org/10.1007/978-3-319-93629-1_5

basis of well-known explicit and implicit "jazz patterns", this kind of process can be viewed as continuously re-designing and re-arranging implicit and explicit procedural patterns based on experiential (implicit) knowledge: they interact based on already known patterns, they will cite other patterns, and by re-designing and re-arranging they also will create a constant flow of new patterns which will be added to their body of experiential knowledge.

Based on this, this paper calls for a truly transdisciplinary format which is integrating natural science, social science and the arts (music, dance, theatre, visual arts) in order to affect the entrepreneurial mindsets and ways of thinking for our current and future leaders, decision makers and entrepreneurs. An experience-based and creative knowing will be able to reveal and teach the tacit knowing patterns we need to develop to go for the next steps toward sustainability.

Contemporary organizations and social systems have to deal with accelerated complexity and growing uncertainties and ambiguities. This paper highlights the contradictive, yet productive tension between rational process thinking and the improvisational field in entrepreneurial processes. It offers a path to develop tools for dealing with complexity and using the strength and the art of improvisation by developing a Performative Pattern Language (PPL). The PPL is based on improvisation theory and aims to allow people to use tacit knowledge in social systems to increase their flexibility, creativity and performance in settings of uncertainty and ambiguity. These kinds of settings provide the culture where innovative approaches emerge.

1 Beyond Bounded Rationality

People in organisations and social systems can act and create new solutions either by analysis and planning, by intuition or by improvisation. However, the vast majority of organizational learning and innovation accepted today still seems to follow one type of procedure: rational planning based upon analysis. For most of last century, and still today, organisational theory and innovation management is based on this rational cognitive mode. Management and engineering are focused upon numbers, influenced by rational industrial thinking, and characterized by measurement and focus on accountability. In the last 20 years, this rather one-sided approach has also been adopted by society at large and professional communities alike, and has infected our everyday way of thinking.

Rational thinking and rational design patterns are based on the assumption that technical and social challenges can be overcome by an objective, step-by-step, rational approach. Although many social scientists, as well as many practitioners, know that this approach captures only a small part of the processes and dynamics existing in both social systems and organizations, it seems to work well for traditional organizations (both profit and non-profit) that are based on the hierarchical model of top-down decision-making and planning. Nevertheless, modern network-type social systems must encourage soft factors like community-building and organizational culture in order to survive in their complex and constantly changing organizational environment. Even organizational systems which rely heavily on "rational" key performance indicators (KPI) based on numbers end up discovering the value of "soft" cultural organizational processes when they are confronted with unexpected dynamics or find that they must use creativity to build corporate cultures of trust and innovation.

The concept of Bounded Rationality (Gigerenzer and Selten 2002) already challenged rationality in decision making based on experimental models. The concept proofed that rational choice is only one part of human choice; non-rational factors are highly influential in everyday- and entrepreneurial decisions. Nevertheless, the majority of entrepreneurs pretend to make rational choices although relying heavily on 'entrepreneurial intuition'. The reason? Complex social systems like modern companies, non-profits, political and informal communities are very often not determined by clearly defined goals and strategies. Most web-based social networks (including facebook, vimeo, cool ideas society, seats2meet, intrinsify.me and many others)[1] and many web-oriented companies are based on the idea of serendipity[2]—that is, they use opportunities that emerge from non-planned networking. Gradually we (re)discover that many settings in which we live and work are governed by unknown situations and ill-defined factors. The ability to be creative, to design innovative environments and to improvise in an ostensibly rational and structured situation may be key factors for survival in a world that is in reality unpredictable and subject to serendipity. Although non-linear and non-deterministic factors should not be ignored in a world of constant change (Looss 2002), they usually are not addressed by a rational approach, and thus are also neglected in engineering and the (social) sciences. Indeed, the dynamic process of organizing (Weick 1995) still is bound to a culture of numbers, results and rationality. The complex network of relations is neither seen nor tackled, since practice and perception are both oriented toward attaining goals, maintain control, and setting strategies. More problematically, entrepreneurs, decision makers, and managers—both in traditional and sustainability-driven organizations—typically lack a language to describe their "tacit knowledge" (Polanyi 1966) (including both: restrictions, fears *and* opportunities, wisdom) in this land of uncertainty.

---

[1] Although facebook and vimeo seem to be well known and linked, smaller and more specialized social networks need links to be seen: http://coolideassociety.com, https://www.seats2meet.com, http://intrinsify.me.

[2] See http://en.wikipedia.org/wiki/Serendipity.

2 Tacit Knowing and the Improvisational Field

The economy of production and organizational technology unfolds in a texture of cooperation of diverse models of partnership—from (a) the small cooperative cell (the team) within an organisation or company, (b) to the organisational design of a company or non-profit-organisation as an entity in itself with explicit structures and implicit knowledge, to (c) the strategic alliance between different types of organisations and stakeholders. These different relations share a common challenge: they all are driven inherently by implicit and "tacit" knowing, and, quite often, emotionally-based decision-making and processes. In contrast, especially in organizational settings, most processes are determined by 'rational planning' which does not grasp the hidden power and potentials of tacit knowing. Nevertheless, the more complex a situation and setting gets, the more planning and rationality loose ground in the process of organizing (Weick and Westley 1996). As shown in a previous papers (Schümmer et al. 2014), up to now in our 'rational world' of buildings, business-statistics and computer/machine-based processes, complexity has not led to the use of emergent and creative processes based on the tacit knowledge of the many. Instead, decision-making still pretends to be built on planned structures that are explicit and too often planned solely using rationality, which try to reduce complexity to a level that is manageable using rational thinking. This limits our ability to cope with the ambiguity and uncertainty that is built into modern organizational dynamics.

In social systems such as like organizations or communities, the improvisational field is the layer beneath planning and acting. It is built upon tacit knowing and experiential wisdom. Dorothy Leonard (Leonard and Swap 2005) calls this phenomenon 'deep smarts'. According to her research deep smarts (i.e. tacit knowing) and the rational field of structures and numbers cannot be separated; instead, they need to rely on each other. Therefore, improvisation and its performative patterns do not replace the rational, cognitive mode: just as the muscle system in the body is needed for and skeleton to move, to balance and to be alert, the improvisational field and performative patterns are needed to balance structures and rules, as well as ambiguities in each situation new to routine and to be alert to innovations and creative opportunities.

To detect the language of tacit knowing we must experiment with new sensorial channels: for instance, if we could "hear" the dynamic processes of organizations, the communicative sensorium in the workplace could be expanded to a new and deeper level which would allow us access both aesthetic and emotional dimensions of processes. Music, as a performing art,[3] can be one key to the "deep level of organizing and innovative processes" which can be used as a reflective tool for both

---

[3]Of course, also other types of performance such as art, dance and theater, as well as modern, performative ways of painting, could be helpful in detecting the potentials of tacit knowledge beyond rational planning (Forsythe 2003). Based on our research at www.micc-project.org, we focus in this paper on music as not only as a metaphor, but also as an analytical tool.

managers and employees but also for people in communities to start a dynamic and creative process of learning for social systems and individuals. To "imagine community processes as a piece of music" (Stark and Dell 2013) opens up social and organizational systems which are often stuck in strategic plans and work-flows, and helps them to creatively re-design the system.

To detect the dynamics of this hidden (implicit and tacit) system inside the visible system, a special form of musical production is necessary in order to foster learning processes in complex and constantly changing settings, which call for the ability for continuous sense-making and serendipity (Weick and Westley 1996): the technology of improvisation has already inspired organizational theory as a metaphor (Weick 1995; Hatch 1999; Barrett 2012); improvisational patterns will be even more important for organizational processes and social systems, if we look at music not only as something that can be received or interpreted, but also as a tool for sense-making. Improvisation then will open up the ability not only to cope with unknown potentials and uncertain processes, but also to redesign patterns and minimal structures in a creative way (Dell 2012; Stark 2014).

This is true not only for music using arts of improvisation such as Jazz, but has also been admired in many pieces of Johann Sebastian Bach (Ruiter-Feenstra 2011), or in the base-lines of Indian music (Kurt and Näumann 2008). It also is common ground for many forms of performative contemporary art such as theater, dance or performing art itself (Johnstone 1987; Forsythe 2003; Fischer-Lichte 2012). In everyday-life, we discover the art of improvisation in many sports activities like modern soccer, sailing, and skiing. One can therefore assume that whenever human creativity and playfulness are triggered, the art of improvisation is one of the keys to joy and self-awareness and furthermore develops skills to cope with ambiguity and uncertainty. Thus, the inventive production of improvisation becomes a norm in itself: challenge and possibility.

Coping with unpredictable processes also is an everyday challenge in organisations and communities. In addition to codified rational procedures, members of social systems usually will develop a set of tacit procedures which prove to be viable (von Glasersfeld 2002). Similar to improvisation in jazz music, where musicians interact on the basis of well-known explicit and implicit "jazz patterns" (Coker et al. 1990), this kind of process can be viewed as continuously re-designing and re-arranging implicit and explicit procedural patterns based on experiential (implicit) knowledge: they interact based on already known patterns, and they also refer to other, already existing or traditional patterns, and by re-designing and re-arranging they also create a constant flow of new patterns which are added to their body of experiential knowledge (Barrett 2012).

3   Improvisation and Sustainable Entrepreneurship: Managing the Unexpected

Research on a technology of improvisation, while using knowledge from the performing arts (especially jazz and contemporary "new" music, but also modern forms of theatre and dance), is one of the back-bones of the 'art of responsible change' and develops practical tools for innovation processes in organisations and social systems. While taking into account that improvisation practice enables us to navigate through new social spaces which are characterised by new dimensions of abrupt change, uncertainty and insecurity, the organization then becomes a transitional place, choreographed by a huge, complex variety of rhythms in which we navigate and perform at the same time. This is exactly the situation in which entrepreneurs have to proof they are 'able to swim'.

Originally, the term improvisation was used to describe an approach to repairing a thing or a situation, a way to correct it in a sloppy way when the original plan was failing. Although improvisation also was inherently associated with high flexibility and mobility, it was only ever meant to be temporary in its use. Now the situation seems to shift: complex social space like modern "fractal" or "fluid" organizations or networks take on the qualities of permanent improvisation. The lifestyle of transition and transformation becomes one of the key features of everyday life, and is also one of the key factors for sustainability. Therefore, entrepreneurial and organizational patterns also, originally being exemplars of linear planning, decision-making and evaluation, must adapt to a situation of complexity and awareness and towards the art (or technology) of improvisation.

Improvisation etymologically descends from the Latin "improvisus", which means unforeseen, unexpected. The term improvisation belongs to the realm of what-is-not-yet (Dell 2002, 2012).

Thus, improvisation cannot be described in itself, but rather can be localized as a continuous readiness and an ability to act-in-an-instant (Scharmer 2009). Everything else will come out of the situation and its processes. Field, network, and variation principles are the categories of action on fluid ground. Therefore, to improvise in situations of ambiguity, alertness and presence will become key features of any organisation or social system. Improvisation positions itself as a technology that also takes into account commitment and trust, self-confidence of actors and their interdependence, as well as autobiographic characteristics of the individual in a group process.

Schön (1983), in describing the "reflective practitioner", refers to the challenge of jazz-musicians to use improvisation in order to create coherence in unpredictable situations: musicians—while collectively trying to develop a creative and inspiring new dynamic of sound—use metric, melodic and harmonic patterns that they are all familiar with to shape the tune or the sound. Musicians most of the time only intuitively grasp the idea of where the tune is developing based on their performance: they will be able to pick up the new sense and adapt their individual playing toward the new goal. Successful improvisations are not only inspiring examples of "reflective practice", says Donald Schön, but *organisational improvisation* can also be seen as the basis of a new praxis of organizing complex systems which are innovative in nature (Johnson 2011).

Improvisation does not differentiate between thinking and acting, but intensifies the movement between the systems of the body (i.e., the systems/components of the organisation). Improvisation therefore acts as a controlling system in the navigation between intersubjective openness and solipsistic moments of subjectivity. Then, intellectual work, social experience and practical-intuitive competence converge—as do the difference between the individual and the collective in social systems, and the difference between the past and the future in time (Scharmer 2009).

4   Improvisational Learning

At first sight, improvisation works in a disorderly fashion and seems to be either unprofitable or ineffective, and therefore seems to describe quite the opposite of entrepreneurship. But, improvisation works because it contains difference, gaps, looseness, and intermediate spaces, which are available for the active interpretative work of the recipients, thus helping to qualify their experience (Hatch 1999). In an improvisational process, the actors develop those sensors that they need in order to grasp directly the ambivalence of a situation, to interpret it, and make it usable—and this is the very core of sustainable entrepreneurship. Cunha (2005) says: "In the improvisational mode, people act in order to learn."

Improvisation thus can be described as a technique which allows its practitioner to integrate serendipity as a learning process and involves proactive learning. Rational analysis is not excluded, rather the opposite: the performative aspect of learning is put into focus. Analysis in the context of improvisation concentrates on the re-arrangement and re-interpretation of material that is gathered through the improvisational process in such a way that it is connectable to new processes in time. The analytic work then relies on qualified experience and the development of complexity-sensors that should lead to a transformation of attitudes and thus enable ecological change. But in order to do this, the improviser needs to develop the abilities needed to recognize change, allow it and help design it.

According to Mintzberg and Westley (2001), people in organisations and social systems can learn by using analysis, intuition or improvisation. Analysis is a structured process that may or may not lead to surprising findings. The analytical mode proceeds from the assumption that an ontological basis is externalized from existing situations. The intuitive mode derives its learning results from establishing connections that were not previously proposed. The improvisational mode is structured very differently: people not only act in order to learn, but they also try to incorporate analytical frameworks into action which itself then becomes a learning laboratory for the "reflective practitioner" (Schön 1983). Graebner (2004) showed that an important source for creating value is a mode of *serendipity*[4] that is caused by exposure to different practices.

Our research practice aims exactly at this mode of serendipity. It aims to trigger its process and to use the fact that the different practices offer different forms of surprise. In this way we try to apply what we analyze to our own process. The only way to do this is to include improvisation in the experimentation and research process itself and to make that visible. Improvisation is *the* mode of action that ensures the independence of the structures it contains as value and precisely in this way maintains the content of design while keeping the process open. That implies that those who practice improvisation also practice recognizing patterns that others overlook, and that they use patterns pragmatically, subtly, or even 'trendily'—as a layer beneath rational planning and sustainability. From this, one important factor emerges: the fact that improvisation does not, as often is expected, need less time and planning. The opposite is the case: a constructive handling of disorder as a cooperative transgression of the plan is potentially more difficult, needs more time for preparation and follow-up. To be too open in the process can also weaken the process and rob it of direction. Therefore improvisation requires high concentration on coordinating activities and interactions.

Improvisation often is avoided because there is no time available for interpreting ambivalent data. Why is it worthwhile to invest time in improvisation, i.e., active interpretation—especially in entrepreneurial processes? Because those who take the time to reflect on situations and their potentials and try to integrate these reflections into open processes of action, are able to accept ambivalence, thus expanding their scope of activity and their effective degree of freedom. Why? Because they are able to recognize when ambivalence is functional and when it is dysfunctional. Here both can be functional on a situation's meta-level. If one's improvisational abilities are expanding, the ability to process ambivalence in a given time frame rises because an ongoing practice of improvisation may enable a person to recognize and play on global time horizons as well as macro-rhythms.

---

[4]*"Serendipity"* means a "fortunate happenstance" or "pleasant surprise". It was coined by Horace Walpole in 1754. In a letter he wrote to a friend Walpole explained an unexpected discovery he had made by reference to a Persian fairy tale, *The Three Princes of Serendip*. The princes, he told his correspondent, were "always making discoveries, by accidents and sagacity, of things which they were not in quest of". The notion of serendipity is a common occurrence throughout the history of scientific innovation such as Alexander Fleming's accidental discovery of penicillin in 1928 and the invention of the microwave oven by Percy Spencer in 1945. (...) Ikujiro Nonaka (2007) points out that the serendipitous quality of innovation is highly recognized by managers and links the success of Japanese enterprises to their ability to create knowledge not by processing information but rather by "tapping the tacit and often highly subjective insights, intuitions, and hunches of individual employees and making those insights available for testing and use by the company as a whole". (...) Serendipity is used as a sociological method in Anselm L. Strauss' and Barney G. Glaser's Grounded Theory, building on ideas by sociologist Robert K. Merton, who in *Social Theory and Social Structure* (1949) referred to the "serendipity pattern" as the fairly common experience of observing an unanticipated, anomalous and strategic datum which becomes the occasion for developing a new theory or for extending an existing theory (Wikipedia-Entrance: Serendipity, 1.11.2014).

5   Improvisation Technology and the Dynamics of Sustainable Entrepreneurship: The Improvisation Lab

In an entrepreneurial mindset for sustainability, patterns of implicit *and* tacit knowing can be used as procedures to cope with as—yet unknown challenges in a creative way in order to find new solutions to given problems. Opposite to instruction manuals or user's guides, performative patterns define the principles that define the solutions that can be adapted to various settings and situations.

Going beyond traditional social science methods such as interviews and participant observation, we elaborate two specific methods in order to detect improvisational patterns in organizations: in a *first step*, the Musical Learning Journey develops a format of a "conversational performance" or "performance lecture", in which a Jazz Combo, working as part of the project group, tries to grasp the artefacts and basic values of the organizational culture in question and to render it as a musical performance as a basis for a cultural discourse about it. This activity serves to open peoples' minds and organizational cultures to musical thought as an alternative way of organizing ones understanding.

In a *second step*, the members of the social system in question (entrepreneurs, organizations, communities) develop a diagrammatical musical notation (score) which is also found in contemporary music in the classical music tradition (sometimes known as "New Music"): the scores consists of a graphic notation which comes to life through its performance instead of through the interpretation of a given piece of music. In the process of the performance, the sound emerges in a creative, rather than an interpretive act.

Thus, non-representational notational (diagrammatic) scores found in contemporary music focus on the relationship between design, recording, repetition, interpretation, reproduction and improvisation.

The goal here is to detect and analyze patterns of intra- and inter-organizational relations and communication processes based on organizational studies and analysis. *Organizational Scores* attempt to use artistic processes and methods of communication to describe and understand temporal and occupational relationships in organizations. In artistic modes of production, notations and diagrams function as working tools that alter the form of art itself, as well as the way it is made and reflected. Organizational Scores use the language of diagrams which has been invented for the notation of Contemporary Classical Music. There, a diagrammatic notation presents the ideas of the composers and also gives plenty of space for interpretation of the musician in a performance.

For organizational scores, entrepreneurs and both organizational and community members are asked to sketch their idea of their challenge/their task as if it would be a piece of music (see Fig. 1). Doing this, we ask for the hidden blueprint of their challenge existing in their minds and bodies based on their individual experience. Based on this peculiar picture of the tacit knowing of their social system, we identify the patterns and principles why they believe their work is successful. By revealing

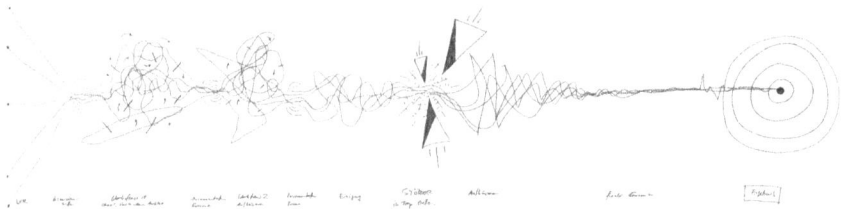

Fig. 1 Organizational score as a musical script of a work-in-progress

their tacit knowing we enable both employees and managers to redesign their work and to elaborate on the art of improvisation in organizations.

The analysis of improvisational (performative) patterns in organizations is based on C. Alexanders (1977) concept of "a pattern language" and its transformational use in organizational contexts (Keidel 1995; Manns and Rising 2005) and social systems (Schuler 2008). According to this concept, patterns of values and principles of organizational cultures unfold and change over time and function as flexible forms of (implicit) tacit problem solving which have proved to be viable and successful in practice.

Therefore, patterns both represent and create sets of relationships which prove to be viable for defined challenges, but, based on the forces they display, also create new patterns and relationships when they are performed. Keidel (1995) analyzed the fundamental forces which have an impact in organisations and determine the types of relationships that occur within their social systems and discovered three of them: "autonomy" describes the phenomenon of individuals and groups acting and deciding in a self-determined manner, while "control" represents the wishes of organisations and individuals to reach their separate and common goals and to balance individual interests and motivations. Finally, "cooperation" describes the integrative power of community and the necessity to collaborate in order to be innovative. Keidel therefore proposes a triadic view of challenges in organisations and social systems instead of the very commonly held dualistic one.

Improvisational patterns therefore exhibit the built-in dynamics of repetitive principles and solutions combined with extreme variability and complex time frames which we also find in fractals (see Fig. 2).

Organizational and community innovation today is based on improvisation which works by asking how and when patterns can create new dynamics and new sets of patterns which themselves create new unexpected combination of patterns as part of their performance process.

 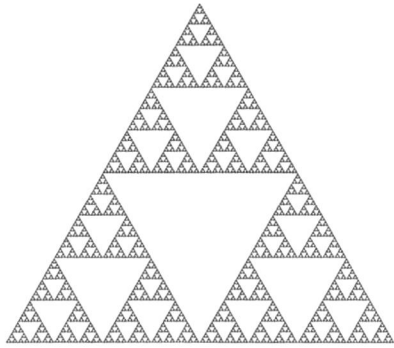

Fig. 2 Mandelbrot set Fractal Picture and the "Serpinski-Triangle" as an example for repetitive creative systems

6 Improvisation in Mode 2: Performative Tacit Knowing to Inspire Sustainability and Innovation in Social Systems

Improvisation follows Latour's (2005) dictum that we cannot externalize. Entrepreneurial processes define not a space where we determine what is already there, but rather a space that is permeated by possibilities. It is a field of possibilities yet to be discovered. In traditional settings we pretend that the epistemological space that we analyse is an objective spatial structure which is only ruptured when something "goes wrong". Then we "fix the situation" and everything continues like expected according to the plan. This way of problem-solving in complex situations might be called '*Improvisation in mode 1*'. Here we improvise in the sense that we react to perceived problems: we repair a perceived lack of order or a problem in the process.

In an entrepreneurial mindset, we reveal the mode of production of current organizational reality while performing, i.e. to view ourselves as performative organizational designers. Therefore, '*Improvisation in mode 2*' aims to transform learned practices, routines and existing patterns into anticipatory concepts. As a permanent experiment and continuous navigational exercise, which sometimes is more and sometimes less in crisis, improvisation in mode 2 tries to transgress planning as a transversal process. This notion of improvisation (mode 2) creates space and opportunities.

Improvisation in this sense is a methodology in the spirit of the word 'techne' (craftsmanship). Just like many forms of art, it cannot be standardised. In this sense, an improvisational technology is a social technology that helps to organize opportunities. But, simultaneously, it also serves to help us perceive reality. It is an ontology of transformation, which shifts our focus while organizing from the object to the relationship and from the relationship to the process. Technology becomes improvisation because on the one hand it generates information from its procedures (in turn providing future orientation for the actors involved) and on the other hand opens up the space for new procedural opportunities.

Today, "production is immediately consumption and recording process (*Enregistrement*), without mediation. The recording process and consumption accord directly with production, although they do so within the production process itself. Hence everything is production." (Deleuze and Guattari 2009). Revealed patterns in the process of improvisation only make sense when shown in their dynamic process, for they are either representations or the bearers of relationships of people (Weick 1995). The diagrams of organization, or organizational scores, are not directly representative. Rather, they are bearers of relationships and distributors of agents. In this respect, the interdisciplinary set-up of research we use is not to interpret the organization or community as music, but is rather an exercise to introduce musical (artistic) thought[5] into the analysis of the organization-as-performative process.

7 Performative Pattern Language: Linking Rationality and Creativity to Produce an Improvisation Technology for Entrepreneurial Processes

If we want to use the patterns and the wisdom they embody as basic elements of complexity and tacit knowing for entrepreneurial processes, we need to develop our ability to cope creatively with ambiguity in a systematic, but not only rational way. We propose, that this ability can be supported by a 'Performative Pattern Language', using the wisdom of implicit and tacit knowing, and the 'art of improvisation'. Performative Patterns will need to go beyond patterns as rational constructs built on pure reason (discussed in more detail in Schümmer et al. 2014). The ability to be creative and to design innovative environments and the art of improvisation in an 'only assumed' rational and structured situation may be a key-factor for survival in a world of unpredictability and serendipity. Patterns of creative action need to interact with rational patterns in order to unleash the potential in their combination. Therefore, identifying patterns of implicit procedures (performative or improvisational patterns)—such as those in music—is important to understanding and to managing codified and documented procedures. This is what we call the 'Improvisational Field' in organisations and social systems. It is a level of action where experience and intuition create emerging structures and movements run in parallel to rational thinking. It may be just as important for social systems as it is for the human body to carry out coordinated contractions and extensions of muscles (Performative Patterns) in order to move the skeleton (Rational Patterns). Performative Patterns create a Performative Pattern Language because they create new situations and results out of a situation—omitting rational planning procedures before acting. Performative Patterns constitute the emerging praxis of communities and organisations; they

---

[5]"thought" in this sense is not to be reduced to a strict cognitive process, but to a conceptual embodiment (Forsythe 2003).

constitute the 'built-in procedural knowledge' of entrepreneurial processes. They are characterized by recursive procedures, viability, performance, and their ability to be re-combined and re-designed creatively and adaptively within given settings. This is what improvisation in music, theatre, dance, and even the world of sports can teach us.

In entrepreneurial processes, *Performative Patterns* can be used as procedures to cope with yet unknown challenges in a creative way and find new solutions to given problems. Opposite to instruction manuals and user's guides, they define the principles of solutions that can be adapted to a wide variety of settings and situations.

The analysis of performative patterns in organisations, although based on C. Alexander's (1977) concept of "A Pattern Language", develops a transformational use of patterns in organisational contexts (Keidel 1995; Manns and Rising 2005) and social systems (Schuler 2008). According to this concept, patterns unfold and change within the values and principles of organisational cultures as flexible forms of (implicit) tacit problem solving which are proven to be viable and successful in practice. It has much in common with the concept of unfolding wholeness, as it was presented in Alexander's more recent works (Alexander 2004; Alexander et al. 2013). Starting from a perceived "minor non-homogeneity in space" (Alexander et al. 2013, p. 430), i.e., a focal point for a new centre, the space is improved by "wholeness-extending transformations" (ibid, p. 428) in order to unfold a new structure triggered by the initial centre. "Morphogenesis then occurs by the repeated application of the 15 transformations on the centers in a configuration" (ibid, p. 430).

A *Performative Pattern Language* needs to support and creatively play with at least three types of transformations: (1) *Basic Action Patterns* which enable people to act-in-an-instant which can be continuously and creatively recombined or in some cases even redesigned in order to develop new processes and results; (2) *Synchronisation Patterns* which orchestrate different "moves and sounds" and develop the "groove" of an organisation or social system, and (3) *Reflection Patterns* which allow people to continuously evaluate the effects of their current actions and the current degree of wholeness of the organization.

Improvisation paves the way from a technology-related world-view toward a view of organizations as social systems for sustainable and performative entrepreneurship.[6] Instead of considering patterns as rules for design, we should consider them as mediators between two different views of design (problem-solving (rational) and performative). This requires patterns to respond both to their dynamic relationships with one another and with the self of the pattern author. Patterns should have both value (dynamic force) (which enables to relate to other patterns) and a subjective component where the pattern author talks about the feelings/values that they create for her or himself.

Improvisation patterns can be the key to understanding the principles of organisations and the deep levels—"the unknown"—of complex modern organizational

---

[6]Like formal organizations, enterprises or public institutions or less formal community of practice which are both based on social relationships.

and entrepreneurial cultures. But patterns and pattern languages in organisations today must go beyond the status quo and must enhance flexibility instead of stability. Further, they must support the creation and detection of new forms of conversation and relationship ("serendipity" acc. to Cunha 2005). They should allow interplay between "move" and "structures", i.e. moves as the creative unfoldings of strong centres triggered by perceived tensions, and structures as the integrative order that connects the different moves in a coherent whole.[7] Therefore the built-in principles of patterns and pattern languages (the "patterns of patterns") must meet the double challenge of providing continuity and variability which is analysed in non-linear systems (Brockman 1995).

References

Alexander, C. (1977). *A pattern language: Towns, buildings, construction.* New York: Oxford University Press.
Alexander, C. (2004). *The nature of order* (Vol. 1–4). Berkeley, CA: Center for Envorimental Structure.
Alexander, C., Neis, H., & Alexander Moore, M. (2013). *The battle for the life and beauty of the earth.* New York: Oxford University Press.
Allan, R. (1998). *Patterns of preaching.* St. Louis: Christian Board of Pubn.
Barrett, F. J. (2012). *Yes to the mess: Surpising leadership lessons from Jazz.* Boston: Harvard Business School.
Brockman, J. (1995). *The third culture. Beyond the scientific revolution.* New York: Simon and Schuster.
Buttrick, D. (1987). *Homiletic moves and structures.* Philadelphia: Augsburg Fortress Publishers.
Coker, J., Casale, J., & Campbell, G. (1990). *Patterns for Jazz: A theory text for Jazz composition and improvisation.* Los Angeles: Warner Bros. Publ.
Cunha, M. P. (2005). *Serendipity. Why some organizations are luckier than others.* Lissabon
Deleuze, G., & Guattari, F. (2009). *Anti-Oedipus. Capitalism and schizophrenia.* New York: Penguin.
Dell, C. (2002). *Prinzip Improvisation (Principles of Improvisation).* Köln: Walther König.
Dell, C. (2012). *Die improvisierende Organisation. Management nach dem Ende der Planbarkeit.* Bielefeld: Transcript.
Feenstra, P. (2011). *Bach and the art of improvisation.* Ann Arbor: CHI Press.
Fischer-Lichte, E. (2012). *Performativität. Eine Einführung.* Bielefeld: Transcript.
Forsythe, W. (2003). *Improvisation technologies. A tool for the analytical dance eye.* Karlsruhe: ZKM.
Gigerenzer, G., & Selten, R. (2002). *Bounded rationality.* Cambridge: MIT Press.
Graebner, M. (2004). Momentum and serendipity: How acquired leaders create value in the integration of technology firms. *Strategic Management Journal, 25,* 751–777.
Hatch, M. J. (1999). Exploring the empty spaces of organizing: How improvisational jazz helps redescribe organizational structure. *Organization Studies, 20,* 75–100.
Johnson, S. (2011). *Where good ideas come from. The natural history of innovation.* New York: Penguin.
Johnstone, K. (1987). *Impro: Improvisation and theatre.* New York: Routledge.
Keidel, R. W. (1995). *Seeing organizational patterns.* San Francisco: Berrett-Koehler Publ.
Kurt, R., & Näumann, K. (2008). *Menschliches Handeln als improvisation.* Bielefeld: Transcript.

---

[7]The distinction between moves and structures has been borrowed from two works in modern homiletic: Buttrick, David: Homiletic Moves and Structures, Augsburg Fortress Publishers, 1987; Allan, Ronald: Patterns of Preaching. Christian Board of Pubn, 1998.

Latour, B. (2005). *Reassembling the social: An introduction to actor-network-theory.* Oxford: Oxford University Press.

Leonard, D., & Swap, W. (2005). *Deep smarts. How to cultivate and transfer enduring business wisdom.* Cambridge, MA: Harvard Business School Press.

Looss, W. (2002). In A. Lenz & W. Stark (Eds.), *Der blinde Tanz zur lautlosen Musik (Dancing Blind to Silent Music).* Tübingen: Empowerment Technologies.

Manns, M. L., & Rising, L. (2005). *Fearless change. Patterns for introducing new ideas.* Boston: Addison Wesley.

Mintzberg, M., & Westley, F. (2001). Decision making: It's not what you think. *Sloan Management Review, 42*(3), 89–93.

Nonaka, I. (2007, July–August). *The knowledge-creating company.* Harvard Business Review Press.

Polanyi, M. (1966). *The tacit dimension.* London: University of Chicago Press Ruiter.

Scharmer, O. (2009). *Theory U – Learning from the future as it emerges.* San Francisco: Berett-Koehler Publ.

Schön, D. (1983). *The reflective practitioner: How professionals think in action.* New York: Basic Books.

Schuler, D. (2008). *Liberating voices: A pattern language for communication revolution.* Boston: MIT Press.

Schümmer, T., Haake, J., & Stark, W. (2014). *Beyond rational design patterns.* In Proceedings of the 19th European Conference on Pattern Languages of Programs (EuroPLoP '14), ACM, New York, NY, USA.

Stark, W. (2014). Implizites Wissen der Improvisation für innovative Organisationskulturen verstehen und nutzen. *Praeview – Zeitschrift für innovative Arbeitsgestaltung und Prävention, 1,* 12 ff.

Stark, W., & Dell, C. (2013). Tuning into the improvisational field. In R. Grossmann, K. Mayer, M. Lenglacher, & K. Scala (Eds.), *Learning for the future in management and organizations.* Charlotte, NC: Information Age Publishing.

von Glasersfeld, E. (2002). *Radical constructivism. A way of knowing and learning.* London: Routledge.

Weick, K. (1995). Organizational redesign as improvisation. In G. P. Huber & W. H. Glick (Eds.), *Organizational change and redesign.* Oxford: Oxford University Press.

Weick, K. E., & Westley, F. (1996). Organizational learning: Affirming an oxymoron. In S. R. Clegg, C. Hardy, & W. R. Nord (Eds.), *Handbook of organization studies* (pp. 440–458). London: Sage.

Wolfgang Stark is founder and director of the Organizational Development Laboratory (www.orglab.org) based at the University of Duisburg-Essen in Germany. He is specializing in community building and empowerment processes in organizations and society, and in organizational/societal learning and organizational culture by linking different disciplines and topics. His research focusses upon empowerment processes, corporate citizenship/corporate social responsibility, value-based management and organizational culture, implicit/tacit knowing and improvisation processes in organisations. He has been founding member and president of the European Community Psychology Association and serves as a regular visiting professor at the 'Instituto Superior Psicologia Aplicada' (www.ispa.pt) in Lisboa (Portugal) and 'University Network for Social Responsibility' (www.netzwerk-bdv.de). Currently he is director of the 'Steinbeis Center for Innovation and Sustainable Leadership' and Visiting Researcher at 'Strascheg Center for Entrepreneurship' in Munich/Germany. His work received various awards—among others a 'Jimmy and Rosallyn Carter-Campus Community Partnership Award 2007', Presidential Award as 'Selected Landmark in the Land of Ideas 2008', and 'Location of the Future' (2014).

X-IDEA: How to Use a Systematic Innovation Method for Social Innovation Projects

Detlef Reis and Brian Hunt

1 Introduction and Theoretical Background

1.1 Introduction

In this chapter, we introduce X-IDEA, a comparatively new innovation process method and related thinking toolbox, and discuss how such a structured innovation

---

Dr. Detlef Reis is the Founder and Chief Ideator of Thinkergy Limited, the Innovation and Ideation Company in Asia (http://www.thinkergy.com). He is also an Assistant Professor at the Institute for Knowledge and Innovation South-East Asia (IKI-SEA), Bangkok University in Bangkok, Thailand, and an Adjunct Associate Professor at the Hong Kong Baptist University.
He is also the creator of four proprietary innovation methods used by Thinkergy: The innovation process method X-IDEA; the innovation people profiling method TIPS; the innovation culture transformation method CooL—Creativity UnLimited; and the creative leadership method Genius Journey. Dr. Reis has written his first two creativity books titled "X-IDEA: The Structured Magic of Playful Innovation" and "Genius Journey. Developing Authentic Creative Leaders for the Innovation Economy", both of which are currently under review with Wiley US (and targeted for publication in Q2.2017 and! 4.2017).

Now retired, Dr. Brian Hunt was formerly Assistant Dean (Quality Assurance) and Assistant Professor at the College of Management, Mahidol University, Bangkok, Thailand. He now researches and writes academic books on management.
® Thinkergy, X-IDEA, TIPS, CooL-Creativity Unlimited, and Genius Journey are (registered) trademarks of Thinkergy.
© The X-IDEA Innovation Method and Toolbox is a registered copyright of Dr. Detlef Reis/ Thinkergy 2009–2016.

D. Reis (✉) · B. Hunt
The Institute for Knowledge and Innovation Southeast Asia, Bangkok University, Bangkok, Thailand
e-mail: dr.d@thinkergy.com; brian.hun@mahidol.ac.th

© Springer International Publishing AG, part of Springer Nature 2018   115
R. Altenburger (ed.), *Innovation Management and Corporate Social Responsibility*,
CSR, Sustainability, Ethics & Governance,
https://doi.org/10.1007/978-3-319-93629-1_6

method can be used in the context of social innovation and corporate social responsibility activities.

We structure this chapter as follows: Section 1 sets the theoretical background by discussing some of the pertinent literature related to structured innovation methods on the one hand, and social innovation, corporate social responsibility (CSR), and sustainability on the other. Next, we introduce the X-IDEA innovation method and thinking toolbox. We describe X-IDEA's comprehensive yet elegant design architecture and special methodological features, and explain why X-IDEA goes beyond existing thinking frameworks for creative problem solving, innovation and design. In the third section, we discuss three case studies that illustrate how an innovation process method like X-IDEA may be used for social innovation projects, and can contribute to corporate social responsibility activities of corporations. In the concluding fourth section, we sum up how corporations may align their innovation initiatives with aspects of sustainability, corporate social responsibility and social innovation.

1.2   Theoretical Background

This chapter is situated at the intersection of three strands of literature: (a) the literature on structured process methods and thinking tools for creativity, innovation and design; (b) the domains on corporate social responsibility and sustainability; and (c) the writings on innovation types (including social innovation). Below, we briefly discuss each of these strands to build-up a theoretical platform for our further discussion.

The literature on problem solving, creativity, innovation and designs abounds with descriptions of structured thinking processes and innovation methods. Popular examples include the Creative Problem-Solving (CPS) model developed by Osborn (1963 [1953]) and Parnes (1967) and, more recently, design thinking (Kelley and Littman 2002; Brown 2008; Kumar 2013). While the different thinking frameworks vary in their detailed design, their common features is that they invite individual innovators or innovation teams to think and work their way through an innovation challenge by passing through different process stages. Other authors offer collections of thinking tools for serious thinking (e.g., de Bono 1992) and creativity and innovation (e.g., Michalko 1991, 2001; VanGundy 2005; Hudson 2007). However, a number of books such as Clegg and Birch (2002), Bragg and Bragg (2005) or, more recently, Kumar (2013) link a collection of thinking tools to an underlying systematic thinking framework.

Innovation facilitators use such innovation process methods and thinking tools for innovation to guide innovation teams through a concrete innovation project. Thereby, innovation projects typically focus on one specific innovation case that relates to a particular innovation type. Classical innovation types that organisations have already pursued for decades or even centuries are process innovation (e.g., Ettlie and Reza 1992; Pisano 1997) and product innovation (or new product development, e.g., Goldenberg and Mazursky 2002; Trott 2002). However, over the past two decades, a wide range of modern innovation types has emerged and complemented the "classical" innovation types. The new innovation types developed and successfully applied by organisations and discussed in the literature include service innovation (e.g., den Hertog and Bilderbeek 1999; Tidd and Hull 2003), customer experience design (e.g., Pine and Gilmore 1998; Diller et al. 2006; Richardson 2010), strategy innovation (Johnston and Bate 2003), or business model innovation (Osterwalder et al. 2010), among others.

One particularly exciting new innovation type that is relevant for our discussion here is social innovation:

- Mulgan (2006) defines social innovation as "innovative activities and services that are motivated by the goal of meeting a social need and that are predominantly diffused through organizations whose primary purposes are social." This differentiation draws a line to typical business innovations created and diffused by for-profit organisations (Mulgan et al. 2007). However, Nichols and Murdock (2012) emphasise that by virtue of proposing a novel, original and meaningful new value proposition, every innovation has also a social dimension organization, regardless of whether it stems from a for-profit or not-for-profit organisation.
- Examples of social innovations include micro-credits first conceived by Muhamad Yunus, the founder of Grameen Bank, the "fair trade" or the open-source software and website development movement, or focused campaigns and activities of non-profit organisations such as Greenpeace.
- As these examples illustrate, and as Nichols and Murdock (2012), highlight, social innovations can happen at three levels: incremental (focusing on improving identified bugs in products and services that the market fails to address adequately); institutional (aiming to reconfigure existing market structures and patterns to create new social value), or disruptive (focusing on starting new social movements that alter the cognitive frames of reference around markets and social issues).

Keeley (2013) integrated ten modern innovation types into one elegant and relevant framework, the "ten types of innovation". Regrettably, Keeley's concept omits a number of important modern innovation types including strategy innovation and social innovation. The first author of this book chapter has personally developed a framework to systematise the modern spectrum of innovation types for his innovation company Thinkergy (Reis 2006, 2014). Labeled the Value-Leverage Innovation Typology, the framework organises the various modern innovation types across four levels (value optimisation, new value creation, value leverage through multiplication, and value leverage through magnification). Moreover, this typology also considers strategy innovation and social innovation as innovation types that may be delivered on any or all four levels. Figure 1 presents this more expansive framework of innovation types with the general innovation focus and the desired impact on each level (Reis 2018).

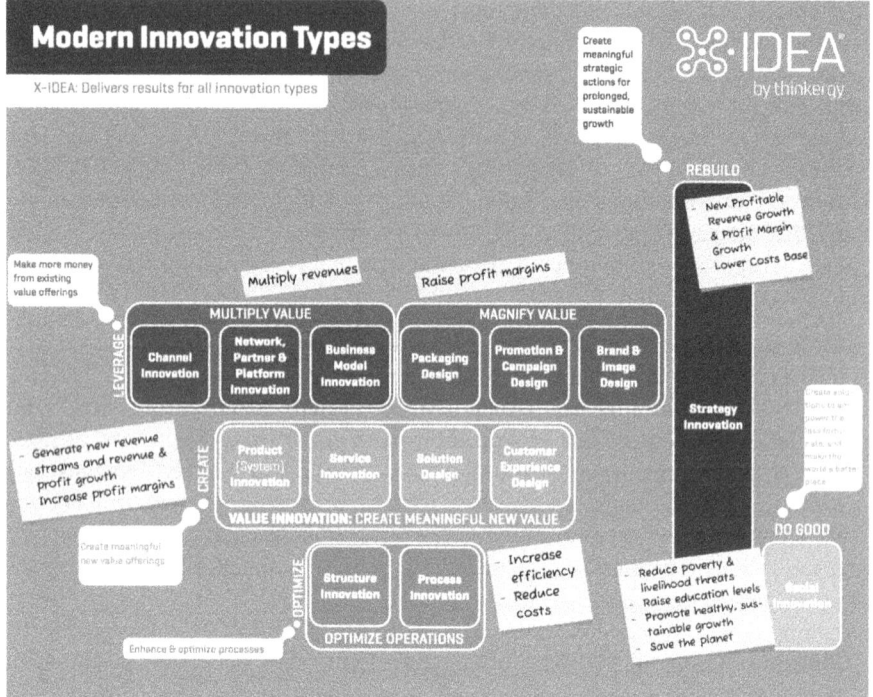

Fig. 1 The modern-spectrum of innovation types

We conclude our discussion of the theoretical background of this chapter by taking a brief glance at Corporate Social Responsibility (CSR). Over the past two decades, CSR has become a more established component of corporate strategic thinking. Organizations have become aware that CSR practices can transform their engagement with local communities as well as adding value through partnerships with local communal ventures. Global consulting firm Accenture reports on a variety of CSR initiatives throughout India, established with the aim of 'shaping India's economic future' (Accenture 2014). Co-prosperity is a core goal of twenty-first century business organizations; in essence, developing the 'business of business is business' ethos that has traditionally informed corporate decision-making and actions. In the process, CSR activities can transform the perceptions of individuals, whether these are employees or members of other stakeholder groups, including consumers (see: Bauman and Skitka 2012; Rupp and Mallory 2015). Not the least attraction for organizations of CSR is its proven valuable contribution to profitability (see, for example, discussions in: Brammer and Millington 2008; Bonini and Swartz 2014).

Inherent is corporate thinking is the belief that 'shared value' brings added value to a broad range of stakeholders, including shareholders and society at large (Porter and Kramer 2011; Rangan et al. 2012). Organizations in many diverse industries devote space to CSR activities in their annual report. Leading edge global companies such as Adidas, Air Canada, British–American Tobacco, Cisco Systems, Johnson & Johnson, McDonalds, Pepsico, Prada, and Procter & Gamble report on-going CSR initiatives, many of these instigated and driven by 'grassroots' employees or external stakeholders. CSR activities in organizations can have a positive influence of employees' creativity. In this case, effective corporate communication is an essential component of managerial discourse as employees need to be aware that their employer is acting for the greater good (Brammer et al. 2015).

2 The X-IDEA Innovation Method and Toolbox: An Introduction

2.1 What Is X-IDEA?

X-IDEA is an innovation process method created by the current first author. The method is marketed and distributed by the innovation company Thinkergy. X-IDEA is a systematic thinking framework designed for innovators and innovation teams to follow while working on an innovation project related to any of the modern innovation types.

Over the past decade, X-IDEA has been used on more than 150 innovation project cases spanning almost all major innovation types. X-IDEA has also been taught to graduate students in master's degree programs at business schools in Thailand, Hong Kong and Finland (Reis 2016b). In a comprehensive longitudinal study, innovation learners confirmed that the use of the structured process flow and integrated application of thinking tools of X-IDEA has led to better thinking and better outputs compared to an unstructured approach (Reis and Hunt 2016).

2.2 How Does X-IDEA Work in General?

The X-IDEA Innovation Method consists of five main process stages: Xploration, Ideation, Development, Evaluation, and Action. Each of the five stages of X-IDEA follows a different objective, requires a different styles of thinking, and focuses on producing a different, yet specific target output. The five main process stages are introduced as follows:

- Stage X—Xploration: In this first process stage, an innovation team thoroughly explores an innovation case to develop a deeper understanding of the related project background. Thereby, the delegates of an innovation project workshop first express their understanding of the challenge, and what they know and do not yet know about the case. Then, they calmly explore the case using four cognitive strategies (check, ask, look, and map). Finally, they extract their "ahas!" (i.e., novel and important insights into their case) as well as a final definition of the challenge.

Practical experiences from more than 150 innovation projects suggest that after a thorough Xploration of the case, the initial perception of the innovation challenge almost always changes. This is because in this stage, participants working on an innovation project uncover knowledge gaps and perceptual blind spots that lead to novel insights into the case, which then allows the teams to uncover their real innovation challenge. As such, this important first stage of X-IDEA ensures that a project team works on and generates ideas for their real challenge, and does not waste scarce resources such as time, effort and capital on what they initially perceive to be the issue.

- Stage I—Ideation: Ideation is the first of two exclusively creative stages with a focus on idea quantity. True to notion of Lateral Thinking, the participants laterally ideate, imagine and incubate raw ideas in this stage. To stress the focus on idea quantity upfront, the innovation facilitator sets an ambitious, yet achievable raw idea quota for the innovation teams to pursue as a target. Depending on the time allotted, an innovation team generates anything between 400–1000 raw ideas with a combination of classic creativity techniques and new ideation-tools developed by Thinkergy. The high number of raw ideas increases the probability of having a sizeable number of original, intriguing ideas.
- Stage D—Development: Development is the second creative process stage of X-IDEA. Now, the objective is to turn idea quantity into quality. At first, the innovation teams are asked to discover intriguing raw ideas within the large pool of raw ideas. Then, they work with this much smaller pool of interesting, original and—at times—wild raw ideas to design and develop these into meaningful idea concepts. This is done by applying the creative principles of elaboration, combination and transmutation and by using special design tools such as Yin And Yang or Get Real (Reis 2016a, 2017).
- Stage E—Evaluation: The fourth stage of X-IDEA, Evaluation, balances the creativity of the previous two stages with realism and pragmatism. Here, the innovation teams evaluate their portfolio of developed idea concepts, enhance promising ones, and finally elect a few top idea concepts that deserve being pitched for real-life activation. As such, the sober Evaluation-stage separates the wheat from the chaff to ensure that time, finances and employee efforts are directed to those few top concepts that are likely to succeed in the market space (high value potential) and can be activated (high implementation feasibility).
- Stage A—Action: Finally, the innovation teams take Action on these top ideas, and turn them into tangible innovation deliverables. Thereby, the participants assess the situation at regular intervals during an idea activation, arrange for the next steps and then activate the planned actions.

Figure 2 illustrates the flow of the five process stages of X-IDEA.

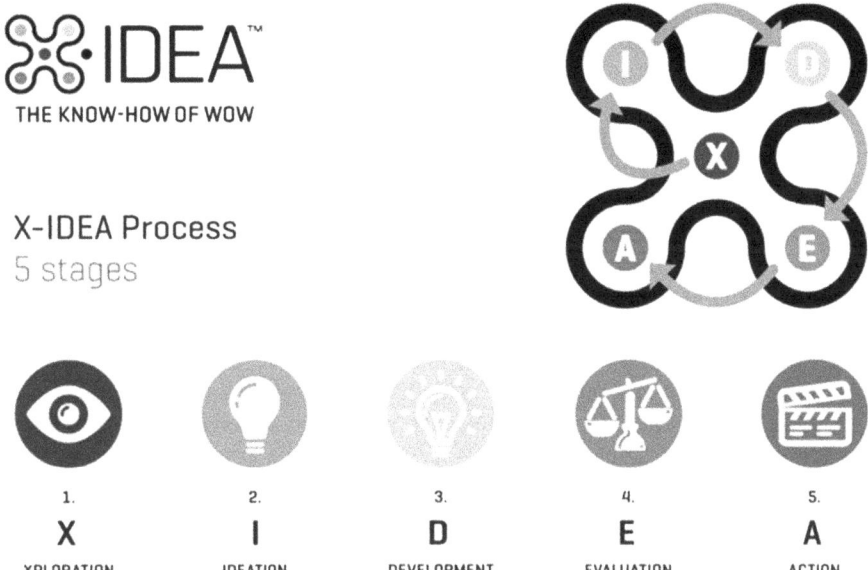

Fig. 2   The five process stages of X-IDEA

2.3   Why Is X-IDEA a Valuable New Addition to the Armoury of Innovation Process Methods?

X-IDEA was created by the current first author with the intent to cure identified flaws and delivery gaps of other innovation methods and creative process methods. What are some of these shortcomings?

- Firstly, many process methods use only one main creative stage. This is problematic because during an idea generation session, most participants tend to immediately judge ideas on their practicality, thus suggesting only "normal", "safe" and "acceptable" ideas. This violates the third ground rules of ideation, which mandates creative thinkers to shoot for wild, crazy and funny ideas.
- Secondly, many innovation project methods also neglect to systematically focus on inputs-throughputs-outputs as an innovation case is taken through the various process stages. Innovation projects tend to be messy and fuzzy, so it's easy for facilitators and participants to lose track of these important questions: What input factors do we need to start a particular step or activity? What interim throughputs are involved in the process? And most importantly, what final outputs do we need to produce in which quantity and quality before we can move on to the next process step?

- Thirdly, while there are many fine books on thinking tools and creativity techniques, comparatively few of these link to a process framework in a systematic way. This is problematic because many inexperienced facilitators and most participants of an innovation project feel overwhelmed having to navigate a rich collection of thinking tools without fully understanding exactly when and where a tool has to be used within a sequence of process steps to accomplish what kind of objective and output.
- Fourthly, a number of popular innovation process methods cater only for one or a limited number of innovation types, such as the Business Model Canvas (Osterwalder et al. 2010) or Blue Ocean Strategy (Kim and Mauborgne 2005). This niche focus typically inhibits those methods from delivering meaningful results for all the many other modern innovation types.
- Fifthly, almost all innovation process methods have been developed—and tend to work fine—in Western cultures. However, most methods neglect the existence of cross-cultural impediments towards certain procedural mechanisms in their framework.

    For example, a common theme in Asian cultures is the fear of losing face. Hence, even when the process instruction during idea generation mandates Asian participants to suggest wild ideas, they rarely do so in order not to "lose face". Another common cultural phenomenon in Asia is to show "consideration to seniority" and "respect to authority." Hence, if a senior participant or superior advocates an obvious idea in a typical "brainstorming" session, most younger participants or subordinates will go along with it although the idea is neither novel, nor original, nor meaningful.
- Sixthly and finally, most innovation process methods do not systematically consider and prevent common cognitive biases and process traps that innovation teams are likely to encounter as they journey through the stages of an innovation project. This oversight leads to the disregard or misinterpretation of project relevant evidence and the production of suboptimal outputs at different process stages, and for the project overall.

Through its integrative design architecture, X-IDEA cures each of these identified ills as follows (Reis 2014):

- One, X-IDEA distinguishes two separate creative stages with different work objectives and target outputs: In the Ideation-stage, your objective is to generate a large pool of raw ideas including wild ones. In the Development-stage, we turn idea quantity into quality by designing and developing a portfolio of realistic, meaningful idea concepts. Because the two creative stages greatly differ from each other in their cognitive activities and output focus, innovation teams are able to move beyond conventional ideas that are usually the result of having only one creative process stage (Reis 2016b).
- Two, X-IDEA has a strong IPO-focus considering inputs, throughputs and outputs on three levels: the overall project, a process stage, and a tool. The IPO focus allows us also to track IPO-related measures on each level. We use these also to set ambitious yet realistic target quota that motivate innovation project teams towards achieving desired outputs and results.

- Three, X-IDEA systematically links every thinking tool to a natural default position within the process flow where a tool is typically used. Currently, the X-IDEA toolbox comprises 150 thinking tools. New tools are added in regular intervals to incorporate new market trends and client needs. The X-IDEA tools are accompanied by related worksheets and stimulus cards that make it easy for innovators to think through a thinking tool, or for innovation facilitators to guide groups through an innovation project.
- Four, X-IDEA is purposefully designed to cater to all modern innovation types (Keeley 2013; Reis 2006, 2017): from process innovation over product and service innovation to customer experience design; from channel over network and platform to business model innovation; from packaging over promotion to brand and image design; and from strategy to social innovation. The comprehensive range of thinking tools in the X-IDEA toolbox allows innovation facilitators to pick those tools that suit the particular nature of, and produce the specific results for, a particular innovation type.
- Five, X-IDEA uses the five X-IDEA Roles of the Xplorer, the Child, the Alchemist, the Judge, and the Champion to overcome intercultural barriers. For example, Asian workshop participants feel comfortable to suggest wild ideas without being afraid of losing face if they know that it's not them suggesting the idea, but them acting in the role of a child.

    Moreover, X-IDEA uses different communication and interaction modes (such as solo brainwriting or pool brainwriting) to circumvent intercultural problems occurring when teams just "brainstorm" for ideas.
- Six, X-IDEA Traps help innovation project teams to systematically avoid cognitive biases and common process traps that are prevalent in any innovation projects.

Figure 3 summarizes how the various X-IDEA features discussed above address the shortcomings of many other innovation methods and problem-solving processes.

3   Application of X-IDEA in Social Innovation: Three Case Studies

X-IDEA is a neutral process method designed to work for all modern innovation types, including social innovation. Below, we present three case studies that illustrate how a systematic innovation method like X-IDEA might be used to support CSR activities and successfully approach social innovation projects:

(a) Case study 1 describes how a company (Merck Thailand) selected an innovation case with a CSR-background to let Thinkergy train their employees and managers in how to use a structured innovation method like X-IDEA (corporate training)
(b) The second case study illustrates how X-IDEA was used in a social innovation project with an external focus for an environmental campaign design with Greenpeace Southeast Asia

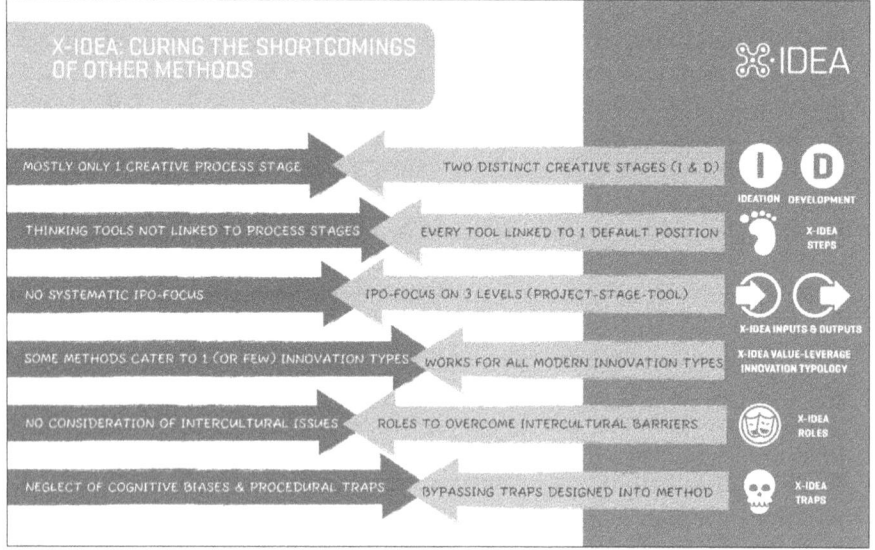

Fig. 3 How the features of X-IDEA counter identified shortcomings of other innovation process methods

(c) The third and final case study features the use of X-IDEA in a social innovation project with a mixed internal and external focus (UNICEF Thailand).

3.1 Case Study 1: X-IDEA in an Innovation Training Using a CSR Case (Merck Thailand)

Merck is a leading science and technology company in healthcare, life science and performance materials. Merck has been active in Thailand since 1991, when the first representation was opened and later expanded into a joint venture.

Merck Thailand has followed a comprehensive stakeholder management approach focusing on customers, shareholders, employees, and society at large. During its first two decades, Merck Thailand achieved double digit revenue growth in four out of 5 years. The former Chairman & Managing Director of Merck Thailand, Mr. Heinz Landau, attributed this standout growth (even at the time of the Asian crisis in the late 1990s) to the four stakeholder approach that he established at the firm. According to Mr. Landau, sustainable leadership coupled with regular CSR activities drove profitable growth and boosted employee engagement in Merck Thailand (Landau 2010a, 2013). Among other corporate social responsibility activities, Merck Thailand established a partnership with the Raks Thai Foundation, an organization that follows the mission to strengthen the capacity of poor and disadvantaged communities in Thailand by analysing root causes of problems, determining suitable solutions and participating in development activities (Panitchpakdi 2010).

In 2005, Merck Thailand approached Thinkergy to train its managers and employees in the use of a structured innovation method. Typically, most innovation training courses use standardised case examples and/or a vanilla innovation case to illustrate the application of an innovation process and a thinking tool, as this approach tends to ensure a smooth, controlled learning experience.

However, Thinkergy agreed to use a neutral CSR case related to Merck's CSR activities to train the delegates on a realistic yet not too stressful case. The case was intentionally framed widely as "How to raise employee engagement for CSR activities in Merck Thailand?"

One idea that came out of the training was to have a joint tree planting day with Merck's staff, lead customers, local shareholders, and CSR partners (such as the Raks Thai foundation) that was later activated with the help of another NGO, the Plant A Tree Today (PATT) Foundation (Landau 2010b).

The Chairman & Managing Director of Merck Thailand, Mr. Heinz Landau, emphasised that the value of a structured innovation method like X-IDEA is enhanced by its playful, energetic delivery style:

> Through his unique style, Dr. Reis was able to energize all workshop participants and to build up confidence in them to deliver creative ideas, not only during the workshop but also for the future.

As this example illustrates, replacing a standardised training case with a real-life CSR-case in an innovation training can be a win-win-win-win for all parties involved:

- Corporations can benefit because the use of a CSR innovation case increases participants' engagement during the training (compared to using an artificial, simulated case);
- Employees get an opportunity to learn how to use a structured innovation method like X-IDEA by working on a realistic, yet at the same time not overly stressful innovation case (compared to using a real case from their own business that tends to increase stress-levels and impairs learning)
- Innovation companies such as Thinkergy typically refrain from using a real innovation case of a corporation in a training. This is because they prefer tackling such a real innovation challenge in the form of an innovation project that has a different emphasis than a training (i.e., producing a tangible innovation deliverable vis-a-vis building up learner's creative competence and confidence), and also sells at a much higher price point compared to an innovation training. However, they tend to be open to replacing a standardised innovation training case with a realistic innovation case with a CSR background, as this promises to enhance the learning experience and gives innovation companies an opportunity to demonstrate their own social responsibility.
- NGOs providing the innovation case can win because they get the resulting ideas for free, some of which they may implement later on with or without the support of their CSR partner.

3.2   Case Study 2: X-IDEA in a Social Innovation Project: Campaign Design at Greenpeace Southeast Asia

In Q4.2011, Thinkergy used X-IDEA to guide Greenpeace Southeast Asia through a social innovation project. Thereby, the focus was on creating a pipeline of impactful environmental campaigns that the NGO may activate in the following 3 years. Thinkergy agreed to conduct a 3-day X-IDEA Social Innovation Project for Greenpeace at third of the normal fee, because as part of our own social contribution, we wanted to support this NGO in the pursuit of the worthy cause of saving our planet from environmental degradation.

The innovation project was integrated in a week-long regional conference that took place at a seaside resort in Chonburi, Thailand. Over 100 delegates from three countries (Indonesia, the Philippines, and Thailand) were split-up into ten teams, each of which worked on one particular innovation challenge that Greenpeace wanted to address in the coming years. As such, Thinkergy helped Greenpeace Southeast Asia to tackle ten different innovation project cases in one event, which was possible thanks to the well-structured approach of X-IDEA.

The innovation challenges addressed five main focus areas identified as important by Greenpeace Southeast Asia: stopping deforestation of rain forest in Southeast Asia; ensuring preservation of maritime resources and oceans; highlighting climate change and promoting sustainable energy solutions; fighting toxic pollutions of environments; and internal talent acquisition.

On the first workshop day, the teams took their innovation case through the initial stage of X-IDEA, Xploration. At first, the teams expressed their initial perception of their challenge in sentences like "How to preserve the population of Sumatra tigers?" or "How to clean-up the dirtiest river in the world?", which is the heavily polluted Citarum river in Indonesia.

Then, the ten teams were sent on the Xplorer's Journey to four different Xploration stations (check-ask-look-map), where each team used a blend of related Xploration thinking tools to thoroughly and deeply Xplore their case. The tools help the teams become aware of perceptual blindspots and knowledge gaps related to their case, and to gain novel insights into what is their real challenge. Finally, the teams stated their deepened understanding in an improved statement of their Final Challenge (e.g. "How to effectively fight toxic industrial pollution of the world's dirtiest river?").

The second day saw an intense morning of outdoor Ideation-activities taking place at ten "Idea Stations", where the delegates used Ideation Tools such as "Star Advisor Board", "Idea Race" and "What If", among others, to generate raw ideas. The teams came up with over 7000 raw ideas for their challenges. In the afternoon, the teams entered the powerful second creative stage of X-IDEA, Development (Reis 2016a). Here, the teams first Discovered the most intriguing raw ideas and then Designed and Developed them into over 250 realistic, meaningful idea concepts.

In the subsequent Evaluation-phase (done on the morning of the third and final day), the ten teams evaluated their portfolios of idea concepts in order to Evaluate each concept, to then Enhance promising ones, and finally to Elect their top five idea concepts.

Finally, each team presented these top five campaign concepts related to their respective innovation challenge to a grand forum. The top ideas that received the best feedback were earmarked for being activated for real as Greenpeace campaigns in the coming 2–3 years. For confidentiality reasons, we cannot present the chosen ideas in greater detail. However, to give you a flavour, one top idea to counter toxic industrial pollution of the world's dirtiest river was the following campaign: A Greenpeace team sets up an aquarium filled with polluted Citarum river-water outside the venue of the annual meeting of a supranational financial institution involved in the funding process for the factories alongside the Citarum river,

Ms. Dawn Gosling, a former Interim Executive Director at Greenpeace Southeast-Asia, summarised her view on the outcomes of the project in 2011 as follows:

> In December 2011 we worked with Thinkergy to bring a new and innovative approach to our campaign planning process. Over the course of 3 days our staff from around the region (approx. 100 people from Indonesia, the Philippines, and Thailand) used the X-IDEA approach to find new ways to tackle some of the most important environmental challenges facing the world today (climate change, deforestation; preservation of oceans and toxic pollutions) plus the organisational challenge of staff/talent acquisition.
>
> After going through the stages of the X-IDEA Innovation Method, the ten teams each presented their top five campaign concepts related to their respective challenge to a grand forum on the third and last day of the event. Over the next month these concepts were developed further and became key elements of our work in 2012.

3.3   Case Study 3: X-IDEA in Social Innovation Projects: Creating the UNICEF of the Future

In October 2016, UNICEF Thailand hired Thinkergy to run a X-IDEA Innovation Project aiming to create a more effective, productive and innovative UNICEF office of the future. The said social innovation project was part of a staff retreat in Hua Hin, Thinkergy had only 2.5 days to guide the 44 UNICEF delegates (split into four project teams) towards the desired project outcomes.

Given the rather short time allocated for the project workshop, Thinkergy held a series of preparation meetings with UNICEF Thailand's management team, and also conducted a series of pre-workshop staff interviews to become more aware of possible innovation challenges (relating to both issues faced by the organization and perceived opportunities to innovate). We used the insights gained from these meetings and interviews to structure the overall flow of activities, and to pick those X-IDEA tools that we believed would help the project teams to become aware of their real challenge.

In the Xploration stage, the innovation teams checked on their core assumptions and on the existence of "rules for fools" (i.e., non-sensical internal guidelines, policies and practices that slow down the teams and prevent conducive actions). They also walked a mile in the shoes of key stakeholders (e.g., donors, delivery partners, government agents, and UNICEF's head office). They looked at their challenge from different scales by examining the big and small picture. They answered powerful—and at times provocative—Xploration questions, too. As a result, the teams realised that compliance to audit guidelines and internal bureaucracy has made the organisation too inside-focused, slow and passive. The teams consensually agreed to reframe their Final Challenge to reflect their real challenge: "How to make UNICEF Thailand fast, flexible and fearless?"

In the subsequent Ideation-stage, the four teams produced over 3000 raw ideas. These formed the ground stock for then designing over 130 realistic, meaningful idea concepts in the Development-stage. In the subsequent Evaluation-stage, each team elected its three top concepts, which they finally pitched in the final Action-stage. Two interesting themes emerged throughout the top idea pitches:

- Several top ideas focused on ways to make internal meetings at UNICEF Thailand shorter, leaner and more output-oriented to free managers' and employees' time and move from talking to action.
- Interestingly, while the Final Challenge was more internally focused, more than half of the pitched top idea concepts made meaning for UNICEF Thailand's external stakeholders (e.g., local and migrant children, partners, and donors).

Mr. Thomas Davin, the Chief Representative of UNICEF Thailand, commented on his take-aways from the social innovation project in 2016 as follows:

> The X-IDEA Innovation Project workshop that we undertook with Thinkergy was an amazing energy boost for all of the UNICEF Thailand team.
>
> The 2.5 days was a real whirl storm of ideation which enabled the team to create over 3000 ideas focused on ways to make UNICEF "Fast, Flexible and Fearless" so as to make an all the more powerful impact in changing vulnerable children lives in Thailand.
>
> Both myself and quite a number of the team were simply amazed at not just the number of ideas we were able to create in such a short time span, but also of the depth and quality of the proposed innovations, as well as the level of energy, buy-in and innovation mindset shift that the workshop created for the vast majority of the staff.
>
> Not the fainthearted given the hard-paced approach and demanding rhythm of the Thinkergy team ☺, but absolutely and totally worth it in terms both of teambuilding impact as well as of the range and depth of innovation concept that the workshop created in a tiny amount of time!

4   Conclusion and Discussion

4.1   Summary and Other Possible Applications of Innovation Process Methods Like X-IDEA for Social Innovation/ CSR/Sustainability

In this chapter, we introduced X-IDEA, an innovation method and toolbox as a framework to guide the thinking of project teams in innovation and more general projects. In particular, we illustrated with the help of three case studies how innovation methods like X-IDEA can add value to individuals, teams and organisations working on projects related to social innovation, corporate social responsibility, and sustainability. Our three case studies showed how the innovation company Thinkergy used X-IDEA to:

- Train businesspeople in how to apply a structured innovation method with the help of a CSR-related innovation case that focused on a partner organization of the corporation undergoing the innovation training (Merck Thailand and Raks Thai Foundation);
- Help an NGO advocating environmental preservation and sustainability to work on campaigns to raise the social awareness for—and affect positive social change for—ten environmental social innovation challenges (social innovation project with Greenpeace Southeast Asia);
- Support a supranational organization caring for child protection and development to create a more agile internal organization as well as more effective external social campaigns for their key stakeholders (social innovation project with UNICEF Thailand).

Abstracting from these three concrete cases, we can envision the following applications for a systematic innovation process methods like X-IDEA to support organisational activities related to sustainability, corporate social responsibility, and social innovation:

- Training companies and their corporate clients might use innovation cases with a social innovation or CSR background to train employees in creativity and structured innovation training courses;
- A corporation might sponsor a social innovation project (conducted by professional innovation experts) for a NGO as part of its CSR activities;
- A NGO may engage a professional innovation company to help them in a social innovation project using a structured innovation method;
- An innovation company (such as Thinkergy) might deliver a social innovation project for a NGO at cost (or do it for free if they're already well-established) as part of its CSR- or social entrepreneurship activities;
- A humanitarian foundation (such as the Gates Foundation) might engage the help of an innovation company (and its structured innovation method) to jointly work on large-scale social innovation challenges.
- A corporation might sponsor a social innovation competition (that is structured following the methodology of a systematic innovation method) as part of its CSR activities.

References

Accenture. (2014). *Organizing for success on corporate responsibility: The path to high performance*. Report by Accenture-FICCI (Federation of India Chamber of commerce and industry).

Bauman, C. W., & Skitka, L. J. (2012). Corporate social responsibility as a source of employee satisfaction. *Research in Organizational Behavior, 32*, 63–86.

Bonini, S., & Swartz, S. (2014). *Profits with purpose: How organizing for sustainability can benefit the bottom line*. New York: McKinsey & Company.

Bragg, A., & Bragg, M. (2005). *Developing new business ideas*. Harlow: Prentice Hall.

Brammer, S., & Millington, A. (2008). Does it pay to be different? An analysis of the relationship between corporate social and financial performance. *Strategic Management Journal, 29*(12), 1325–1343.

Brammer, S., He, H., & Mellahi, K. (2015). Corporate social responsibility, employee organizational identification, and creative effort: The moderating impact of corporate ability. *Group & Organization Management, 40*(3), 323–352.

Brown, T. (2008). Design thinking. *Harvard Business Review, 86*, 84–92.

Clegg, B., & Birch, P. (2002). *Crash course in creativity*. London: Kogan Page.

de Bono, E. (1992). *Serious creativity*. London: Harper Collins.

den Hertog, P., & Bilderbeek, R. (1999) Conceptualising service innovation and service innovation patterns. In *Research Programme on Innovation in Services (SIID) for the Ministry of Economic Affairs*. Utrecht: Dialogic.

Diller, S., Shedroff, N., & Rhea, D. (2006). *Making meaning. How successful businesses deliver meaningful customer experiences*. Berkley, CA: New Riders.

Ettlie, J. E., & Reza, E. M. (1992). Organizational integration and process innovation. *Academy of Management Journal, 35*(4), 795–827.

Goldenberg, J., & Mazursky, D. (2002). *Creativity in product innovation*. Cambridge: Cambridge University Press.

Hudson, K. (2007). *The idea generator. Tools for business growth*. Crows Nest: Allen & Unwin.

Johnston, R., & Bate, D. (2003). *The power of strategy innovation*. New York: AMACOM.

Keeley, L. (2013). *Ten types of innovation. The discipline of building breakthroughs*. New York: Wiley.

Kelley, T., & Littman, J. (2002). *The art of innovation*. London: Harper Collins Business.

Kim, W. C., & Mauborgne, R. (2005). Blue Ocean strategy: From theory to practice. *California Management Review, 47*, 105–121.

Kumar, V. (2013). *101 design methods. A structured approach for driving innovation in your organization*. Hoboken, NJ: Wiley.

Landau, H. (2010a). *Providing meaning and purpose*. [Blog] Care guys. Retrieved January 20, 2017, from http://www.thecareguys.com/2010/03/01/providing-meaning-and-purpose/

Landau, H. (2010b). *Leading with a green heart*. [Blog] Care guys. Retrieved January 20, 2017, from http://www.thecareguys.com/2010/05/05/leading-with-a-green-heart/

Landau, H. (2013). *The 4 stakeholders approach: A great recipe for sustainable leadership success* [online]. Retrieved January 20, 2017, from http://www.yespeoplematter.com/ypm-whitepaper-jan-2013.pdf

Michalko, M. (1991). *Thinkertoys*. Berkley, CA: Ten Speed Press.

Michalko, M. (2001). *Cracking creativity*. Berkley, CA: Ten Speed Press.

Mulgan, G. (2006). The process of social innovation. *Innovations, 1*(2), 145–162.

Mulgan, G., Tucker, S., Ali, R. , & Sanders, B. (2007) *Social innovation: What it is, why it matters and how it can be accelerated*. SAID Business School, Working Paper.

Nicholls, A., & Murdock, A. (2012). The nature of social innovation. In Social innovation, 1–30. In A. Nicholls & A. Murdock (Eds.), *Social innovation: Blurring boundaries to reconfigure markets*. London: Palgrave Macmillan.

Osborn, A. F. (1953). *Applied imagination*. New York: Scribner's.

Osterwalder, A., Pigneur, Y., Smith, A., et al. (2010). Business model generation, Self published.

Panitchpakdi, P. (2010). *Raks Thai Foundation and Merck Thailand: Evolving CSR – partnerships*. [Blog] Care guys. Retrieved January 20, 2017, from http://www.thecareguys.com/2010/01/22/raks-thai-foundation-and-merck-thailand-evolving-csr-partnerships/

Parnes, S. J. (1967). *Creative behavior guidebook*. New York: Charles Scribner.

Pine, B. J., & Gilmore, J. H. (1998). Welcome to the experience economy. *Harvard Business Review, 76*, 97–105.

Pisano, G. P. (1997). *The development factory: Unlocking the potential of process innovation*. Boston: Harvard Business Press.

Porter, M. E., & Kramer, M. R. (2011). Creating shared value. *Harvard Business Review*, (January–February), 63–70.

Rangan, K., Chase, L. A., & Karim, S. (2012, April 5). *Why every company needs a CSR strategy and how to build it* (Working Paper 12-088). Harvard Business School.

Reis, D. (2006). *Business education in the age of value creation*. The Asia Forum on Business Education, Hanoi.

Reis, D. (2014). *X-IDEA: The structured magic of systematic innovation*. ISPIM Asia-Pacific Innovation Forum 2014, Singapore.

Reis, D. (2016a, June 19–22). *Ideation vs. development in X-IDEA: How to move beyond conventional ideas in an innovation project?* In XXVII ISPIM Innovation Conference 2016, Porto.

Reis, D. (2016b). *Teaching business creativity to young professionals: Course design, pedagogy, and methodologies*. In Proceedings of the International Conference "Creative industries in Asia: Innovating with constraints".

Reis, D. (2018, forthcoming). *X-IDEA. The structured magic of playful innovation*.

Reis, D. & Hunt, B. (2016, December 4–7) *Training businesspeople in structured innovation: Uncovering the innovation Learner's experience*. In ISPIM Innovation Summit 2016, Kuala Lumpur.

Richardson, A. (2010, November 15). *Using customer journey maps to improve customer experience*. HBR.

Rupp, D. E., & Mallory, D. B. (2015). Corporate social responsibility: Psychological, person-centric, and progressing. *Annual Review of Organizational Psychology and Organizational Behavior, 2*, 211–236.

Tidd, J., & Hull, F. M. (Eds.). (2003). *Service innovation: Organizational responses to technological opportunities & market imperatives*. London: Imperial College Press.

Trott, P. (2002). *Innovation management and new product development*. Harlow: Pearson Education.

Vangundy, A. B. (2005). *101 activities for teaching creativity and problem solving*. San Francisco: Pfeiffer.

Detlef Reis is the Founder of the Asian innovation company Thinkergy (www.thinkergy.com). He is also the inventor of Thinkergy's four proprietary innovation methods: X-IDEA, an award-winning innovation process method and related thinking toolbox; Genius Journey, a creative leadership development method; TIPS, an innovator people profiling method; and CooL—Creativity UnLimited, an innovation culture transformation method.

Dr. Reis is also an Assistant Professor at the Institute for Knowledge and Innovation Southeast Asia, Bangkok University, Thailand, and an Adjunct Associate Professor at Hong Kong Baptist University. He graduated with a doctorate in international management from Saarbruecken University, Germany.

Brian Hunt earned his Ph.D. from the University of Technology Sydney (UTS). He has held academic positions in UK universities and training functions in organizations in the Middle East and Asia. He is co-author of four books on business and management, and has published articles in international academic journals.

Innovation, Business Models, and Catastrophe: Reframing the Mental Model for Innovation Management

Friedrich Glauner

1 Innovation and Global Collapse: The Paradox of Destructive Wealth Creation

This paper explores the mental model that informs our understanding of innovation and innovation management. Its purpose is to challenge the fundamental conceptual framework that underlies this model and our thinking about the economy, about the aim and scope of business in general, and about innovation and innovation management in particular.

The following definition should help us distinguish the terms "innovation" and "innovation management" from each other: By "innovation", we mean a substantial product, service, or process that constitutes the core of a newly invented or substantially refined business model or value proposition. Innovations can be either disruptive and radical or incremental. Whereas incremental innovations follow the logic of an existing business model or value proposition, radical or disruptive innovations either change the logic and rules of an existing business model or value proposition or replace the entire business model and value proposition in itself. An example for the latter is "Voice over Internet Protocol" (VoIP), the technology underlying e.g. Skype. This invention jeopardized the traditional business models of present day telecommunication companies such as Verizon, AT&T, or the German TELEKOM, whose business models depended on the value proposition of selling access to their communication channels. The latter companies invested heavily in all kinds of hardware and software for regional, national, and transnational transmission systems, including phones, switchboards, or wired and wireless networks. With VoIP, all of these investments into the infrastructure and services of communication channels have

F. Glauner (✉)
Cultural Images, Grafenaschau, Germany
e-mail: friedrich.glauner@culturalimages.de

© Springer International Publishing AG, part of Springer Nature 2018
R. Altenburger (ed.), *Innovation Management and Corporate Social Responsibility*,
CSR, Sustainability, Ethics & Governance,
https://doi.org/10.1007/978-3-319-93629-1_7

become essentially obsolete, as users of VoIP can at most be charged only for access to the internet. The consequence was a dramatic erosion of charges for national or international calls, including roaming fees, a perennial favorite story of newspaper editors during holiday season. The invention of VoIP should thus be regarded as a radical game changer for the telecommunications market. An example for more incremental innovation is the development of electric bicycles. The invention of e-bikes was not a game changer in the bike market, but more an extension of already existing business models and value propositions. It only expanded the traditional bicycle market with a new product, i.e. a bicycle with an additional electrical drivetrain. This substantially refined type of bicycle is capable of opening up completely new markets (for example handicapped or elderly persons who are one of the prime target audiences for the new e-bikes). To draw on the metaphor of games: incremental innovations seek to invent new moves within a game, whereas radical or disruptive innovations either change the rules of the game (radical inventions) or the game itself (disruptive inventions).

While innovations are about inventing, changing, or enhancing products, services, business models, or value propositions, "innovation management" deals with processes and people. It concerns how an organization can foster a climate in which innovation can take place. Innovation management deals with issues of human systems development, such as the question of how to improve team motivation or flexibility, how to implement diversity without raising transaction cost for settling the possible conflicts triggered by diversity, or how to promote corporate creativity and focus by sharing knowledge and mutual visions. The relationship between innovation management and innovation thus has to be regarded as a relation of means (innovation management) and ends (innovation as a primary source of corporate success). In this relationship, innovation management serves as a tool used to develop the key driver and core substance for gaining competitive advantages.

The following will analyze how this key driver for corporate success has to be redesigned if innovation is to keep its primary role in sustaining corporate and societal prosperity and wellbeing. The paper applies a conceptual approach to innovation and innovation management. This implies that the arguments unfolded here need to be expanded on further in subsequent papers. These would need to address the material consequences for corporate strategies and innovation management which flow from the arguments presented here.

Innovation is regarded as the golden bullet for corporate success. By contrast to efficiency, cost cutting, and organizational or process design measures, such as TQM, EFQM, lean management, Kaizen, or Six Sigma, innovation and innovation management promise quantum leaps for profitability. Apple's invention of the iPhone tells the story: this invention was not only a game changer in the market for mobile communication, but propelled Apple within a mere 2 years into one of the most profitable and, for several years, most highly valued enterprise on Earth. By establishing genuine competitive advantages, innovation and innovation management secure market leadership and above-average cash flow, revenue, and greater corporate value. However—as this paper will argue—*the inherent logics of technology-induced innovations has a dark side to the bright shining moon of corporate success.* This dark side can be called the *paradox of destructive wealth creation* (Glauner 2018). It lies not only at the heart of disruptive innovations that are fuelling winners-take-all markets (Seba 2006, 2014), but also in the corporate pursuit of incremental innovations that similarly lead to developments which, by and large, can be highly successful for individual firms, but highly destructive for the surrounding systems these firms are living in and living from. This paradox consists in the fact that the unbroken success of today's technology-driven business models continues to fuel a historically unprecedented period of new opportunity and wealth creation not just for the happy few, but indeed for the great many—while also eroding the very ecological and social foundations of that wealth and prosperity. This erosion of the basis of today's golden age of wealth creation can be seen in three, co-dependent and mutually reinforcing dimensions:

On the *political level,* the destructive nature of wealth creation can be seen in the constant shifting of power and influence between the spheres of economy, politics, and society. With the new consolidation of influence, increasingly apparent monopolies of knowledge, and restricted access to market,[1] this shift is not only undermining the concept of free markets, but subverting democratic decision making processes. The threat lies in the fact that trans, multi, and supranational corporations are increasingly beyond the reach of official authorities, political parties, or legislators, as the agenda is set by the forces of the market and the interests of market actors. In their most advanced form—Shoshona Zuboff speaks of the advent of surveillance capitalism (Zuboff 2016)—the algorithms of Google and Facebook even influence the behavior and choices of the greater public, feeding through into elections and political processes. Companies like Apple, Google, Facebook etc. get that data with which they create new semblances of reality that determine our political discourse from us, their very users and customers. A Facebook experiment in 2013 revealed

---

[1]The pursuit of dominance in the market to gain control over market access can be called the pursuit of *inverse monopolies* (Glauner 2016). The market strategies and negotiating power of leading retailers show the mechanisms behind such inverse monopolies. According to a report published by German anti-trust officials in April 2014, 85% of the German food retail industry is dominated by a mere four companies: Aldi, Edeka, Lidl/Kaufland, and REWE. They use their power to force suppliers in increasing dependence, while also engaging in a strategy of flooding the market with discount to premium store brands. Traditional brands are under pressure from three sides, which only reinforces the monopoly-lite power of the big four retailers: First, small to mediums-sized producers are forced to accept often ruinous discounts and bonus payments, and even subsidize the shelf space and marketing campaigns for their products. Second, since the retailers do not expect the same discounts or payments from their own store brands, the independent producers are placed at a natural disadvantage. This is made worse by a third aspect of the strategies used by the major retailers: The creation of store brands often goes hand in hand with more vertical integration, with retailers acquiring or establishing own production and processing facilities. This in turn gives the retailers even more leverage. A recent comparative advertising campaign by Lidl pitted well-known brands against the discount retailer's own brands, representing a type of price war against products on offer by Lidl itself: A radio advert compared the sound of opening a bottle of Lipton's ice tea with the sound of a Lidl ice tea, with a voice over asking: "What's the difference? Lipton's ice tea, €1.50. Lidl ice tea, €0.49".

that every user's behavior can be changed and controlled by a random change in the algorithms. This gives the algorithms the power to not only influence markets and product preferences, but even affect the outcome of elections, as the discussion about the impact of social media on the US presidential elections shows (Lobe 2016).

On the *social level,* the danger of destructive wealth creation lies in the increasing disconnect between the winners and losers, made worse in today's winners-take-all markets. The world is increasingly falling apart into a handful of people creating and enjoying their access to excessive wealth and exclusive information and opportunity and the masses disenfranchised from these opportunities despite the apparently substantial growth in wealth.[2] As Boston Consulting puts it in its research paper "Global Wealth 2015: Winning the Growth Game" the index of private wealth has increased by 11.9% in 2014, after having grown by 12.3% in 2013—2 years with historically low interest rates. For comparison: In the same time, the global GDP grew by 3.41% in 2013 and 3.39% in 2014. Thus the increased wealth recorded by Boston Consulting is the product of a disproportionate increase in the value of highly unequally distributed wealth (BCG 2015a, b). These findings are called to mind again in the study "Work crisis—a divided tale of labor markets" of Aleksandar Kocic, Managing Director Research Deutsche Bank in New York (Kocic 2015). Confirming a McKinsey report by Lowell Bryan and Diana Farrell published in 1996 (Bryan and Farell 1996), the Deutsche Bank assumes that, for the first time in economic history, the significant rate of growth is not leading to a trickle-down effect, adding to the common good or creating new jobs. "For the first time since the industrial revolution, new technology is destroying more jobs that it is able to remobilize. And as ever less labor is needed to produce the same output, it is becoming clear in some countries that growth is now possible without rising employment and wages. Such a profound change is bound to have immense economic and social implications" (Bryan und Farell 1996, p. 47). This analysis is again confirmed by a study of ING Diba Bank that explores how the technology-driven, disruptive business models of today are widening the gap between the winners and the losers. This wealth creation is fueling a process that—according to the authors Erik Brynjolfsson, Carl Frey, Andrew McAfee, and Michael Osborn—is not only

---

[2]"In 1990, 1.9 billion were living in extreme poverty. By 2015 that number had been cut by more than half, to 830 million, while in parallel the global middle class had almost tripled. And most citizens of advanced economics today command goods and services that were beyond the reach of even kings and emperors only 200 years ago"(Stuchtey et al. 2016, S. 8). Even if these figures sound extremely positive in the abstract, they need to be taken with a pinch of salt: As the Pew Research Center authors Rakesh Kochhar and Russ Oats stress in their study "A global middle class is more promise than reality", poverty is defined as an income of less than $2 a day, average and high middle-class income as $10.01 to $20 or $20.01 to $50 per day, and high income as $50 per day. People above the global poverty line and up to those defined as the global middle classes therefore have an income of between $730 and $7300 per year: "Even those newly minted as middle class enjoy a standard of living that is modest by Western norms. As defined in this study, people who are middle income live on $10–20 a day, which translates to an annual income of $14,600–29,200 for a family of four. That range merely straddles the official poverty line in the United States—$23,021 for a family of four in 2011" (Kochhar und Oats 2015, p. 6).

putting almost half of all jobs at risk, but actually eroding the very basis of self-sustaining economic prosperity (Brynjolfsson and McAfee 2014; Frey and Osborne 2013).

On the *ecological level,* the erosion of the basis of prosperity is immediately evident in the phenomenon of global resource depletion and the resulting destruction of the foundations of our natural existence. Before and beyond all ideologically charged debates about global resource consumption, climate change, or man's role in the sixth great extinction of life on Earth (Wilson 1992, p. 30), it is plain to see that these forces are also endangering the economic foundations of growth. This is due to the simple fact that economic growth is owed to the factors of labor, capital and, especially, access to resources which, as the McKinsey writers Stuchtey et al. (2016) stress, is cannibalizing itself at the current rate of exploitation.[3] It can also be seen at work in how the exploitation of resources is threatening to undermine the political and social foundations of our nations and societies. Consider the phenomenon of climate change: "Global warming is likely to uproot hundreds of millions, perhaps even billions of people, forcing them to leave their homes and creating severe global political and economic challenges. . . . Of the 6 billion-plus humans . . . nearly a fifth are threatened directly or indirectly by desertification. China, India, Pakistan, Central Asia, Africa, and parts of Argentina, Brazil, and Chile all have areas with low rainfall and high evaporation that account for more than 40% of Earth's cultivated surface. . . . Because such a large portion of the Earth's people live near sea level, a significant rise, even by a foot or two, could cause forced migrations of tens or even hundreds of millions of people. . . . The Christian development agency Tearfund has estimated that there will be as many as 200 million climate refugees by 2050 and as many as 1 billion by the end of the century if global warming and its impact continue" (Watts 2007, p. 101).

At its core, this triple paradox of destructive wealth creation forces us to acknowledge that our economic means and ends are collapsing under their own weight, because our—per se rational and apparently extremely successful—behavior on the

---

[3]Referencing Ayres and Warr (2005, 2009), Stuchtey et al. (2016) have revealed that the exponential growth since the industrial revolution has for the most part been due to the exploitation of fossil fuels and other natural resources: "Taming wind and hydro energy, and inputting them into the economy, once allowed mechanization of grinding, pumping, sawing, irrigation and many other laborious tasks . . . Taming coal and vastly increasing the amount of energy put into the economy was crucial for the first industrial revolution. While our modern economy has of course moved on from horses and steam engines, it is still striking how many industries continue to depend heavily on natural resources: food, transport, construction and all primary material production, for instance" (Stuchtey et al. 2016, p. 59). The idea "that the success [of modern day wealth creation, FG] is largely built on transforming natural capital, the economist's word for natural resources, into other forms of capital" (Stuchtey et al. 2016, p. 9) leads them to conclude that today's untenable rate of resource depletion will lead to an inevitable collapse in the accepted model of constant economic growth: "Since the mid-1980s and with ever-increasing speed, environmental depletion has reached a global scale and scope where it actually starts to threaten the viability of our model of wealth creation itself. Our economy has grown so big, so fast, that it is quickly depleting the very same natural capital on which it thrives. In a way, it is falling victim to its own success" (Stuchtey et al. 2016, p. 11).

individual level is creating highly destructive forces on the level of the total system. It is evident when we see this mechanism from the perspective of a potentially impending global collapse. As Safa Motesharrei, Jorge Rivas, and Eugenia Kalnay argue in their seminal NASA study "Human and Nature Dynamics (HANDY): Modeling Inequality and Use of Resources in the Collapse or Sustainability of Societies" (Motesharrei et al. 2014), societies will falter with mathematical certainty either if the usage of resources exceeds the replacement rate of natural growth, or if the degree of inequality in distribution and participation leads to upheaval and turmoil (Motesharrei et al. 2014; Williams 2013). Since our economies as well as our social discourse seem to be bound to the concept of growth,[4] and since this growth is increasing inequality rather than reducing it, the collective push for growth speeds up resource depletion, thus leading to the destructive dynamics which place individual businesses and our societies as a whole in existential jeopardy. For the future viability of companies, this means that they have two options. Either they try to play the Great Game of Strategy (Glauner 2016, 2017a, b), i.e. the game of innovations which accelerate the downward-spiral of disruption, concentration or resource depletion, or they must find a different way to secure their existence in present and future markets. However, in order to do so, our visions of truly viable innovations, venturing and business management first need to free themselves from the shackles of existing economic mental models. Only then can we develop innovative business models that could cut through this spiral and become sustainable enough to enable wider participation in wealth by creating and augmenting resources sustainably. If fiction informs reality, future viability needs the strategic development of new fictions to undergird the models and markets of tomorrow.

To overcome the paradox of destructive wealth creation, we thus not only need to change our understanding of innovation and innovation management, aligning both towards more sustainable and CSR-related goals. We also need to change the basic mental model that forms our understanding of innovations and innovation management, channeling both into a rationality which, at its heart, fuels the paradox described here. To this end, this paper will analyze, in a first step, the mental model which informs the perspective of our present games of innovation. In a second step, the paper will outline the parameters of a new mental model for a more viable concept of enterprise and economy, a filter by which corporations can install innovation processes which will lead to corporate excellence by cutting through the downward cycle of destructive wealth creation at the same time.

---

[4]One is the human pursuit for differentiation, analyzed by Pierre Bourdieu (1982). It triggers our continuous search for new products, which we want to consume more often than not simple in order to show that we can do it. The other two are the economic spiral of prices and innovations and the need to yield interest on invested capital (Glauner 2016).

2 Scarcity, Competition, Value Creation, and Growth: The Mental Model of Economy and Innovation Management

2.1 Challenges and Driving Forces of Innovation Management

What is the crucial driver of innovation? The blunt answer is: our human pursuit of progress and success. This is based in our understanding that "the better is the enemy of the good", which fuels a mechanism that Joseph Schumpeter labeled the principle of "creative destruction" (Schmupeter 1942). This principle leads to the corporate pursuit of innovation alongside three mantras which, in combination, determine the logics of modern enterprise. The first prays "Outwit, outperform, outsmart". This mantra addresses the belief that it is self-serving human creativity which must be activated if one wants to excel in the games of competition. The second mantra states "Be better or vanish". It sings the high notes of ceaseless progress which, if practiced with single-minded determination, will lead to success in one's markets. It addresses the mechanisms that shape the fields of innovation. Finally, the third mantra is the solemn bass voice of "Be different or die". With the deep tones of faith, this mantra addresses with utter conviction the belief that any presence in any market comes from a self-centered notion of differentiation, in which all others—Michael Porter calls them the five forces of rivals, suppliers, customers, new entrants, and substitutes (Porter 1980)—have to be viewed more as threats to one's own success than as possible partners for more mutual and longer-lasting forms of benefit creation. If we interpret innovation in terms of these three mantras, innovation is understood as the stealth weapon in the fight for corporate success. It secures by winning in a market in which only some, but never all players can win.

However, if we really want to understand the crucial challenges that accompany innovation and innovation management, we have to look at our games of innovation from a different angle. To capture this perspective, we need to ask about the crucial effects present day innovations are triggering. These effects consist of four global megatrends that, on the macro and supra level of markets and societies, form the core challenges and frames of reference which innovation and innovation management are bowing to. By contrast to most other megatrends, such as urbanization, health, the aging society, digitalization, population growth, or the gender shift and female empowerment, these four megatrends will affect all and every corporation everywhere in the same manner (cf. Fig. 1): The first of these is the *acceleration of all corporate and societal processes*. This megatrend affects all parameters of any business: its direction, its structure, its strategy, and its substance. By raising their ability to innovate and change themselves, all companies are forced to find ways to keep up with a market that is constantly speeding away from them. The second megatrend consists in the *loss of established boundaries in markets and services*. This second megatrend means that the impact of globalized information and production processes is merging and blurring formerly separate markets and services both horizontally and vertically. The unbounded nature of modern business forces

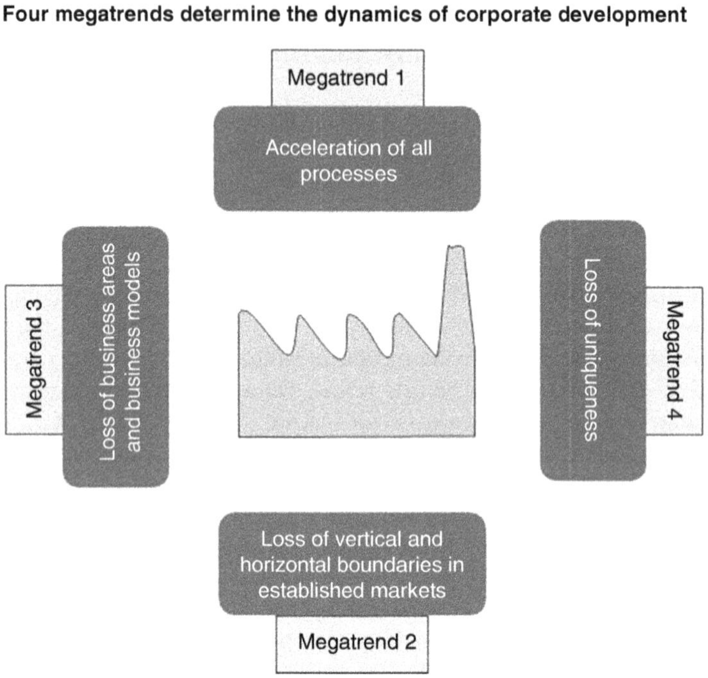

Four megatrends determine the dynamics of corporate development

Megatrend 1

Acceleration of all processes

Megatrend 3

Loss of business areas and business models

Loss of uniqueness

Megatrend 4

Loss of vertical and horizontal boundaries in established markets

Megatrend 2

Fig. 1 The megatrends squeezing businesses

companies to understand how ever greater transparency, global competition, and the blurring of old business models and ways of doing business affect their current and future models and how they can respond to this blurring. The *loss of established business areas and business models* represents the third megatrend. It affects the shape of current and future business models. The constant availability of information everywhere poses a new question: How can new media and organizational forms be used to accommodate the loss of old business areas and business models? Finally, all corporations must face the challenge of *loss of uniqueness*. This fourth megatrend is a consequence of the first three, as products and services are becoming more easily replaced in a globally homogenized economy and older price policies are harder to justify. Companies need to know how they can develop unique and inimitable competences to counter the easy replication and substitution of products and services and the downward pressure on prices it causes.

With a view to these megatrends, all corporations are faced with two quite divergent challenges: first they must become ever more flexible; second, they must become ever more distinct. In order to stay competitive, companies need to establish structures that allow them to become more flexible, adaptable, and changeable, and they need to establish a benefits-oriented profile that protects their unique identity for the future. To bridge the gap between these divergent challenges, innovation management tries to walk down two distinct roads: First by establishing corporate cultures which help breech the negative effects and constraints that flexibility places on individuals and their ability to be motivated to change (Sennett 2006). Second, they seek to win within the games of innovation by possibly inventing new business models, processes, or products which yield either long-lasting competitive advantages or, even better, the disruption of entire markets. However, it is exactly this latter road which fuels these four megatrends in the first place, thus leading not only to the paradox of destructive wealth creation, but also, on the micro level of corporations, to ever more obvious existential jeopardy challenging individual firms. As Clark Gilbert, Mathew Eyring, and Richard Foster show, the average lifespan of all corporations listed in Standard and Poor's has dropped from 61 years in 1958 to 25 years in 1980 to 18 years in 2012 (Gilbert et al. 2013, p. 44). As McKinsey Consultants Richard Foster and Sarah Kaplan predicted already in 2002, this pattern of discontinuity will continue. In their view, the average life expectancy of the companies listed on the S&P500 index will be 10 years by 2020 (Foster and Kaplan 2002, p. 13).

To overcome this development, innovation management cannot proceed as usual in its attempt at keeping the spiral of acceleration and disruption spinning ever faster. But how could this get achieved? Our answer is: by reinventing the mental frame which drives our actions in economic reasoning.

2.2   A Philosophical Remark on Mental Frames and Their Role for Human Cognition and Action

Before analyzing why and how the binding logics of our present model of economics is triggering the paradox of destructive wealth creation, we need a careful look at the role of mental models. Designing models is one of the academic's preferred ways of gaining an understanding of complex phenomena. To this end, we reduce the phenomenon at stake to its basic elements, i.e. the elements which seem to us to be crucial. Once they are identified, we then arrange and combine them like parts of an engine, a machine, or a building in order to analyze how they interact and work together. As is already the case for the selection of the elements, this usually rather mechanistic analysis of causes and effects is carried by a set of background assumptions (hypotheses about laws, driving forces, dependencies, and causal relations) which embed the model into a broader realm that gives the model its specific aim. Thus scientific and real-world modeling—for example in physics, carpentry, economy, or architecture—is the process of reducing multidimensional complex phenomena into less complex forms that depict only 'the relevant' aspects which then can be analyzed, tested, and proven or disproved in reality.

The ideal of this type of modeling can be found in physics and mathematics, which is the reason why the study of economics seems so committed to clothing its models in mathematical terms and equations, even though most economists know that they are dealing with beliefs and mindsets, not laws of nature, which make the markets turn.[5] However, in this rather mechanistic process of modeling, not only our models often seem to behave badly (Derman 2011), but so too does reality, which quite often does not bow to the expectations we cast upon it. This leads to such phenomena as black swans, i.e. the completely unexpected (Taleb 2008). However, even more critical is the fact that a crucial blind spot remains at work in all modeling endeavors. It consists in the primary assumptions about the object in the first place. For our models in economics as in all other social sciences, this becomes a problem, because all of our reasoning about reality is bound to quite often subconscious systems of belief and prior assumptions forming specific theories, which serve as filters when modeling the world.[6] This holds true for natural sciences and even more so for all forms of social sciences whose laws, by contrast to the laws of nature, are only present in our self-fulfilling descriptions of the world.[7] Following Voltaire's quip that cause can follow effect like a doctor following a patient's coffin, our perception of the world in the social sciences depends on the narratives that we use to describe the reality of our social institutions. Our beliefs make reality and not reality our beliefs.

A simple example can show this: As Juan Elegido found in his study "Business Education and Erosion of Character", referencing Simon (1985), Jensen and Meckling (1994), and Tetlock (2000), our mental models and attributions of what human beings are and how they act have major consequences. If people understand

---

[5]This can be shown not only with a view to the formation of bubbles, but even more so for one of the core inventions of economic transactions: the invention of money. Money, whether it be a thousand dollar bill or the Rai stone money of the Micronesian people of Palau and Yap weighing up to 5 tons, always carries its wealth only by force of the trust people put into it and into its owners. Such trust is a social, not a physical construct. As vast numbers of bursting bubbles show, it can vanish in seconds, leaving nothing else behind but waste paper and stones.

[6]Following Pierre Duhem (1908) and Thomas Kuhn (1962) on the paradigmatic nature of scientific theory formation and Heinz von Foerster (1972, 1987) and Ernst von Glasersfeld (1995) on the cybernetic construction of reality, we have to admit that the objects of scientific enquiry are essentially constituted by the theories and observation / measurement processes we bring to the equation. But our perception of the world is not just dependent on the theories we espouse. It is also un- or underdetermined by its objective nature (Quine 1960). What we experience as existent has always been 'ontologically relative' (Quine 1969). The world is changing in the wake of new theories coming to the fore—our many scientific revolutions—in its basic form, function, and make-up. This now implies that the world can be described in many different ways and that no single way can lay claim to absolute truth. As Paul Feyerabend argues against the adherents of a positivist science and logical positivism and against the deceptively enlightened methods of rational science, such as Popper's fallibilism (Popper 1935), all descriptions of the world are allowed. There is no one method with a rightful claim to universal validity: "anything goes" (Feyerabend 1975, 32).

[7]As John Searle argues with good reason, the forms and laws of our social world, that is, the world of our man-made institutions and belief systems, are made by our declarations. Declarations are statements which bring about the fact when uttered, as the statement "You are fired!", said by a boss to his or her employee, brings about the very fact it states. As Searle concludes with a view to our social practices and belief systems, "all of institutional reality, and therefore, in a sense, all of human civilization, is created by speech acts that have the same logical form as declarations "(Searle 2010, 12f; vgl. 1995, 59ff und 1969, 50ff und 175ff). These declarations are mirrors of the culturally informed hopes and aspirations that we carry into the world.

themselves as rational maximizers of self-interested benefits, they will change their behavior to match the expectations triggered by our economic concepts of scarcity and competition: Whenever a win–win outcome is impossible, the mission is to maximize one's gains at the expense of the others involved. Since we cannot ever be absolutely certain that the other side will be cooperative, every economic relationship suffers from an inherent lack of trust. This is made worse by the mental models of economics, such as the principal-agent problem in contractual relations put forward by Jensen and Meckling (1976). Almost all facets of competitive thinking are governed by the idea that human beings will follow their own good first. Selfishness abounds. In the case of the economics undergraduate: "Students will come to expect that other people will act that way [i.e. selfishly, FG]. This has clear practical consequences because it is well established in prisoner dilemma experiments that most subjects will defect if they are told that their partners are going to defect (Dawes 1980). In other words, the mere fact that people expect that others will behave selfishly will tend to make them behave selfishly (Miller 1999)" (Elegido 2009, p. 18).[8]

Coming back to the insight that our beliefs make reality and not reality our beliefs, the cautious remarks on mental models and modeling in science and reality can be applied in two directions. First, and seen from an epistemic point of view, we have to become aware of the fact that all our models of reality are ineluctably affected by initial background assumptions which cannot be reflected by the model itself, since they function as founding principles for inventing the model in the first place. This epistemic fact I shall call the intrinsic values bias in modeling reality. Second, and seen from an action-oriented point of view, we must understand that in social reality every mental model sets up a set of expectations upon which we act, thereby triggering a self-fulfilling feedback loop which brings about the very reality we expected. This fact I shall call the self-fulfilling prophecy of mental models in the social realm.[9] The danger of mental models thus consists in the fact

---

[8]This explains why even Michael Jensen nowadays prefers to talk about integrity and morality, instead of how to control misconduct. As one of the founders of the principal-agent approach, which was built on the concept of mistrust, his recent thinking on integrity and morality as "powerful access to increased performance for individuals egoism, groups, organizations, and societies" (Erhard et al. 2009, see also Erhard et al. 2016) acknowledges that stick and carrot systems designed to address defection and non-cooperative behavior do not serve to raise trust. Referring to integrity and morality in this sense is an alignment to the humanistic insight that only mutual trust and faith serve to raise what Elinor Ostrom and Richard Sennett call the social capital that is needed if a business wants to excel (Ostrom 2000; Sennett 2006).

[9]This self-fulfilling prophecy becomes most apparent in our mental model of who and what man is. We can see opposite ideas of the nature of humankind: on the one side Thomas Hobbes' homo homini lupus, assuming that our bio-psycho-social conditioning makes us inherently evil and our actions selfish, uncivilized, and self-interested (Duerr 1988; Dawkins 1976) and, on the other side, the emancipated and enlightened idea of Jean-Jacques Rousseau and Immanuel Kant, who considered man to be empathetic and good by nature, capable of behaving in a civilized and cooperative fashion in normal circumstances (Elias 1939; Bauer 2006, 2008). In fact, and on well-established biological and historical grounds, we have to acknowledge that we are both (de Waal 2005; Harari 2015). Even though this is true, it still remains the case that individual people tend to think one way

that fiction determines reality: our mental models of the world determine our actions and our actions determine our social reality. The only way out of this circle is to confront a specific model with critical and open reflection about its intrinsic values bias, its inherent binding logics, and the consequences which flow from both in reality.

2.3   Innovation Games: The Mental Model of Economy

Applying these cautious remarks on modeling reality, we need to distinguish between three levels of bias when modeling the economic realm that informs our understanding of innovation as the core driver for corporate success.

At the most basic level, we must ask about the role of human nature in economic decision-making. Here, the bias of economic theory-building consists in the concept of homo oeconomicus, i.e. the understanding that humans act in the economic realm as rational maximizers of their own good. On a second level, we ought to ask about the purpose of the economy and the purpose of organizations which facilitate economic activity. Here, the bias of economic theory consists in the understanding that business is a game of competition devoted to raising value and profit by allocating scarcity. At the top level, the third and most obvious one, we want to model best practices and rational procedures, for example in designing processes and organizations, building strategies or developing human resources. Here, the bias consists in the assumption that raising value can be calculated and modeled in rather mechanistic terms according to clear-cut input-output ratios that would lead the way to competitive advantages, which in turn lead to certain results like more market share, profits, or share value.

Taking a step back when analyzing this threefold bias of modern economic reasoning, four central concepts stand out: scarcity, competition, value creation, and growth. While adherents and critics of the market economy quarrel over how to interpret and deal with competition, value creation, and growth, they seem to share the same view on scarcity. They differ only insofar as critics of market capitalism reference scarcity with a normative claim on just distribution and sustainable ventures, while the adherents of market capitalism approach the problem of scarcity in functional terms.

However, if we found the mental model of economics and business upon the notion of scarcity, we cannot escape the fly bottle of the initially described paradox of destructive wealth creation. This is due to two facts. The first consists in the already outlined understanding that economic reasoning evolves from the self-fulfilling feedback loops of our expectations, which are themselves determined by our mental model of the subject. The second fact is a direct consequence of the first.

---

or the other. And this mere belief alone already triggers an expectancy loop which most often leads to its own fulfillment.

If we base our mental model of economics on the concept of scarcity, it becomes difficult to bridge the normative and the functional views which are at the heart of the conflict on how to deal with the paradox of economic wealth creation. This is due to the fact that the rationale of competition on and about scarce resources necessarily pushes the competing parties into a race about relative gains,[10] a race that leads on the level of the complete system to the paradox described above.

If we assume that the homo oeconomicus is a rational self-interested actor pursuing the greatest good for himself, and if we then combine this concept with the concept of free markets distributing scarce goods in the best possible manner and producing individual profit at the same time, this scarcity-driven competitive model of individual wealth creation cannot but lead to a spiral of all market relations being increasingly subject to actors trying to outsmart each other in the pursuit of greater gain. More and more aspects of our world are being pulled into this spiral of economization, and the forces of concentration in collective competition are beginning to undermine the very basis of our existence.

This becomes plain when we combine the consequences of technological evolution with the key premises of the economic worldview in a causal chain:

Premise 1: People and enterprises are rational maximizers of their own good.
Premise 2: Goods and resources are scarce.
Premise 3: In the fight for scarce goods and resources, people and enterprises are in competition.
Premise 4: The optimal fora for managing scarce goods are competitive, liberalized markets governed by the free rules of supply and demand.
Premise 5: An advantage in the profit-driven competition for scarce goods and resources generates greater profit, which can again be used to gain more competitive advantage.
Premise 6: Creativity and intelligence can overcome scarcity in means, goods, and equity, so that the invention of novel, disruptive business models can make long-established "top dogs" obsolete.

Combining these premises of the economic mental model with the above-described paradox of modern wealth creation, we can see the following feedback spiral forcing the hand of the economic actors and accelerating the economization of our world:

---

[10]Robert Frank understands this race about relative gains to be the basic principle of the present day "Darwin economy". Its internal logics consists in "the fact that individual and collective incentives diverge when relative income" or gains matter (Frank 2011, p. 44). Since in most economic games of competition, relative gains of individuals matter more than absolute gains of the group or the whole system, these divergent interests of individuals and groups trigger behaviors and "arms races" that not only cause enormous harm to the group, but tend to lead systematically to market failures which, in the end, offset any lasting advantages even for the most successful individuals in the system (Frank 2011, 30ff).

– Competition for scarce goods in an iterative process of profit and loss is leading to the concentration of assets in the hands of a few selected players, which is globalizing and accelerating technological evolution.

  – The concentration creates additional competitive advantages for the big players (economies of scale and horizontally or vertically integrated value chains), which are reinforced by dominance strategies which close off markets.

    – Competitors in markets closed off by dominance strategies are forced to find disruptive business models as the only way forward for smaller or less established players.

      – Globally disruptive business models (such as Industrie 4.0) are leading to new concentration.

        – This wave of concentration is putting countless jobs and a broad and diverse enterprise landscape at risk.

          – The erosion of a broad, small-scale employment and enterprise landscape is leading to increased pressure on prices for the products and services of remaining actors.

            – The increasing pressure on prices for products and services can only be balanced with new disruptive business models or even more extreme cost-reduction measures in production (reducing labor and resource costs, outsourcing the costs down the supply chain, production in cheap labor countries with low or no social or environmental standards)

              – The increasing pressure on prices echoes around the world and places the resources and income of a broad landscape of suppliers and their employees under pressure.

                – The growing pressure on resources exacerbates the spiral.

Seeing this downward spiral, all corporations are confronted with the task to innovate and transform themselves and their business models in ever quicker cycles. However, it is exactly this pressure to innovate which keeps this spiral spinning. Innovation thus takes a Janus-faced shape: it turns out to be the engine that will give individual firms the lead in the collective race for unbridled gains, a race that more and more destroys the economic, social, and ecological resources individual firms and our societies as a whole are living on.

3   Reframing Innovation and Innovation Management: Ethicology and the Natural Art of Yielding Benefits

As the last section has shown, we cannot overcome the present paradox of economic wealth creation being triggered by our games of innovation if we stick to the concept of scarcity as a core driving force of economic activity and business management.

But is there a genuinely sound concept or idea that could replace the fact of scarcity on which the mental models of economics are built? And if yes, how does it alter the three mantras of business management which guide our understanding of innovation and innovation management? The answer to the first question will guide us to the answer to the second. It can be found if, for a moment, we turn away from economic reasoning and look into the probably most efficient, dynamic, and innovative form of, in a sense, value creation which ever took place on earth: Nature. Seen in terms of exchange, the living world has to be understood as the most dynamic, innovative, and efficient system of beneficial transactions. However, even as every individual living being and all ecosystems are finite entities, stocked with only limited means and scope, nature does not operate in a way that would yield the destructive effects our economically informed innovation games are triggering. To understand this fact, we need to turn to the processes which guide natural development and growth.

3.1 The Principles of Natural Resource Creation

Let us consider the environment: All ecological processes are grounded in cyclical systems operating freely in accordance with five natural principles: locality, freedom, sense of scale, diversity, and value creation. The basic resources that are available to the cyclical system and the sum total of the value creation cycles in the system determine the system's potential for growth and differentiation. In light of these principles, nature has to be understood as a highly dynamic feedback loop whose growth feeds on its own substance. This can be seen if we look for a moment into the extinction rate of species: As a matter of fact, a vast number of formerly stable ecosystems collapsed and 99% of all species that ever lived became extinct over the course of the last 3.5 billion years of life on earth (Otto et al. 2007). Usually, we interpret this fact with the idea that the fittest win in the game of selection and adaptation. According to this, the development of a given species is a continuous fight over scarce resources. We explain the principles of nature with man-made ideas of scarcity and competition, arguing that surviving in this fight requires fitness, i.e. relative competitive advantages in adaptation. But this is not the central principle of nature. Not competition, but symbiosis is the driving force of all that lives (Capra 1996). Whether we look into our body or the plant systems of any forest, it is the symbiosis of many different types of microbial organisms and bacteria that makes them viable in the first place.[11]

---

[11] As Bernhard Kegel says with a nod to Gilbert et al. (2012), the biological concept of individuals should be dropped in favor of the idea of holobionts, that is, of symbiotic systems that engage in value-adding interactions with other similarly symbiotic systems (Kegel 2015, p. 309). Even though this statement is highly problematic if we see it from the human cognitive perspective—i.e. if we take into account that most of our ethical, philosophical, psychological, religious, social, and economic concepts, theories and world views and thus our basic understanding of rights, duties, liabilities, and responsibilities depend on the concept of a conscious, self-reliant and free-willed person—the cognitively necessary concept of an individual can be challenged in biological terms.

If we describe nature with economic principles, we get stuck in what could be called the Darwin trap. This trap lies in our tendency to describe nature according to our concept of human competition and interpret it mechanistically with the economic concepts of competition and scarcity. But: The principle of nature is not scarcity, but abundance—abundance understood as the holistic feedback cycle of continuous and substantive surplus and added value creation which in turn feeds the dynamics of selection, differentiation, adaptation and natural growth. What has enabled certain species to survive? The answer is not that they were fitter and quicker in adaptation and the race of competition, but that they create greater value for the system around them than what they take out of it. Bees and mycorrhizae show us how this works: Both live in symbiosis with their hosts and the environment. They create added value for that environment that far exceeds what they get from it in return. Symbiotic surplus and added value creation is the basic principle of nature. Natural growth is based on systemic added value creation, in which the value that a sub-system contributes to the greater whole is greater than the benefits it takes from it.

What enables a species to survive in its ecosystem? Not flexibility or adaptive fitness in a competition for scarce resources, but the creation of added value that keeps the greater system intact and helps it to grow and differentiate. With their forces of constant differentiation by adaptation and change, natural cycles are thus governed by two general laws. The first systemic law of ecology says that only those subsystems of ecosystems will survive in the long run that contribute added value for the total system, value that goes beyond what the subsystems take out of it. This principle of value contribution leads to the second systemic law of ecology: Added value cycles are cycles of exchange whose basic pool of resources keeps growing in accordance with the five principles of natural resource creation.

To escape the Darwinian trap, we need therefore to open up to the basic laws of natural cycles. Instead of either seeing nature in terms of the economy, interpreting it with the mental models of scarcity and competition, or in a misconstrued understanding of sustainability, namely in terms of mere efficiency and the careful use of resources, we need to bring our economy into alignment with the resource-creating growth principles of nature. This means kick-starting value-creating cycles of surplus. By contrast to the zero emissions ideals of the blue economy proposed by Gunter Pauli (2010), the efficiency or nutrition cycles of von Weizsäcker et al. (1995) and Braungart and McDonough (2013), or the model of closed, self-organized dynamic living systems (Capra and Luisi 2014), the key here is the open organization of surplus, added value, participation, and growth processes that protect their own survival by creating new resources. This type of growth is not just a qualitative, but

---

This is due to the fact that all species of a higher order, such as mammals, birds, insects or fish, have to be understood as complex systems of living entities (cells, bacteria . . .) which need to interact symbiotically if those smaller entities (cells, bacteria . . .) and the higher system they constitute want to stay alive. This holds true for all living systems of a higher order, be it a rabbit, a bee, a wolf, or indeed homo sapiens. Another example of this would be the complex root systems of forests. We now know that trees not only live and thrive in symbiosis with mycorrhizae, but actually use that symbiotic network to communicate with other trees to protect each other's survival (Hachtel 1998).

also a quantitative process that stops today's spiral of disruption, concentration, and resource exploitation. It does so by establishing business models and structures that work with the natural and man-made social systems to form differentiated, regionally decoupled, and decentralized, autonomous value creation cycles, thereby growing and enriching the very stock of resources sustaining the system as a whole.

Resource-creating business models that emulate the five natural principles are therefore squarely opposed to the basic ideals that lie at the heart of economic reasoning: the ideas of absolute dominance over the market, the value chain, and the revenue stream are replaced with the idea of cooperative value-adding cycles for holistic and substantive value creation. This idea is therefore the counterpoint to the short-term profit logics of exploitative economics that flood with their ever quicker innovations already saturated markets with more products and services, whose sense is often limited to the surrogate experience of meaningless consumption. In their focus on enabling, enhancing, and enriching all links of the value chain, resource-creating business models aim to achieve value-adding participation, i.e. an economy that applies the natural principles to the micro, meso, macro, and supra-levels of economy, society, and nature. The micro-level concerns the relationships between companies and people (employees, customers, suppliers, and business partners); the meso-level concerns the relations between companies (suppliers, clients, or competitors); the macro-level covers the relationships between companies and the systems surrounding them, external stakeholders, politics, and society at large; and the supra-level concerns the relations between companies and the environment.

However, in order to engage successfully in this kind of mutual value creation, business management in general and innovation and innovation management in particular need a different guiding line than the three mantras of "Outwit, outperform, outsmart", "Be better or vanish", and "Be different or die" which shape the self-referential understanding of competition, advantages, and success in the games of allocating scarcity. This other guiding line is provided by the values of reciprocity, fairness, truthfulness, and partnership. These values are at the heart of the concept of the Global Ethos as analyzed by Hans Küng (2012) and reflected in the concept of humanistic management (Dierksmeier 2016, Pirson 2017). The Global Ethos includes a set of rules and values that are shared by all cultures and great religions, thus fostering mutual engagements which can bridge conflicts of interest. The principle of humanity means that every human being has inalienable and inviolable dignity and deserves to be treated humanely. The second principle is the principle of reciprocity as it is stated in the *Golden Rule* "Do unto others as you would have do to you." These two basic principles are expressed in the four basic global ethos dimensions: First, by the values of peace and respect for life; second, by the values of justice and solidarity; third, by the values of truthfulness and tolerance; and fourth, by the values of mutual respect and partnership. As universal principles of human exchange, they can serve to foster corporate cultures which adapt to the mental model of added value creation, sustaining more future-viable forms of business than the mere pursuit of added value.

Applying these values to the natural laws of serving benefit and resource creation, we thus reach a paradigm which can actually generate future-viable business models, i.e. business practices which cut through the paradox of destructive wealth creation.

This is the paradigm of ethicology. Ethicology means the application of the Global Ethos values and the five principles of natural resource creation to the realm of economics and business management. It bridges the paradox of economic wealth creation insofar as it serves the human need for differentiation by cutting through the downward spiral of present-day economic value creation. According to this view, sound economic activity consists not so much in the just allocation of scarce goods as in the implementation of natural, i.e. mutually symbiotic wealth-creation processes. This shift in paradigm generates a variety of different niche strategies which not only allow individual corporate success, but also enable the overall system to grow and feed upon itself in a way which is economically, socially, and ecologically sound. The case of Schamel might serve as an example. Managed by the fifth generation of the family, **Schamel Meerrettich GmbH & Co.KG** from Baiersdorf in Franconia employs around 50 members of staff. To protect its future viability, Schamel formed the "Protection Initiative Bavarian Horseradish". With around 100 local radish farmers, the company has managed to win EU certification for Bavarian horseradish. Its thinking was to protect the Bavarian culture of horseradish cultivation, with the regionally typical smallholder structures and with a unique variety in taste and style. The entire production chain benefits from this, because the price of certified Bavarian horseradish is about twice that of comparable produce elsewhere in the global market. With its commitment to quality and value creation, Schamel was awarded a TOP brand award in 2014. The company had already been included as one of Germany's "brands of the century" in a 2007 book on the subject, joining the likes of Mercedes, Lufthansa, Nivea, Duden, Tempo, Miele, or Persil. The business model and dedicated niche strategy of Schamel have excellent value creation scores in the areas of regional decoupling, the formation of cycles of value creation, and ecologically sound production processes. It rests on the understanding that only mutual beneficial strategies in designing regionally decoupled, symbiotic added value chains will secure lastingly viable revenue streams, i.e. economic success that breaks with the paradox of destructive wealth creation.

3.2 "Be Valuable or Die": The Mantra of Lastingly Viable Innovation Management

As these brief statements regarding the principles of natural resource creation have shown, nature aligns its evolutionary mechanisms with a set of values that is squarely opposed to the values which inform and guide our economic behavior. By contrast to the economic rationale which bows to the belief that size and dominance matter,[12] nature never went into size nor into dominance, but into the

---

[12] A good example is Wincor Nixdorf. With its branch of automated teller machines, it competed on a global level with only two rivals, Diebold and NCR. After years of fruitless effort to step up from global number three to either number two or number one, Wincor decided in 2015/2016 to sell its

development of highly diverse and free floating systems of mutuality, cooperation, and symbiotic dependencies. If we abstract from this difference between natural and economic forms of value creation, the difference between natural and economic exchange systems has to be viewed as a difference in the core values according to which each system operates. Within nature, these core values are, on the level of species interacting (species perspective), the values of symbiosis, i.e. mutuality, reciprocity, and symbiotic dependency, and on the level of ecological systems functioning (systemic perspective), the principles of locality, freedom, sense of scale, diversity, and added value creation. Within the mental model of economy, it is competition, scarcity, and dominance which informs the three fundamental mantras of innovation and business management, i.e. our self-centered conceptions of "Outwit, outperform, outsmart", "Be better or vanish", and "Be different or die".

If we analyze this difference regarding the values schemes that guide the exchanges and transactions within natural and economic systems, a first and crucial conclusion can be drawn: All forms of value creation rest in prior forms of values creation. It is this insight which leads the way to an answer regarding the question of how the mental model of natural value creation would alter the three mantras of our economic reasoning. If we follow the mental model of natural value creation, the self-centered momentum of the three mantras "Outwit, outperform, outsmart", "Be better or vanish" and "Be different or die" is changed into a systemically oriented pursuit of holistic value creation. The mantra of nature could thus be stated as "Be Valuable or die". This natural principle of "being valuable" shifts the focus away from the maximization of one's own good towards the maximization of the good of one's adversary, i.e. the alter which needs to grow and foster if the ego wants to gain (this is the deeper sense of symbiotic dependency), and of the system in which both alter and ego can exist. Transformed into the language of corporations, "being valuable" neither means just increasing revenue or corporate value nor just raising shareholder or even stakeholder value, but organizing complex and holistic values and value chains which lead into innovative business models that foster the multidimensional growth of resources and added value, thereby cutting through the spiral of globalized destructive wealth creation.

Companies as diverse as the global market leaders Interface Inc., Schmalz GmbH, Hilti AG, or companies like HiPP GmbH (a quality leader in baby food products) and the merino wool apparel producer Icebreaker Ltd. show how this shift in business focus can be achieved. Let us take the case of Interface. **Interface Inc.** is the world market leader for modular floor coverings in the office and public facilities market. Their floor coverings are woven from synthetic fabrics. Producing the material needs enormous amounts of oil and energy. A few years ago, Interface introduced a completely new strategy for commodity sourcing. This is not just meant to introduce sustainable recycling processes. It also tries to establish truly viable, resource-creating economic cycles on the local level. For this purpose, Interface

---

business to the market's number two, Diebold, as they believed that remaining number three would not suffice to reach the planned turnover targets that could sustain Wincor in the long run.

launched the "Net-Works" project. Net-Works is a sub-project of the long-term strategy Mission Zero developed by the company's founder Ray Andersen. Mission Zero wants to source all resources needed by Interface from recycled or renewable sources. As part of this program, Interface bought up the discarded nets of local fishermen and the general population, which might cause massive damage if left afloat in the ocean. With its call for collecting, selling, and thus returning used nets back into the global supply chain, the company is not just preventing ecological damage, but putting the nets to a new use in a second life. Later, this might be followed by a third or fourth life, as the floor coverings themselves are again recycled. What is more important is that Interface is creating an eco-friendly source of income that is helping the global eco-systems by making the local society wealthier. Thus Interface not only follows the sustainability mantra of establishing resource cycles according to circular economy and "cradle to cradle" principles (Braungart and McDonough 2002), but indeed exemplifies the idea of setting up complex resource growth and value added cycles which foster qualitative and quantitative growth in line with the social, economic, and ecological demands of a lastingly viable economy. The business model of Interface, as well as all other business models which could be labeled to follow the principles of ethicology, are thus squarely opposed to the self-centered monopolistic ideals of economic strategy development: to secure corporate success by the invention of products, services, value propositions, and business models which facilitate, if possible, absolute dominance over the market, the value chain, or the revenue stream. Instead, they focus on fostering cooperative value-adding cycles for holistic and substantive value creation. They are therefore a counterpoint to the short-term profit logics of exploit- ative economics that flood already saturated markets with more products and services, whose sense is often limited to the surrogate experience of meaningless consumption. In their focus on enabling, enhancing, and enriching all links of the value chain, resource-creating business models aim to achieve value-adding partic- ipation, i.e. an economy that applies the natural principles of value creation to the micro, meso, macro, and supra-levels of the firm, the economy, society and nature.[13]

To become valuable in this sense, corporations need to foster a climate which aligns the understanding of its business purpose with the values scheme of ethicology, i.e. the principles of natural value creation on the one hand and the principles of an ethic that fulfills at least the basic principles defined above as the values of the Global Ethos on the other hand (Glauner 2016, 2017a, b). Guided by these values schemes, the focus of corporate value creation will go into innovations of what I call future viable niche strategies for fostering resource growth and added value cycles. For a hint as to how such future-viable niche strategies could be established, we can apply the following taxonomy of ethicological business models. To develop such models,

---

[13]The micro-level concerns the relationships between companies and people (employees, cus- tomers, suppliers, and business partners); the meso-level concerns the relations between companies (suppliers, clients, or competitors); the macro-level covers the relationships between companies and the systems surrounding them, be they external stakeholders, politics, or society at large; and the supra-level concerns the relations between companies and the environment.

we apply two effect-oriented indices which work together in line with the ethical concepts of the Global Ethos values and the five principles of natural value creation.

The first index looks at the participatory potential of a business model. The criterion of participation asks: Who is part of the system, and who is outside the system? Looking at a typical customer relationship, we can phrase this question as follows: Does the company make its products for its customers or with its customers? In the former case, the customers are outside the system; they are a means to an end, specifically, a means of generating revenue and profit for the company. In the latter case, the company makes its products with its customers, which makes them part of the purpose—the purpose of generating shared value and benefits in a cooperative network. We can apply the same criterion to relationships with suppliers, business partners, or any other type of stakeholder relationship. Instead of seeing these with the eyes of Michael Porter as possible threats, they become the crucial partners for business models which focus on mutual, holistic, and future viable value creation.

The second index we can use to assess ethicological business models is their added value creating potential. The key here is the question: Where, how, and for whom does the business model create substantive beneficial value? These potential beneficiaries can be identified on the micro-level of the company-to-people relationships, on the meso-level of company-to-company relationships, on the macro-level of company-to-other-systems relationships, and on the supra-level of the company-to-environment relationship. We can calculate the value creating potential both in material and in ideal terms by using indicators of empowerment, expansion/integration, and enrichment. This gives us a control system which includes indicators for ethicological performance on top of the financial indicators already proposed by the enterprise cockpits of EFQM or Balance Score Card models (Kaplan and Norton 1996; Müller-Stewens and Lechner 2003) and the corporate cultural indicators of the values cockpit (Glauner 2017a). These can include the participatory potential, the networking/integration potential, diversity, regionality, resource creating potential, degree of regional decoupling and many more.

Ethicological business models are therefore based on a vector analysis of the company's effectiveness in the areas in which it creates value added and systemic surplus by fostering new forms of participation, empowerment, or resource creation. These should not only look at the economy, the environment, and wider society as three distinct areas, but also include the systemic levels (micro, meso, macro, and supra) and consider the scope and degree of regional spread. By scope, we mean the number of individual elements of the total system (people, companies, and other affected parties) who benefit from the added value or resources created by the company.

When assessing ethicological business models with this taxonomy, we need to remember: The potential for creating resources is defined by how a company creates added value with business models that are in line with the values of the global ethos and the five principles of nature. The more added value a business model creates on more levels (micro, meso, macro, supra) and in more spheres (economy, society, environment), the greater its potential for resource creation will be and the more profitable, effective, and viable for the future it will be.

To break through the paradox of economic wealth creation, companies thus need to develop innovative business models, value propositions, and organizational structures that try to establish viable added value chains. The formation of these chains can be compared to the evolution of self-sustaining ecosystems. If companies engage in ethicological forms of business development, they can be viewed as actors in a new kind of civil society. This new form of civil society views corporations not as forces opposing an otherwise fair and responsibility-driven society, but as possible allies who generate an economy that follows the path of nature, developing forms of diversity and growth which add to the great ledger by serving all who are part of the system.

References

Ayres, R., & Warr, B. (2005). Accounting for growth: The role of physical work. *Structural Change and Economic Dynamics, 16*(2), 181–209. https://doi.org/10.1016/j.strueco.2003.10.003.

Ayres, R., & Warr, B. (2009). *The economic growth engine: How energy and work drive material prosperity*. Cheltenham: Edward Elgar.

Bauer, J. (2006). *Prinzip Menschlichkeit: Warum wir von Natur aus kooperieren*. Munich: Heyne (2008).

Bauer, J. (2008). *Das kooperative Gen. Abschied vom Darwinismus*. Hamburg: Hoffmann & Campe.

BCG. (2015a, June). *Global wealth 2015: Winning the growth game*. Boston Consulting Group. https://www.bcgperspectives.com/content/articles/financial-institutions-growth-global-wealth-2015-winning-the-growth-game/

BCG. (2015b, April). *Industry 4.0 the future of productivity and growth in manufacturing industries*. Boston Consulting Group. https://www.bcgperspectives.com/Images/Industry_40_Future_of_Productivity_April_2015_tcm80-185183.pdf

Bourdieu, P. (1982). *Die feinen Unterschiede. Kritik der gesellschaftlichen Urteilskraft*. Frankfurt/Main: Suhrkamp (4th ed., 1987).

Braungart, M., & McDonough, W. (2002). *Cradle to cradle: Remaking the way we make things*. New York: North Point Press.

Braungart, M., & McDonough, W. (2013). *The upcycle. Beyond sustainability – Designing for abundance*. New York: Melcher/North Point Press [deutsch: *Intelligente Verschwendung. The Ubcycle: Auf dem Weg in eine neue Überflussgesellschaft*. (oekom) Munich].

Bryan, L., & Farrell, D. (1996). *Market unbound: Unleashing global capitalism*. New York: Wiley.

Brynjolfsson, E., & McAfee, A. (2014). *The second machine age. Work, progress, and prosperity in a time of brilliant technologies*. New York: Norton.

Capra, F. (1996). *The web of life: A new scientific understanding of living systems*. New York: Anchor/Random House [deutsch: *Lebensnetz. Ein neues Verständnis der lebendigen Welt*. (Schwerz) Bern, Munich, Vienna].

Capra, F., & Luisi, P. L. (2014). *The systems view of life. A unifiying vision* (3rd ed.). Cambridge: Cambridge Univeristy Press.

Dawes, R. H. (1980): Social dilemmas. *Annual Review of Psychology, 31*, 163–193.

Dawkins, R. (1976). The selfish gene. Oxford: Oxford University Press (30th Anniversary ed., 2006).

de Waal, F. (2005). *Our inner ape. A leading primatologist explains why we are who we are.* New York: Riverhead.

Derman, E. (2011). *Models behaving badly. Why confusing illusion with reality can lead to disaster, on wall street and in life.* New York: Free Press.

Dierksmeier, C. (2016). *Reframing economic ethics. The philosophical foundations of humanistic management.* New York: Palgrave MacMillan.

Duerr, H.-P. (1988). *Der Mythos vom Zivilisationsprozeß – Vol. 1: Nacktheit und Scham.* Frankfurt/Main: Suhrkamp.

Duhem, P. (1908). *Ziel und Struktur der Physikalischen Theorien.* Hamburg: Meiner (1978).

Elegido, J. (2009). Business education and erosion of character. *African Journal of Business Ethics, 4*(1), 16–24.

Elias, N. (1939). *Über den Prozeß der Zivilisation. Soziogenetische und psychogenetische Untersuchungen.* Frankfurt/Main: Suhrkamp (1976).

Erhard, W.H., Jensen, M.W., & Zaffron, S. (2009). *A new model of integrity: An actionable pathway to trust, productivity and value* (PDF File of Keynote Slides). Barbados Group Working Paper No. 07-01, Harvard NOM Working Paper No. 07-01, 1st IESE Conference on "Humanizing the Firm and the Management Profession" Presentation. Retrieved June 16, 2016, from http://papers.ssrn.com/sol3/papers.cfm?abstract_id=932255

Erhard, W.H., Jensen, M.W., Zaffron, S. (2016). *Integrity: A positive model that incorporates the normative phenomena of morality, ethics, and legality – Abgridged.* Harvard Business School NOM Working Paper No. 10-061, Barbados Group Working Paper No. 10-01, Simon School of Business Working Paper No. 10-07. Retrieved June 16, 2016, from http://papers.ssrn.com/sol3/papers.cfm?abstract_id=1542759

Feyerabend, P. (1975). *Against method: Outline of an anarchist theory of knowledge.* London: Verso (4th ed., 2010).

Foster, R., & Kaplan, S. (2002). *Creative destruction. Why companies that are built to last underperform the market – and how to successfully transform them.* New York: Currency.

Frank, R. H. (2011). *The Darwin economy. Liberty, competition, and the common good.* Princeton: Princeton University Press.

Frey, C.B., & Osborne, M.A. (2013). The future of employment: How susceptible are jobs to computerisation? Retrieved September 17, 2013, from http://www.oxfordmartin.ox.ac.uk/downloads/academic/The_Future_of_Employment.pdf

Gilbert, S. F., Sapp, J., & Tauber, A. I. (2012). A symbiotic view of life: We have never been individuals. *The Quarterly Review of Biology, 87,* 325–341. https://doi.org/10.1086/668166.

Gilbert, C. Eyring, M., & Foster, R.N. (2013). Duale transformation. *Harvard Business Manager,* S. 34–44.

Glauner, F. (2016). *Future viability, business models, and values. Strategy, business management and economy in disruptive markets.* Heidelberg: Springer.

Glauner, F. (2017a). *Values cockpits. On measuring and steering corporate cultures.* Berlin: Springer.

Glauner, F. (2017b). Compliance, global ethos and corporate wisdom: Values strategies as an increasingly critical competitive advantage. In J. D. Rendtorff (Ed.), *Perspectives on philosophy of management and business ethics, series ethical economy. Studies in economic ethics and philosophy 51* (pp. 121–137). Berlin: Springer.

Glauner, F. (2018). Redefining economics: Why sharing value is not enough. Accepted to be published in the *Competitiveness Review* Special Issue Call "Creating Shared Value: Restoring the Legitimacy of Business and Advancing Competitiveness".

Hachtel, W. (1998) Mykorrhiza vermittelt Stofftransfer zwischen Waldbäumen. In *Spektrum der Wissenschaft 4/1998,* 25. http://www.spektrum.de/magazin/mykorrhiza-vermittelt-stofftransfer-zwischen-waldbaeumen/824505

Harari, Y. N. (2015). *Sapiens. A brief history of humankind.* New York: HarperColins.

Jensen, M. C., & Meckling, W. H. (1976). Theory of the firm: Managerial behaviour, agency costs and ownership structure. *Journal of Financial Economics, 3*(4), 305–360.

Jensen, M. C., & Meckling, W. H. (1994). The nature of man. *Journal of Applied Corporate Finance, 7*(2), 4–19.

Kaplan, R. S., & Norton, D. P. (1996). *The balanced scorecard: Translating strategy into action.* Boston, MA: Harvard Business Review Press.

Kegel, B. (2015). *Die Herrscher der Welt. Wie Mikroben unser Leben bestimmen.* Cologne: Dumont.

Kochhar, R., & Oats, R. (2015, July 8). *A global middle class is more promise than reality.* Pew Research. Retrieved October 4, 2016, from http://www.pewglobal.org/2015/07/08/a-global-middle-class-is-more-promise-than-reality/

Kocic, A. (2015, June). Work crisis—A divided tale of labour markets. In *Deutsche Bank Konzept. Reflections on unusual issues* (pp. 46–53). https://www.dbresearch.de/PROD/DBR_INTERNET_DE-PROD/PROD0000000000357626/Konzept+Issue+05.pdf

Kuhn, T.S. (1962). The structure of scientific revolutions (50th Anniversary Edition). Chicago: University of Chicago (2012).

Küng, H. (2012). *Handbuch Weltethos. Eine vision und ihre Umsetzung.* Munich: Pieper.

Lobe, A. (2016, April 29). Wird Facebook Donald Trump verhindern? *Frankfurter Allgemeine Zeitung*, 17.

Miller, D. T. (1999). The norm of self-interest. *American Psychologist, 54*(12), 1053–1060.

Motesharrei, S., Rivas, J., & Kalnay, E. (2014, May). Human and nature dynamics (HANDY): Modeling inequality and use of resources in the Collaps or sustainability of societies. In *Ecological Economics* (Vol. 101, pp. 90–102). doi:https://doi.org/10.1016/j.ecolecon.2014.02.014

Müller-Stewens, G., & Lechner, C. (2003). *Strategisches Management. Wie strategische Initiativen zum Wandel führen. Der St. Galler General Management Navigator.* Stuttgart: Schaeffer-Poeschel (2nd revised ed.).

Ostrom, E. (2000). Social capital: A fad or a fundamental concept. In D. Partha & I. Serageldin (Eds.), *Social capital. A multifaceted perspective* (pp. 172–214). Washington: The World Bank.

Otto, K.-S., Nolting, U., & Bässler, C. (2007). *Evolutionsmangagement. Von der Natur lernen: Unternehmen entwicklen und langfristig steuern.* München: Hanser.

Pauli, G. (2010). *The blue economy. 10 years, 100 innovations, 100 million jobs.* Taos, NM: Paradigm.

Pirson, M. (2017). *Humanistic management: Protecting dignity and promoting well-being.* Cambridge: Cambridge University Press.

Popper, K.R. (1935). *Die Logik der Forschung.* Tübingen: J.C.B. Mohr (Paul Siebeck) (8th ed., 1984).

Porter, M. E. (1980). *Competitive strategy: Techniques for analyzing industries and competitors.* New York: Free Press.

Schumpeter, J. A. (1942). *Capitalism, socialism, and democracy* (p. 1994). London: Routledge.

Searle, J.R. (1969). *Speech acts. An essay in the philosophy of language.* London: Cambridge University Press (11th ed., 1984).

Searle, J. R. (1995). *The construction of social reality.* New York: The Free Press.

Searle, J. R. (2010). *Making the social reality. The structure of human civilization.* Oxford: Oxford University Press.

Seba, T. (2006). *Winners take all. The 9 fundamental rules of high tech strategy.* San Francisco, CA.

Seba, T. (2014). *Clean disruption of energy and transportation. How Silicon Valley will make oil, nuclear, natural gas, coal, electric utilities and conventional cars obsolete by 2030.* San Francisco, CA.

Sennett, R. (2006). *The culture of the new capitalism.* New Haven: Yale University Press.

Simon, H. A. (1985). Human nature in politics: The dialogue of psychology with political science. *The American Political Science Review, 79*(2), 293–304.

Stuchtey, M., Enkvist, P.-A., & Zumwinkel, K. (2016). *A good disruption. Redefining growth in the twenty-first century.* London: Bloomsbury.

Taleb, N. N. (2008). *Der Schwarze Schwan. Die Macht höchst unwahrscheinlicher Ereignisse.* München: Hanser.

Tetlock, P. E. (2000). Cognitive biases and organizational correctives: Do both disease and cure depend on the politics of the beholder? *Administrative Science Quarterly, 45*, 293–329.

Van Orman Quine, W. (1960). *Word and object.* Cambridge, MA: The MIT Press.

Van Orman Quine, W. (1969). *Ontological relativity and other essays.* New York: Columbia University Press.

von Foerster, H. (1972). Bemerkungen zu einer Epistemologie des Lebendigen. In ders. (1993) *Wissen und Gewissen. Versuch einer Brücke.* Frankfurt/Main: Suhrkamp (8th ed., 2011, pp. 116–133).

von Foerster, H. (1987). Kybernetik. In idem (1993) *Wissen und Gewissen. Versuch einer Brücke.* Frankfurt/Main: Suhrkamp (8th ed., 2011, pp. 72–76).

von Glasersfeld, E. (1995). Radikaler Konstruktivismus. Ideen, Ergebnisse, Probleme. Frankfurt/Main: Suhrkamp (7th ed., 2011).

von Weizsäcker, E.U., Lovins, A.B., & Lovins, L.H. (1995). *Faktor Vier. Doppelter Wohlstand – halbierter Verbrauch. Der neue Bericht an den Club of Rome München: Droemer Knaur* (Engl. Factor four: Doubling wealth, halving resource use). The new report to the Club of Rome. London: Earthscan (1998).

Watts, R. G. (2007). *Global warming and the future of the earth. Synthesis lectures on energy and the environment: Technology, science, and society #1.* San Rafael, CA: Morgan & Claypool.

Williams, J. N. (2013). Humans and biodiversity: Population and demographic trends in the hotspots. *Population and Environment, 34*, 510–523. https://doi.org/10.1007/s11111-012-0175-3).

Wilson, E.O. (1992). *The diversity of life.* London: Penguin (2001).

Zuboff, S. (2016, March 5). The secrets of surveillance capitalism. *Frankfurter Allgemeine Zeitung.* Retrieved March 7, 2016, from http://www.faz.net/aktuell/feuilleton/debatten/the-digital-debate/shoshana-zuboff-secrets-of-surveillance-capitalism-14103616.html

Friedrich Glauner links long-term business-practice with interdisciplinary academic expertise at the intersection of philosophy, economy, and knowledge transfer. He worked for 18 years as an entrepreneur, CEO and manager, and enjoyed research and teaching stints at the Technical and the Free University Berlin, at the European Business School, and at the University of California, Berkeley. He currently lectures future viable business models, strategy development, business management, business ethics, and leadership skills at the Weltethos-Institut/Global Ethic Institute (University of Tübingen), at the Bundeswehr University Munich and at other Universities. Dr. Glauner majored in Philosophy with an emphasis on epistemology, linguistic philosophy and philosophy of science with minors in Economics, Religious Studies, History and Semiotics in Berlin, Cologne, London, and as a Fulbright Scholar at Berkeley.

CSR Behavior: Between Altruism and Profit Maximization

Klaus Kotek, Alina M. Schoenberg, and Christopher Schwand

1 Introduction

While Corporate Social Responsibility (CSR) is a very broad concept, it mostly refers to firms' activities that account for the interests of all stakeholders such as customers, employees, shareholders, society (community) and environment and go beyond legal obligations. In other words, firms should actively consider and improve the impact they have on society and drive change towards a sustainable development of their business. The concept of CSR has therefore two important dimensions: (1) the (measurable) integration of social and environmental needs in the firms' business operations and (2) the voluntary nature of CSR activities. According to the PWC Global CEO survey (2016) 64% of the CEOs see CSR as a core part of their business, 59% of them believe social values are important to attract top employees and 37% agree that CSR attracts investments. Furthermore, the importance of CSR is expected to rise within the next 5 years. However, according to CECP (2016) 53% of the surveyed companies did not increase total CSR spending between 2013 and 2015 (47% decreased total giving, 8% did not change the total sum) leading to an increase in overall spending of only 1%, while 87% of the companies measure societal outcomes and became aware of the strategic dimension of societal outcome measurement. Managers are often caught between the expectations of ethical consumers and the profit-maximizing expectations of the investors, and therefore often use CSR in order to promote their image or their brand. Corporate philanthropy is therefore also used as an instrument to uphold the image of companies (Porter and Kramer 2002). This raises the question whether CSR behavior of firms is more driven by

K. Kotek (✉) · A. M. Schoenberg · C. Schwand
Department of Business, IMC University of Applied Sciences Krems, Krems, Austria
e-mail: klaus.kotek@fh-krems.ac.at; alina.schoenberg@fh-krems.ac.at;
christopher.schwand@fh-krems.ac.at

© Springer International Publishing AG, part of Springer Nature 2018   159
R. Altenburger (ed.), *Innovation Management and Corporate Social Responsibility*,
CSR, Sustainability, Ethics & Governance,
https://doi.org/10.1007/978-3-319-93629-1_8

profit maximization strategies that might concentrate on communicating instead on enhancing CSR activities than by altruism and philanthropy and whether altruism is necessary in order to increase CSR behavior.

In this paper we give a brief economic overview on CSR considering the companies motivations for engaging in CSR activities and discussing the optimal level of CSR. In the second section we will discuss different theoretic reasons behind CSR activities of firms offering insight on whether strategic considerations might overweigh corporate philanthropy. In the third section we will discuss the significance of CSR activities for the overall provision with public goods. The fourth section concludes.

2 CSR Behavior

2.1 General Considerations

Many economic agents like firms and households are involved in pro-social activities opposed to the common assumption of self-interested decision-making in economic theory. However, engaging in philanthropy does not automatically imply pure altruistic motives of the giver, since pro-social behavior might be driven by the desire for social acclaim or by the prevention of public shame. Fehr and Gächter (2000) show in their economic experiment that people also care for fairness. Furthermore, a strategic component in the decision-making of the firms, that facilitates companies to use philanthropic actions to increase profits, suggests that self-interest and pro-social behavior do not necessarily exclude each other. Since the motivation behind CSR actions do influence the economic outcome of the firm's stakeholders in different ways, Kitzmueller and Shimshack (2012) suggest a typology of CSR activities that accounts for the preferences of stakeholder and shareholder (Fig. 1).

Shareholder Preferences

		social	purely monetary
Stakeholder Preferences	social	altruistic behavior (mixed effects on profits)	strategic behavior (profit maximization)
	purely monetary	altruistic behavior (negative effects on profits)	self-interested behavior (profit maximization, no CSR)

Fig. 1 Typology of CSR activities. Source: Kitzmueller and Shimshack (2012)

The difficulty of determining if CSR is mostly altruistic motivated lies in the variety of stakeholders involved in the decision making process of the firm such as shareholders, customers, investors, employees, competitors and government. According to the PWC 19th Annual Global CEO Survey (2016) all these stakeholders have a high impact on the firm's strategy. A stakeholder-oriented perspective of CSR motivation is therefore important, because pro-social activities at the firms' level imply that the stakeholders (consumers, employees, investors) and the shareholders (owners) of businesses are (partly) handing over their own donation decision to the firm.[1] CSR activities are meant to provide (impure) public goods, reduce negative externalities of production or generate positive externalities. In order to do so, firms spend the shareholders money for social goods, deciding instead of the shareholder whether and which cause to support (Friedman 1970). According to Kotchen (2006) firms often add characteristics of a public good to a private product in order to facilitate a joint consumption of both, the private and the public good. The firm offers the consumer in exchange for a price premium such joint products consisting in a private good (e.g. coffee) and a public good if the product is produced in a certain socially responsible way (e.g. reduction of pollution if the coffee is organic). The consumer therefore donates to a public good in addition to buying the private good. This raises the question whether stakeholders and shareholders would not be better off by choosing type and extent of the pro-social activity themselves.

2.2 Altruism and Impure Altruism

In the context of CSR, altruism implies sacrificing profits in order to provide a public good or to internalize negative externalities of production (Bénabou and Tirole 2010). Altruism refers here to voluntarily giving for the good of others, whereas the individual utility only depends on the provided level of the public good. In other words, economic agents are altruistically motivated if benefits to others increase their own utility.

According to economic theory public goods are often underprovided because individuals have an incentive to free ride due to the non-excludability from using the public good. In order to deal with the undersupply of public goods both, governmental and non-governmental solutions are possible. The private (non-governmental) provision of a public good (e.g. CSR) leads to positive externalities, allowing everybody to benefit from personal cost. Philanthropic activities that improve the competitive context of firms such as factor market conditions, demand conditions, the presence of supporting industries or the regulatory context might lead other firms to free ride because of the shared benefits of CSR (Porter and

---

[1]Philanthropic activities mostly align with the personal beliefs of firms executives and do not necessarily represent the pro-social behavior of shareholders or stakeholders (Friedman 1970; Porter and Kramer 2002).

Kramer 2002). Due to the non-excludability characteristic of public goods, the optimal level of public good won't be achieved through voluntary action (e.g. CSR) (Becker 1974; Bergstrom et al. 1986). Nevertheless, public goods are provided on markets due to altruism. According to Arrow (1972) altruistic behavior could decrease the welfare loss of the undersupply of the public good due to the higher tendency of altruistic economic agents to reveal their willingness to pay for public goods.

However, altruism might not be the only driver of pro-social behavior since individuals might benefit from their social reputation or self-respect as well (Bénabou and Tirole 2006). The perspective of philanthropic reputation may enhance (impure) altruistic behavior of firms and therefore their CSR activities. In this case, besides the overall provision level of the public good, economic agents also care about their own contribution to the public good. In other words, one reason why private provision of public goods takes place in spite the incentive to free ride is because economic agents are not indifferent between own and public (governmental) pro-social activities (Andreoni 1990). While pure altruism refers to the benefit of making someone else better off, impure altruism implies a selfish component. This might be "motivated by the desire to win prestige, respect, friendship, and other social and psychological objectives" (Olson 1965). Thus, social signaling is an important motivation behind philanthropic behavior, which might explain why less than 1% of donations are anonymously (Bénabou and Tirole 2003; Glazer and Konrad 1996). Furthermore, the warm glow effect of giving, which, other than the desire for prestige, etc., is purely intrinsic, also leads to the impurely altruistic pro-social behavior. Thus, even though firms are primarily profit oriented (selfish) they might gain benefit from (impure) altruistic behavior if they interact in a context where reciprocation is rewarded. The degree of the altruistic motive in CSR is therefore different for economic agents creating disbelieve about the true philanthropic intent of giving.

If the stakeholders and shareholders preferences coincide in their (impure) altruistic motivation, firms are going to realize CSR projects more or less regardless their impact on profits. According to Kitzmueller and Shimshack (2012) ethical consumers that engage in pro-social activities of their own might experience a decrease in utility due to an inconsistence in their behavior if the social responsibility does not reflect in their purchasing habits. Furthermore, consumers without preferences for pro-social activities but for high preferences for their own reputation might choose to purchase ethical goods due to benefits from signaling. Both cases would align with a purely altruistic behavior of the firm without necessarily causing a loss of profits. According to Habel et al. (2016) customers experience a warm glow effect when buying products from philanthropic firms only if they believe in the altruistic motivation behind the pro-social behavior of the firm but do not accept price premiums without questioning price fairness. Firms therefore need to emphasize that they carry the cost of pro-social activities in order to improve the credibility of the altruistic motivation. Altruistic behavior is also assumed if firms engage in CSR activities in spite of purely monetary preferences of customers, since loss of profits is most likely to occur in this case.

2.3   Strategic CSR

Since pure altruism implies a negative effect on profits, the choice of pro-social behavior of the firm needs to consider also a strategic dimension. If opposed to the socially motivated stakeholders, the shareholders' only goal is motivated by monetary incentives the firms might use CSR activities as a strategy to avoid the loss of ethically consumers and other stakeholders and to increase the competitive advantage of the firm. According to Friedman (1970) CSR is either a possibility to increase profits or a conflict of interests between shareholders and manager, which chooses to spend the shareholders money for social goods that could actually be provided by the shareholder himself in a more efficient way. In this case corporate philanthropy might be understood as an transfer of business interests from shareholder to stakeholders. Porter and Kramer (2002) argue that, opposed to Friedman's assumption that economic and social objectives are separate, strategic CSR might improve the competitive advantage of the firm when pro-social activities of the firm are used to enhance the competitive context of the company. The competitive context refers to the "quality of the business environment in the location or locations where they operate" and improving it through CSR activities might provide higher social benefits than individual donors could achieve. Therefore, the economic benefits of CSR activities will be higher, the more the pro-social behavior relates to the company's business. Recognizing the right competitive context and yet the right strategic pro-social behavior implies a thorough understanding of the stakeholders' preferences. Hence, there is a variety of reasons to engage in CSR activities (Maxwell et al. 2000):

– Increase in consumer demand
– Enhance in employer satisfaction
– Attracting Social Responsible Investments (SRI)
– Avoiding government regulation

In other words, there are some reasons increasing the incentive of firms to act strategically in their decision to provide social goods. In this section, we present certain motivations for providing CSR activities considering the stakeholders' preferences for pro-social behavior and their implications for the firms:

2.3.1   Consumer Related CSR

Empirical evidence and several surveys suggest that consumers consider CSR activities of firms as well as their willingness to contribute to society on the expense of profit maximization for their buying decision (Environics 1999; MORI 2000; Nielsen 2014). Philanthropic activities of firms influence therefore both, the consumption and the willingness to pay of customers (Kitzmueller and Shimshack 2012). In other words, ethical consumers are willing to pay a price premium for social goods while they might punish unethical firms by decreasing their willingness to pay for their products or by not buying at all. In addition to the consumers' expectations towards the ethical behavior of firms, they expect transparency and communication of pro-social behavior. The consumer response to CSR is therefore depended on the consumers' awareness and their trust in the companies' philanthropy. Due to an information asymmetry between customer and firm, the CSR motivation of the firm is difficult to observe which might lead to free riding and/or greenwashing behavior, e.g. to higher CSR advertising effort than the actual CSR spending. According to Navarro (1988) the purpose of CSR is the signaling of the firm's quality. However, consumers might be able to distinguish between ethically motivated CSR or CSR activities induced by self-interest or social pressure and perceive this kind of behavior as greedy (Baron 2009; Becker-Olsen et al. 2006; Kitzmueller and Shimshack 2012).

2.3.2 Employee Related CSR

There are many theories relating CSR with the labor market: Social identity theory suggests that employees experience an improved self-perception when working for a philanthropic employer. Signaling theory indicates that CSR activities signal information on business ethics and working conditions (Albinger and Freeman 2000). Thus, the reputation of a firm's pro-social behavior can affect the employees' choices in favor of CSR providing firms and induce competitive advantages on the labor market for those firms. There are two dimensions that need to be considered to this regard: the impact of CSR on employees' motivation and work ethic as well as the effects on labor cost of the firm. Both dimensions can benefit the firm and enhance strategic CSR. Stigler (1962) argued that firms donating to employee welfare attract better applicants due to the signal of better working conditions. According to Brekke and Nyborg (2008) philanthropic behavior of firms attract morally motivated workers that are likely to put more effort in their tasks and less likely to shirk. This increases the firm productivity and reduces expensive fluctuation of employees. Furthermore, if motivated employees have non-pecuniary interests in the success of the company, monetary incentives are less needed (Besley and Gatak 2005; Bowles et al. 2001). This implies lower wages for employees that value CSR activities of their employers. According to Besley and Gatak (2005) intrinsic motivation of employees might substitute extrinsic motivation such as incentive payments if the employer is "mission oriented".

2.3.3 Investor Related CSR

Opposed to other stakeholders, firms have a clear financial obligation towards their investors/shareholders, thus financial performance of the firm along with the strategic element of CSR do have an important role. In addition to Friedman's (1962, 1970) arguments that "few trends could so thoroughly undermine the very foundations of our free society as the acceptance by corporate officials of a social responsibility other than to make as much money for their stockholders as possible", Atkinson and Galaskiewicz (1988) argued that firms with high CEO ownership are less charitable than firms with low CEO ownership. Furthermore Dewatripont et al. (1999) see charitable activities as detrimental for the accountability of CEOs. Due to the increasing relevance of CSR for stakeholders, investors began to recognize the necessity of strategic CSR due to a positive effect on financial performance (Ambec and Lanoie 2008; Dixon-Fowler et al. 2013) as well as a driver for innovation activities and operational efficiency (Porter and Van der Linde 1995).

Chatterji et al. (2009) categorize investors' motives for Social Responsible Investments (SRI) as financial (environmental performance increases financial performance), deontological (profits from unethical companies are avoided), consequentialist (unethically firms are punished while ethical firms are rewarded through lower cost of capital) and expressive (association with investment in good causes is important). All of these reasons for investment induce strategic behavior of the firm that might determine the corporate philanthropy. Once again, reporting of CSR activities and social ratings serve as a signal of the firm's pro-social behavior and its credibility.

2.3.4 Government Related CSR

In order to counteract market failures such as negative externalities or the undersupply of public goods, government actions are taken to impose a certain behavior on firms. Firms that intend to avoid future regulations or to enhance a smother adjustment in case of new regulatory actions might strategically implement CSR activities. Due to the lack of compliance control or punishment mechanisms such as fines or proceedings in the case of self-regulation (e.g. CSR), which is by definition voluntarily, pro-social activities are appealing for firms. According to Maxwell et al. (2000) firms can preempt regulations if they succeed in meeting consumer expectations regarding regulatory actions. This might be the case when influencing the political process is costly for the consumers since a low level of self-regulation will stop consumers from entering the political process. The incentive for self-regulation therefore increases with decreasing cost of the consumer to influence the political process. Self-regulation also allows firms to avoid competitive disadvantages in the case of unforeseen regulations due to an easier adjustment process (Kitzmueller and Shimshack 2012). In case of existing regulations, an over-compliance through CSR might reduce regulatory monitoring and scrutiny (Maxwell and Decker 2006).

To sum up, corporate philanthropy increases the competitive advantage of firms in terms of attraction of ethical consumers, investors and employees and also in terms of a better compliance with government regulations. Furthermore, CSR might help avoid social activism and boycotts.

3   Optimal Level of CSR

As previously seen there are several motives behind corporate philanthropy, suggesting that CSR is not only a product of altruistic attitude of the firm and its stakeholders, but also an important part of the business strategy. However, firms need to fulfil expectations of stakeholder with different preferences toward social responsibility, which implies that a (social) responsible firm might not be valued as such by different stakeholders. Since opposed interests with respect to the degree of altruism of stakeholders and shareholders might occur, an optimal level of CSR needs to be determined in order to fulfill as many expectations as possible without disregarding the financial obligation towards shareholders. This raises the question whether there is such an optimal level of pro-social behavior and whether CSR affects the level of total giving in society.

As shown in Sect. 2 the discussion about on optimal level of CSR and on whether CSR is used in a strategic way to maximize profits or is purely altruistic relates best with the optimal (public or private) provision of (impure) public goods. Despite the incentive to free ride, private agents supply (impure) public goods due to altruistic or strategic reasons. According to economic literature, private provision of a public good creates a positive externality leading to an undersupply on markets. In order to correct the undersupply with public goods, public provision occurs. This leads to a complete crowding-out of private provision of the public good due to a decreased incentive of private agents to supply public goods living the overall level of the public good provision unchanged (Roberts 1984; Bergstrom et al. 1986; Andreoni 1988). However, the degree of crowding-out of CSR activities depends on the warm-glow effect as well as on the strategic motivation behind the pro-social behavior. Assuming that economic agents are not indifferent between own pro-social behavior and the donation behavior of other economic agents, Andreoni (1990) suggests that public provision of public goods does not entirely crowd out private philanthropic behavior.

Strategic CSR implies somehow the delegation of pro-social behavior from stakeholders to the firm, which might imply a changed behavior of those stakeholders and shareholder towards charitable giving. According to Baron (2009) CSR crowds out personal giving if shareholder perceive a warm glow effect and therefore a higher market value of the firm than its financial return. Bénabou and Tirole (2006) argue that the crowding out effect might reduce total supply of charitable giving when high rewards reveal extrinsic motivation of the firm and therefore spoil the reputational motive for good behavior. For Morgan and Tumlinson (2013) corporate provision of public goods reduces the free riding behavior among shareholders and increases the overall supply.

In another line of arguments, consumers tend to behave opportunistic if they cannot observe pro-social behavior when buying goods (Darby and Karni 1973), which implies that firms might need to engage in social signaling through communication strategies in order to increase the overall level of pro-social activities. Furthermore, market competition is an important determinant of the optimal CSR level. A high market competition for ethical consumers does lower the possibilities to charge higher prices and therefore lowers pro-social behavior and CSR incentives of the firm (Bagnoli and Watts 2003; Shleifer 2004). According to Besley and Ghatak (2007) the CSR level corresponds the (suboptimal) private provision equilibrium when consumers self-select among ethical and non-ethical products according to their valuation of the public good. For the case of an impure public good that bundles a private with a public good the provision of the public good also corresponds the private provision equilibrium, which is below the social optimum (Kotchen 2006). To conclude, CSR enhances a Pareto improvement only if governments fail to implement the Samuelson provision of the public good (first best solution) and is therefore economically justified in this case (Kitzmueller and Shimshack 2012).

4 Conclusion

In this paper, we analyze the motivation behind CSR activities of firms. Furthermore, we consider the effect of self-motivated and altruistic CSR on the overall philanthropic outcome. We considered different preferences of stakeholders and shareholders in order to delimit altruistic behavior from strategic pro-social activities. This allows for three main conclusions: (1) even though CSR activities are often motivated by strategic considerations in order to meet the stakeholders' expectations, philanthropic behavior of firms does contribute to well-being of society, (2) the lack in CSR does not necessarily imply unethical behavior of the firm, since it might be the result of fierce market competition and (3) under certain circumstances CSR might improve the overall provision with public goods, however a first-best-solution cannot be achieved without government intervention.

Current surveys show that the willingness to further increase CSR spending is limited, which can be seen as consolidation process in that reflection and strategic decision is key to optimizing CSR efforts. It is important for all stakeholders—consumers, employees, investors and their agents, government to understand CSR strategies in order to reflect and adjust own (re-)actions.

References

Albinger, H. S., & Freeman, S. J. (2000). Corporate social performance and attractiveness as an employer to different job seeking populations. *Journal of Business Ethics, 28*(3), 243–253.
Ambec, S., & Lanoie, P. (2008). Does it pay to be green? A systematic overview. *Academy of Management Perspectives, 22*(4), 45–62.
Andreoni, J. (1988). Privately provided public goods in a large economy: The limits of altruism. *Journal of Public Economics, 35*(1), 57–73.
Andreoni, J. (1990). Impure altruism and donations to public goods: A theory of warm-glow giving. *The Economic Journal, 100*(401), 464–477.

Arrow, K. J. (1972). Gifts and exchanges. *Philosophy and Public Affairs, 1*(4), 343–362.

Atkinson, L., & Galaskiewicz, J. (1988). Stock ownership and company contributions to charity. *Administrative Science Quarterly, 33*(1), 82–100.

Bagnoli, M., & Watts, S. G. (2003). Selling to socially responsible consumers: Competition and the private provision of public goods. *Journal of Economics and Management Strategy, 12*(3), 419–445.

Baron, D. P. (2009). A positive theory of moral management, social pressure and corporate social performance. *Journal of Economics and Management Strategy, 18*(1), 7–43.

Becker, G. S. (1974). A theory of social interactions. *Journal of Political Economy, 82*(6), 1063–1093.

Becker-Olsen, K. L., Cudmore, A., & Hill, R. P. (2006). The impact of perceived corporate social responsibility on consumer behaviour. *Journal of Business Research, 59*(1), 46–53.

Bénabou, R., & Tirole, J. (2003). Intrinsic and extrinsic motivation. *Review of Economic Studies, 70* (3), 489–520.

Bénabou, R., & Tirole, J. (2006). Incentives and prosocial behaviour. *American Economic Review, 96*(5), 1652–1678.

Bénabou, R., & Tirole, J. (2010). Individual and corporate social responsibility. *Economica, 77* (305), 1–19.

Bergstrom, T., Blume, L., & Varian, H. R. (1986). On the private provision of public goods. *Journal of Public Economics, 29*(1), 25–49.

Besley, T., & Ghatak, M. (2005). Competition and incentives with motivated agents. *American Economic Review, 95*(3), 616–636.

Besley, T., & Ghatak, M. (2007). Retailing public goods: The economics of corporate social responsibility. *Journal of Public Economics, 91*(9), 1645–1663.

Bowles, S., Gintis, H., & Osborne, M. (2001). Incentive-enhancing preferences: Personality, behaviour and earnings. *American Economic Review, 91*(2), 155–158.

Brekke, K. A., & Nyborg, K. (2008). Attracting responsible employees: Green production as labor market screening. *Resource and Energy Economics, 30*(4), 509–526.

CECP. (2016). *Giving in numbers*. 2016 Edition.

Chatterji, A. K., Levine, D. I., & Toel, M. W. (2009). How well do social ratings actually measure corporate social responsibility? *Journal of Economics and Management Strategy, 18*(1), 125–169.

Darby, M. R., & Karni, E. (1973). Free competition and the optimal amount of fraud. *Journal of Law and Economics, 16*(1), 67–88.

Dewatripont, M., Jewitt, I., & Jean, T. (1999). The economics of career concerns, part I: Comparing information structures. *The Review of Economic Studies, 66*(1), 183–198.

Dixon-Fowler, H. R., Slater, D. J., Johnson, J. L., Ellstrand, A. E., & Romi, A. M. (2013). Beyond "Does it pay to be green?" A meta-analysis of moderators of the CEP–CFP relationship. *Journal of Business Ethics, 112*(2), 353–366.

Environics. (1999). *International Environmental Monitor*.

Fehr, E., & Gächter, S. (2000). Fairness and retaliation: The economics of reciprocity. *The Journal of Economic Perspectives, 14*(3), 159–181.

Friedman, M. (1962). *Capitalism and freedom*. Chicago: The University of Chicago Press.

Friedman, M. (1970). The social responsibility of business is to increase its profits. *New York Times Magazine, 32*(13), 122–126.

Glazer, A., & Konrad, K. A. (1996). A signaling explanation for charity. *The American Economic Review, 86*(4), 1019–1028.

Habel, J., Schons, L. M., Alavi, S., & Wieseke, J. (2016). Warm glow or extra charge? The ambivalent effect of corporate social responsibility activities on customers' perceived price fairness. *Journal of Marketing, 80*(1), 84–105.

Kitzmueller, M., & Shimshack, J. (2012). Economic perspectives on corporate social responsibility. *Journal of Economic Literature, 50*(1), 51–84.

Kotchen, M. J. (2006). Green markets and private provision of public goods. *Journal of Political Economy, 114*(4), 816–834.

Maxwell, J. W., & Decker, C. S. (2006). Voluntary environmental investment and responsive regulation. *Environmental and Resource Economics, 33*(4), 425–439.

Maxwell, J. W., Lyon, T. P., & Hackett, S. C. (2000). Self-regulation and social welfare: The political economy of corporate environmentalism. *The Journal of Law and Economics, 43*(2), 583–617.

Morgan, J., & Tumlinson, J. (2013). *Corporate provision of public goods.* Academy of Management Proceedings

MORI. (2000). *European attitudes towards corporate social responsibility. Research for CSR Europe.* London: Market & Opinion Research International.

Navarro, P. (1988). Why do corporations give to charity? *The Journal of Business, 61*(1), 65–93.

Nielsen. (2014). *Doing well by doing good.* Global Survey of Corporate Social Responsibility.

Olson, M. (1965). *The logic of collective action. Public goods and the theory of groups.* Cambridge: Harvard University Press.

Porter, M. E., & Kramer, M. R. (2002). The competitive advantage of corporate philanthropy. *Harvard Business Review, 80,* 56–68.

Porter, M. E., & Van der Linde, C. (1995). Toward a new conception of the environment-competitiveness relationship. *The Journal of Economic Perspectives, 9*(4), 97–118.

PricewaterhouseCoopers. (2016). *Redefining business success in a changing world.* In 19th Annual Global CEO Survey.

Roberts, R. D. (1984). A positive model of private charity and public transfers. *Journal of Political Economy, 92*(1), 136–148.

Shleifer, A. (2004). Does competition destroy ethical behavior? *The American Economic Review, 94*(2), 414–418.

Stigler, G. J. (1962). Information in the labor market. *Journal of Political Economy, 70*(2), 49–73.

Klaus Kotek is a professor at the University of Applied Sciences (IMC FH Krems/Austria) since 2009. He teaches marketing (especially strategic planning, budgeting and performance management), corporate communications and project management in bachelor and graduate programs. Klaus Kotek has more than two decades of professional experience in marketing and corporate communications, with more than 18 years at two MNE (GE and Zurich Financial Services). His research concentrates on employer branding, strategic brand management as well as evaluation of communication efforts.

Alina Schoenberg is a professor for economics at the University of Applied Sciences (IMC FH Krems/Austria) since 2016. She received her degree in economics from the Ludwig Maximilians University and obtained her PhD degree at the Bundeswehr University in Munich. Her primary research interests focus on public finance/economics and regional science.

Christopher Schwand is programme director of the Export-Oriented Management study programme at IMC University of Applied Sciences Krems. Previously, his most important stations working as researcher in industrial technologies and business-to-business markets include Coopers&Lybrand, PWC, University of Vienna and Technical University of Vienna. At present, digital transformation processes are his key research interest.

Sustainability in Fashion: An Oxymoron?

Doris Berger-Grabner

1 Initial Situation

The last few years have witnessed an increased interest in sustainable fashion and ethical practices within the fashion industry (Henninger et al. 2016). Sustainability is emerging as a so-called "megatrend" (Mittelstaedt et al. 2014) and in order to stay competitive more and more companies start to use catchwords like "sustainable", "eco-friendly", "social" or "ethical" in their marketing communications. But this overdosing of sustainability claims has the effect that consumers very often mistrust such claims, especially when they can't verify the credibility.

Moreover, although most of the consumers are already aware of unsustainable practices of fast fashion retailers, ethically made clothes make up only 1% of the one trillion global fashion industry (2013). What are the reasons behind this reluctant purchase decisions for sustainable fashion? Why do consumers still buy fast fashion items although they know about critical incidents, like the collapse of the Rana Plaza factory building in Bangladesh in 2013 in which more than 1000 garment workers had to die. McNeill and Moore (2015) have found several reasons for this reluctance: a higher price level of slow fashion items, a perceived lack of social acceptance for sustainable fashion and, still, unawareness.

On the other side, many fashion companies are still unable or unwilling to take on sustainability issues such as waste and energy reduction, the payment of higher labour and production costs, traceability, slower turnover or a smaller range of products. Sometimes fashion companies don't even know whether their clothes are made in safe factories due to the fact that they use a complex web of suppliers, mainly in emerging countries, which often engage sub-contractors. Therefore shared

D. Berger-Grabner (✉)
Department of Business, IMC University of Applied Sciences Krems, Krems, Austria
e-mail: doris.grabner@fh-krems.ac.at

© Springer International Publishing AG, part of Springer Nature 2018   171
R. Altenburger (ed.), *Innovation Management and Corporate Social Responsibility*,
CSR, Sustainability, Ethics & Governance,
https://doi.org/10.1007/978-3-319-93629-1_9

responsibility along the whole supply chain—from designers, manufacturers, retailers to buyers—is necessary. In addition, it needs government standards to remove clothing with the most negative impact from the market.

The purpose of this article is to contribute to a better understanding of sustainable fashion business and practices, ethical consumer behavior and its influencing factors on the fashion industry. Last but not least, it should be discussed whether fashion can ever be sustainable or stays full of contradictions—an oxymoron?

2 Ethical Consumer Behaviour in Fashion

This chapter examines literature and secondary studies on ethical consumer behaviour and its influencing factors on the fashion industry, and summarizes the findings.

Since the 1980s, consumers are becoming increasingly aware of the social and environmental consequences of apparel they purchase. Still, the fast fashion retail market experiences a rapid growth, although most of the consumers know about the criticism of sweatshop labour by fast fashion companies like Primark, etc. Consumers don't always show ethical behaviour when buying apparel and are not always willing to pay a price-premium for sustainable fashion. Potential causal factors behind this behaviour, attitudes and perceptions towards sustainability and influencing factors for ethical consumer behaviour are identified.

2.1 Ethical Consumer Behaviour and Influencing Factors

Previous research has shown that consumers' beliefs about ethical fashion are based on their perceptions of a company in terms of the reputation in the fashion industry. Moreover their beliefs influence their support for what they perceive as socially responsible and eco-friendly businesses. Consequently, consumer education is essential to raise consumers' awareness of ethical fashion issues (Shen et al. 2012).

McNeill and Moore (2015) found in their study that fashion consumers can be categorized in three different consumer groups. Each group has a different view in regard to fast fashion and perceptions of fashion, and different fashion consumption behaviour. In general, ethical fashion purchase behaviour is determined by "their general level of concern for social and environmental well-being, their preconceptions towards sustainable fashion and their prior behaviour in relation to ethical consumption actions" (McNeill and Moore 2015, S. 220). Other influencing factors are the influence of peer groups and the consumer's knowledge about sustainable fashion products and practices (Barnes and Lea-Greenwood 2006). Self-enhancement and openness to experience personal values are also regarded as influencing factors when it comes to engagement in ethical fashion consumption (Manchiraju and Sadachar 2014).

With our clothes we would like to make a statement. Most of the consumers want to wear fashionable clothes and sustainable clothing is very often perceived as less fashionable. Another conflict potential is the so-called "fashion obsolescence". The fashion industry, especially the fast fashion sector, uses planned obsolescence a business strategy so that the consumer feels a need to purchase new, fashionable products which replace the old ones. Fast fashion is often assessed as "disposable" fashion due to short product life cycles and to the fact that unfashionable items are commonly thrown away with the household waste. Emotionally durable designs ("slow" design) through customization of clothing (user-centred fashion), reuse and upcycling of clothing and collaborative consumption might work against this fashion obsolescence.

Last but not least, consumers are often missing a true added value which justifies the price premium of sustainable fashion. Therefore, to make consumers care about ethical behaviour in general and, in particular, purchase more sustainable clothing, marketing communication and consumer education are supposed to be the most successful tools which result in an ethical consumer behaviour.

2.2 Attitudes and Perceptions Towards Sustainability

Henninger et al. (2016) examined what sustainable fashion means to consumers. They conducted a qualitative study and concluded that "the interpretation of sustainable fashion is context and person dependent" (Henninger et al. 2016, S. 400). Most of them do care about unethical behaviour, but this attitude does not always result in sustainable purchase decisions (McNeill and Moore 2015).

Joergens (2006) states that personal needs motivate consumers primarily when they buy apparel and have priority over ethical issues. Although consumers have a gaining interest in ethical fashion, they are very often unable to make ethical judgements due to a lack of comprehensive information. Consumers only care about ethical issues which influence them directly.

Consequently, a truthful label on clothes by a reliable and independent institution or organization, containing important ethical information, might be a helpful tool to facilitate customers' purchase decisions and to generate greater awareness. Awareness campaigns, in general, but also the increasing involvement of consumer activist groups and the availability of ethical products will help to encourage ethical consumer behaviour and change consumers' consumption (Teah and Chuah 2015).

3 Sustainable Brands and Ethical Practices

This chapter provides information on sustainable brands and their ethical practices. There do already exist a lot of fashion companies that have a sustainable performance and others, mainly fast fashion companies, are constantly trying to improve their performance in regard to sustainability.

3.1 Sustainable Brands: Opportunities and Barriers

The fashion industry faces a lot of opportunities in regard to sustainable business practices, concerning environmental, social and ecological issues, but it also face some barriers. Black (2008) identified the following barriers as the biggest ones:

- The fact that fashion is regarded as disposable and is turned into waste as soon as an item gets old or unfashionable and
- "the basic tenet of consumption for the sake of consumption" (Black 2008, S. 63).

Opportunities are many and varied. Digitalization in general and new technologies in particular will have an important impact on sustainable fashion. For instance, 3-D body scanning and automated custom-fitting systems have the potential to move parts of clothing production from mass fashion to individualized fashion. It can be assumed that well-fitted clothing, well-aligned with personal tastes, give more enjoyment and have longer lifecycles than ready-to-wear clothing (Ashdown 2008).

Furthermore, mass customization is a promising way for creating more customer-centric products that perfectly fit customers' needs. Mass customization can provide environmental benefits in all phases of a product's life cycle (Piller and Steiner 2008):

- During the manufacturing process: no storage of finished products is needed and there exists no overproduction;
- During the use of the product: customized products provide better fulfilment of customer needs; products will be handled with greater care;
- At the end of a product's life cycle: efficient recycling and reuse of product components due to a stronger customer-manufacturer relationship ("crade-to-cradle");

Literature research has shown that there do already exist various websites which help consumers to identify whether a brand is sustainable or not and to which degree. For instance "The sustainable fashion directory" (http://sustainablefashiondirectory.com/about/) provides information on ethical practices by fashion companies so that consumers can shop responsibly, without spending a lot of their time doing research. Engagement of the customers is necessary to create a sustainable fashion industry. But, what is more, the supply side and the demand side have to work together on ethical issues within the whole value chain and make fashion more sustainable.

3.2 Competitive Advantage for Companies Adopting Sustainable Practices

Fashion brands have to develop competitive advantages to be successful in the long run. Competitive advantages could be generated by adopting various sustainable practices, for example by developing new sustainable business models or by introducing innovations in fabrics, like new methods of creating fabrics and enhanced fabric materials or smart fabrics.

3.2.1 Innovations in Fabrics

Some retail and manufacturing companies, within the apparel industry, have already started to use **sustainable fabric materials** for producing their products.[1] Their products are made of bio-based synthetic fibers from renewable resources (e.g. sugar cane, corn sugars, agricultural waste, etc.), instead of petroleum-based synthetic fibers, which leave a lower carbon footprint during the production process, and new materials like synthetic spider silk (Weinswig 2016b). They also use water- and stain-resistant cotton fabrics which help to reduce water and energy necessary for the washing process (Weinswig 2016b).

Another innovation field are so-called "**smart fabrics**". Smart fabrics are textiles with embedded technology that enables clothing made from them either to perform functions that regular garments cannot perform or to have special characteristics that regular clothing does not have (Weinswig 2016a). Innovative companies[2] have developed nanofabrics and integrate silver nanoparticles into garments to make them stain-, odor-resistant and waterproof. Another category are connected fabrics which have embedded digital, wearable technology (Weinswig 2016a). Smart apparel offers the possibility to connect to other devices and collect and analyse data or offer additional features like music storage, file-playing features, charging batteries, tracking and tracing etc.

All in all, smart apparel provides a huge potential and new business opportunities for the apparel industry, especially for sportswear and accessories manufacturer and retailers, although innovations in fabrics involve higher investment in R&D and usually higher production cost which means that customers would have to pay a premium price for these products. Therefore many of these innovations are not relevant for the mass market, at least at the moment (Weinswig 2016b).

Another competitive advantage can be generated by developing new methods of creating fabrics. Examples are biomimicry, spray-on fabricant, enzyme-grown fabrics or recycled materials.

The core idea behind **biomimicry** is that nature has already solved many of our problems and has developed sustainable solutions by emulating nature's time-tested patterns and strategies (Biomimicry Institute 2016). For example, by mimicking the way color is produced in the Morpho butterfly's wing,[3] the fibre appears colored without any dying. Iridescence in the butterfly's wing is formed through structural or physical color, rather than chemical color (Donna Sgro 2012).

---

[1]For instance www.spiber.jp/en/endeavor, www.nanexcompany.com or www.basf.com

[2]For instance www.myzone.org, www.owletcare.com and www.clothingplus.com

[3]One of the largest butterflies in the world, with wings that span 5–8 inches.

Spray-on fabricant creates an instant sprayable non-woven fabric. This liquid suspension is applied via spray gun or aerosol on to any surface. The fabric is formed by the cross linking of fibers. It helps to create more customer-centric products that perfectly fit customers' needs and personalize the wardrobe (Fabrican Ltd 2011).

Enzyme-grown fabrics.

3.2.2 New Sustainable Business Models

Another competitive advantage for companies willing to adopt sustainable practices can be generated through innovative business models, in particular **online apparel and accessory rentals,**[4] especially suitable for higher-end designer merchandise. These services allow customers to borrow fashion items for a limited period of time at lower costs than buying these items (Weinswig 2017). This business model is especially suitable for millennials (born between 1980 and 2000) because they put far more emphasis on corporate responsibility and sustainability in their purchase decisions than other generations (Weinswig 2016b).

Online collaborative consumption for apparel and accessories helps to expand an item's lifecycle, decreases waste production, declutters people's homes and reduces, in general, overconsumption of fashion items.

Virtual fashion in general and digital prototyping in particular is a business model which helps to reduce the amount of waste produced during the sampling stage of designing a fashion line and to promote a zero-waste fashion industry (Fibre2fashion 2016) but also to reduce shipping transactions. Through virtual fashion platforms, e.g. My Virtual Model,[5] customers can make use of virtual models that can be personalized to look like the customer, and virtually try on clothes before they buy them (Calderin 2009). One of the main advantages is that customers don't have to order clothes to try them on physically or to visit a brick-and-mortar store which helps to reduce the amount of returns and the carbon footprint, and provides the advantage of customizing products (user-centric fashion).

4 Sustainable Marketing Communication

This section discusses sustainable marketing communication in general and communication strategies to encourage ethical consumer behaviour. This chapter also takes up the issue of overstating "green" communication and the negative consequences.

---

[4]A well-known example is Rent the Runway (https://www.renttherunway.com/).

[5]http://myvirtualmodel.com/

4.1 Effective Marketing Communication and "Greenwashing"

When consumers are asked what comes up to their mind when they hear the term "ethical fashion", it is associated with "fair working conditions, a sustainable business model, organic and environmentally friendly materials, certifications and traceability" (Henninger et al. 2016, S. 400). As a consequence, these are important issues to be addressed in an effective marketing communication strategy.

Moreover, marketing communication is most effective when sustainable consumption issues are linked to values that are harmonious and congruent to the values the target group holds important, for instance family health, attractiveness or economic value (Martin and Schouten 2014). The advertisements should include social and environmental messages about ethical fashion (Shen et al. 2012).

But if such "green" claims and sustainability issues are exaggerated, misleading or not verifiable, companies might be accused of so-called "greenwashing", which leads to a negative brand image or company's reputation.

In order to avoid "greenwashing", misleading marketing communications and not proper use of "green" words, it is essential to define sustainable fashion in general and to identify key sustainable fashion criteria in particular, which provide assurance for the customers that fashion items are produced with social, ethical and eco-friendly aspects in mind (Henninger et al. 2016). Shen et al. (2012) have learned in their study that consumers' confidence in a fashion company can be enhanced by increasing transparency of production and manufacturing processes.

4.2 Communication Strategies to Encourage Ethical Consumers Behaviour

Manchiraju and Sadachar (2014) suggest that companies should develop communication strategies that focus on self-enhancement values, such as hedonism, power and achievement, because these values to intentions to engage in ethical consumer behaviour. They also suggest that older populations, especially females, should be targeted in ethical marketing campaigns, as women and baby boomers, for instance, are more likely to engage in ethical consumption.

When it comes to the younger age group (e.g. Generation Y), which are normally very concerned with how they are perceived by their peers, targeting opinion leaders is likely to be efficient when it comes changing a certain behaviour (McNeill and Moore 2015). Especially social media is very important for this age group. Therefore social media campaigns with role models and opinion leaders, promoting ethical consumer behaviour, might increase interest in sustainable fashion.

Sustainable marketing communication has to have certain aspects to be successful (Martin and Schouten 2014, S. 209–212):

- Accountability and transparency: a company should be openly accountable for its actions and expose business and marketing practices to outside observers;
- Credibility: consumers should regard companies' claims as credible, for instance by third-party verification and reporting;
- Consumer education: consumers should be well-informed about the need for sustainability, the consumption of green products and services and its benefits and ethical business practices;
- Value congruence: a marketing message has to be consistent with a target consumer's values by providing accurate information;

To summarize, the relationship between consumer education (knowledge and awareness) and support from the supply side (designers, manufacturing and retailing), but also from the government and other external stakeholders, are the most important aspects of establishing sustainable business and business practices.

5   Conclusion

Based on the theoretical fundamentals and practical examples described in the chapters above, this chapter briefly summarizes the paper, states implications and gives recommendations for the fashion industry.

5.1   Summary

In the last few years ethical behaviour and sustainable business practices have become vital for a company's reputation and profitability. There are already a lot of fashion companies that have a sustainable performance. Also fast fashion companies are constantly trying to improve their performance in regard to sustainability.

Although consumers have a gaining interest in ethical fashion and "green" issues, they very often don't buy sustainable fashion. One of the reason is that they are very often unable to make ethical judgements due to a lack of comprehensive information. Other reasons are that sustainable clothing is very often perceived as less fashionable and that consumers don't see a true added value which justifies the price premium.

Therefore, to make consumers care about ethical behaviour in general and, in particular, purchase more sustainable clothing, marketing communication and consumer education are essential to achieve these objectives.

To conclude with, the most important aspect of establishing sustainable business practices is that all stakeholders involved in the fashion industry collaborate with each other and try to make the fashion industry more sustainable.

5.2   Implications and Recommendations

New technologies and innovations might help fashion companies to adopt sustainable practices and gain competitive advantages, for example by developing new sustainable business models or by introducing innovations in fabrics, like new and enhanced fabric materials or smart fabrics. Moreover, 3-D body scanning and automated custom-fitting systems have the potential to move parts of clothing production from mass fashion to individualized fashion and extend a garment's lifecycle. Also mass customization is a promising way to create more customer-centric products that perfectly fit customers' needs and therefore won't be thrown away after a short period of time.

Furthermore, a truthful label on clothes by a reliable, independent institution or organization, containing important ethical information, might be a helpful tool to facilitate customers' purchase decisions and to generate greater awareness.

Last but not least, sustainable fashion has to become more fashionable and widely socially accepted. Many fashion companies and designers already head in the right direction, but they still have a lot of work to do.

References

Ashdown, S. (2008). Bodyscanning and fit: The impact of new technologies on sustainable fashion. In S. Black (Ed.), *The sustainable fashion handbook* (p. 289). London: Thames & Hudson.

Barnes, L., & Lea-Greenwood, G. (2006). Fast fashioning the supply chain: Shaping the research agenda. *Journal of Fashion Marketing and Management, 10*, 259–271.

Biomimicry Institute. (2016). *What is biomimicry.* Retrieved from Biomimicry: https://biomimicry.org/what-is-biomimicry/

Black, S. (2008). *The sustainable fashion handbook.* London: Thames & Hudson.

Calderin, J. (2009). *Form, fit, fashion: All the details fashion designers need to know but can never find.* Beverly, MA: Rockport.

Donna Sgro. (2012). Morphotex dress. Retrieved from donnasgro: http://donnasgro.com/Morphotex-Dress

Fabrican Ltd. (2011). *What is fabrican?* Retrieved from http://www.fabricanltd.com/

Fibre2fashion. (2016, November 24). *'Designers should adapt digital sample to reduce waste'.* Retrieved from http://uk.fashionnetwork.com/news/–Designers-should-adapt-digital-sample-to-reduce-waste-,756985.html#.WKL-2GaFN9A

Henninger, C. E., Alevizou, P. J., & Oates, C. J. (2016). What is sustainable fashion? *Journal of Fashion Marketing and Management: An International Journal, 20*(4), 400–416.

Joergens, C. (2006). Ethical fashion: Myth or future trend? *Journal of Fashion Marketing and Management, 10*(3), 360–371.

Manchiraju, S., & Sadachar, A. (2014). Personal values and ethical fashion consumption. *Journal of Fashion Marketing and Management, 18*(3), 357–374.

Martin, D., & Schouten, J. (2014). *Sustainable marketing* (Pearson New International Edition). London: Pearson Education.

McNeill, L., & Moore, R. (2015). Sustainable fashion consumption and the fast fashion conundrum: Fashionable consumers and attitudes to sustainability in clothing choice. *International Journal of Consumer Studies, 39*, 212–222.

Mittelstaedt, J. D., Schultz, C. J., Kilbourne, W. E., & Peterson, M. (2014). Sustainability as megatrend. *Journal of Macromarketing, 34*(3), 253–264.

Piller, F., & Steiner, F. (2008). Mass customization: A strategy for sustainability in the fashion industry. In S. Black (Ed.), *The sustainable fashion handbook* (pp. 287–288). London: Thames & Hudson.

Shen, B., Wang, Y., Lo, C. K., & Shum, M. (2012). The impact of ethical fashion on consumer purchase behaviour. *Journal of Fashion Marketing and Management, 16*(2), 234–245.

Teah, I. P., & Chuah, J. (2015). Consumer attitudes towards luxury fashion apparel made in sweatshops. *Journal of Fashion Marketing and Management, 19*(2), 169–187.

Weinswig, D. (2016a, November 21). *Smart fabrics: The future of apparel?* Retrieved January 4, 2017, from Fung Global Retail & Technology: https://fungglobalretailtech.com/wp-content/uploads/2016/11/Smart-Fabrics-Report-November-21-2016-2.pdf

Weinswig, D. (2016b, December 9). *Innovations in fabric materials.* Retrieved January 8, 2017, from Fung Global Retail & Technology: https://fgrtresources-rm31c3pfhm5geadr1.netdna-ssl.com/wp-content/uploads/2017/01/Innovations-in-Fabric-Material-December-9-2016.pdf

Weinswig, D. (2017, January 16). *Millenial lifestyles drive growth in apparel rental.* Retrieved January 18, 2017, from Fung Global Retail & Technology: https://fgrtresources-rm31c3pfhm5geadr1.netdna-ssl.com/wp-content/uploads/2017/01/Millennial-Lifestyles-Drive-Growth-in-Apparel-Rental-January-16-2017.pdf

Doris Berger-Grabner is professor at the IMC FH Krems in the department business. Her specialization is marketing, retail management, market research and scientific work. She is also a part-time lecturer at the Danube University Krems (Lower Austria).

She holds a master's degree in Business Education and a PhD in Business Administration from the Business University of Vienna. She is the author of various books and journal publication.

The Bosch Group's Approach to Innovation and Sustainability Communication

Bernhard Schwager

1 Company Founder and Corporate Foundation

Robert Bosch founded his company in 1886 in Stuttgart as the "Workshop for Precision Mechanics and Electrical Engineering." Today, the Bosch Group comprises Robert Bosch GmbH and around 450 subsidiaries and regional companies in some 60 countries. If its trading and service partners are included, Bosch is present in around 150 countries. Its global development, manufacturing, and sales network shapes the foundation of the company's further growth. The ownership structure of Robert Bosch GmbH ensures the Bosch Group's financial independence. It enables the company to plan for the long term and invest in the future. For the Bosch Group, sustainability and a long-term strategic orientation are integral parts of corporate culture.

In 1921, Robert Bosch made the following statement in the Bosch Zünder associate newspaper: "I would rather lose money than trust". The company founder's word and faith in his products were always more important to him than short-term profit. Today, the Bosch Group continues to observe its founder's values (see Fig. 1).

One particular characteristic that sets Robert Bosch GmbH apart from most other companies of comparable size is its ownership structure. 92% of the company's share capital is held by Robert Bosch Stiftung GmbH, a charitable foundation. Robert Bosch Industrietreuhand KG, an industrial trust, holds the majority of voting rights. It thus acts as the executive shareholder. The Bosch family holds the remaining shares; the family continues to have an interest in the company's success and wishes to hold a lasting stake in it (see Fig. 2).

B. Schwager (✉)
Corporate Health, Safety, Environmental and Fire Protection as well as Sustainability – Sustainability and Think Tank (C/PSS), Robert Bosch GmbH, Stuttgart, Germany
e-mail: Bernhard.Schwager@de.bosch.com

© Springer International Publishing AG, part of Springer Nature 2018   181
R. Altenburger (ed.), *Innovation Management and Corporate Social Responsibility*,
CSR, Sustainability, Ethics & Governance,
https://doi.org/10.1007/978-3-319-93629-1_10

Fig. 1 Robert Bosch
(1861–1942)

Fig. 2 Current ownership
structure of the Bosch
Group

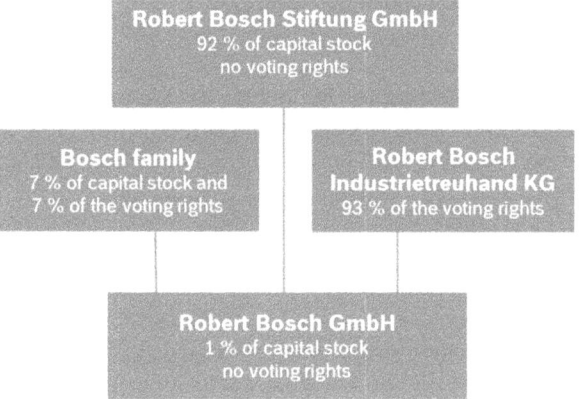

It is important to note that the owners are not focused on quick profit. Rather, the aim is to develop high-quality, beneficial products and services. Today, this aim is reflected in the slogan "Invented for life". In everything it does, Bosch strives for responsible management, which has been a defining characteristic throughout the company's history. On the occasion of its 125th anniversary, Christoph Bosch, the company founder's grandson, said: "I think the term sustainability best describes him." If the term had been used outside of forestry at the time, Christoph Bosch believes that his grandfather would have used it.

Robert Bosch Stiftung spends the dividends it receives exclusively on charitable initiatives—to the tune of 700–800 projects each year. To reach its targets, the Stiftung supports third-party projects in addition to developing and carrying out its own programs. Since it was founded more than 50 years ago, the Stiftung has spent some 1.4 billion euros on approximately 20,000 of its own and other projects. For instance, the Stiftung owns Robert Bosch Hospital and the Institute for the History of Medicine.

2   From Patent to Innovation

Thanks to its entrepreneurial spirit and corporate foundation, Bosch is better able to focus its long-term activities on striking the right balance between economic, social, and ecological interests. In some instances, contradictions between these spheres can only be removed with innovation. Developing innovative products that meet market needs is decisive to the company's success. In 2017 alone, Bosch filed around 5800 patents, or 24 patents per working day. In total, Bosch has over 100,000 active property rights. The aim is to develop inspiring solutions for its customers that not only meet quality, reliability, and cost requirements, but that also serve as benchmarks in their respective market segments.

The automotive segment is a strong example of this. Bosch has filed patents for products in all parts of the car, among them lighting systems, starters, horns, windshield wipers, diesel injection pumps, servo brakes, direction indicators, car radios, and car heating systems.

For a long time, the problem of wheels locking when a car was braking appeared to be insurmountable. It wasn't until the 1960s, when electronics made sufficiently fast and reliable brake regulation possible, that Bosch began developing an anti-lock braking system (ABS) for series-produced vehicles. 15 years past from the initial stages of development to start of series production in 1978 (see Fig. 3), and it took another 5 years for ABS to pay off for Bosch. Today, the system is mandatory for every newly registered vehicle in the European Union. This example shows that long-term thinking must sometimes go beyond a company's usual medium- and long-term planning. This also clearly demonstrates that patent registrations only lead to innovations when researchers' findings can be turned into new products and services.

Bosch innovations make life safer and more eco-friendly, in line with the legacy of company founder Robert Bosch. Today, Bosch is expanding its activities to meet the demands of modern human life. For instance, the company is coming up with innovative mobility concepts and smart infrastructure that are making decisive contributions to enhancing quality of life in a sustainable manner, especially for the world's growing urban population.

3   Attitudes and Traditions

Franz Fehrenbach, who succeeded Hermann Scholl as chairman of the board of management on July 1, 2003, upheld the strategy of systematically reducing the company's dependence on the automotive industry by striving for strong growth in

Fig. 3 Bosch has been making brakes safer for the past 30 years with its anti-lock braking system

other business sectors. In so doing, Fehrenbach placed great importance on globalization, environmental protection, resource conservation, and energy efficiency. In Automotive Technology, the largest and oldest Bosch business sector, the company stepped up research on hybrid and electric drives. The first hybrid car with Bosch technology was launched in 2010.

Bosch has always considered that corporate responsibility involved striking a balance between business success and social concerns. In recent years, the company has expanded its definition of responsibility to include environmental protection. In 2007, Franz Fehrenbach said: "Our top priority is without question to secure the company's long-term future, but we are also equally committed to doing so by achieving a balance between ecology, economy, and corporate social responsibility." This statement is based on the belief that a company can only be successful in the long term if it pursues a policy of sustainable business management and does not infringe on social and ecological interests.

Bosch practices corporate social responsibility and a sustainable approach to doing business out of a firm belief that it is the right thing to do. In so doing, the company always strives to ensure that it meets the needs of internal and external stakeholders, and in some instances it assumes a pioneering role. The Bosch commitment to entrepreneurial responsibility must be equally reflected both in the product portfolio and in the pursuit of sustainability targets in the company's value-added processes. Strategies, targets, and opportunities are outlined in the "We are Bosch" mission statement, which replaced the "House of Orientation" at the start of

2015. The statement presents a new framework of reference upon which the Bosch Group and its business sectors base their strategic orientation. All of the focal points comprised in "We are Bosch" are based on influential factors such as megatrends, changes in the competitive environment, innovations, customer expectations, resource scarcity, and political developments. Training and continuous professional development also play a very important role.

As a result of the shift from a seller's to a buyer's market, companies must more than ever offer products that are tailored to the current needs of customers. To this end, Bosch takes advantage of the innovative potential of global developments at some 120 locations. The company sees change as a major opportunity, especially in the areas of energy efficiency, electrification, automation, emerging markets, and connectivity. To seize these opportunities, Bosch benefits from its strengths, such as its corporate culture, long history, innovative strength, the quality of its products and services, and its broad global set-up. The Bosch values continue to shape the foundation of the company's strategy and approach to business: a clear future and profit orientation, responsibility and sustainability, initiative and consistency, openness and trust, fairness, reliability and credibility, legality, and diversity.

4 The Innovation Process

For more than 125 years, the name Bosch has been associated with forward-looking technology and groundbreaking inventions that have made history. Today, the Bosch Group is a leading global provider of technology and services with 402,000 associates worldwide (status of December 31, 2017). In fiscal 2017, Bosch generated sales of some 78 billion euros in the four following business sectors: Mobility Solutions, Industrial Technology, Consumer Goods, and Energy and Building Technology. The Bosch Group aims to improve the quality of life with innovative, beneficial, and inspiring products "Invented for Life". The board of management's commitment to responsible corporate management is closely linked to these goals (Bosch 2017). This responsibility has played an important role throughout the company's history.

Bosch innovations are by no means the product of chance. Rather, research and development work is systematic and targeted. This has not changed over the course of Bosch history. The only thing that has changed is the company's focus. Today, the subject of energy efficiency is more important than ever. However, development activities not only focus on creating energy-efficient products. Bosch also puts a great deal of effort into optimizing its own value creation. Global increases in energy demand have driven this development, as have ever-stricter climate protection standards and the finite nature of fossil fuels. These include the increasingly demanding emissions standards and regulations concerning the fuel consumption of vehicles. Today, Bosch already generates around 40% of its sales with products that contribute to energy efficiency, environmental protection, and resource conservation. What is more, Bosch currently spends half of its research and development expenditure on these products.

For instance, Bosch household appliances have made everyday life easier and improved quality of life. In developing these products, Bosch has long paid close attention to energy efficiency and protecting the environment. Inspired by its company founder's organic farming activities, Bosch continues to lead the way in green technologies, making certain that all appliances lower water and energy use. For instance, the EcoLogixx 7 heat-pump dryer consumes up to 55% less energy than previous models and reflects the Bosch commitment to continuously offering sustainable, efficient technologies.

Developing innovations for a connected life is another strategic aim that reflects the "Invented for life" leitmotiv. These include, for instance, the "Home Connect" software, which allows people and machines to come closer together. Connected ovens and dishwashers have been available since December 2014, and refrigerators, washing machines, dryers, and coffee machines followed in the fall of 2015. These devices have made it possible to do housework from anywhere, and this gives users more flexibility and thus also more free time. For instance, via a smart phone app, the washing machine can let its user know when it has finished the laundry, or the dishwasher can warn its owner that it is running out of rinse. This can prove practical when a user happens to be at the supermarket.

In strategic terms, electrification is a significant trend for the automotive industry in general, and for the Bosch Mobility Solutions business sector in particular. According to experts, the electric car will not see a breakthrough for some time yet. By 2020, Bosch expects electric vehicles to account for 2.5 million out of more than 110 million new cars. The same year, the company expects to see some three million plug-in hybrids and 6.5 million hybrid vehicles on the world's streets (see Fig. 4).

For the product portfolio, the "Invented for life" strategic leitmotiv calls for technical solutions that contribute to protecting the environment. The leitmotiv thus aims to make renewable sources of energy even more profitable, and mobility even safer, cleaner, and more efficient, and products more eco-friendly. Ecology is

Fig. 4 The future is electric

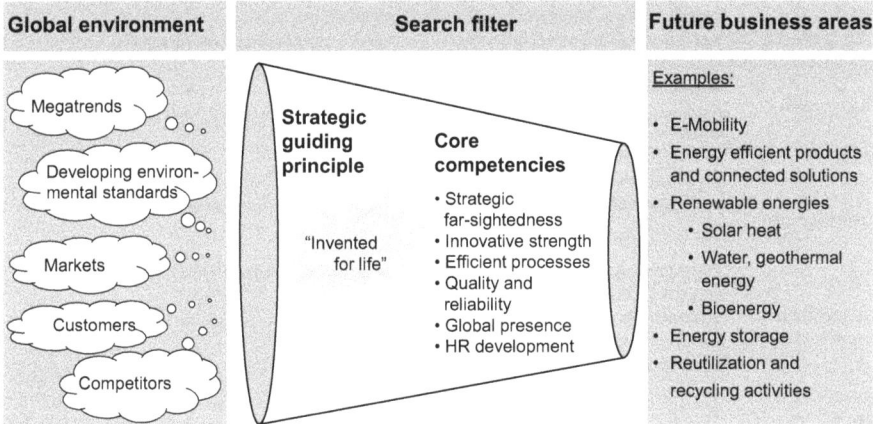

Fig. 5 Focus on sustainable sectors that show promise for the future

an engine of innovation at Bosch, and the Bosch Group sees environmental protection as a business opportunity.

The leitmotiv is a strategic filter for the further development of Bosch business with sustainable products. This applies both to the further development of existing fields of business, and to the creation of new ones. What is more, to secure their success, product and business ideas must reflect the company's core competencies. Figure 5 shows examples of strategic products in areas that show promise for the future, and in which Bosch is already active today.

Over the past 5 years, the Bosch Group has spent more than 20 billion euros on research and development that aims to make life safer, more comfortable, and more eco-friendly. Of its 402,000 associates around the world, in 2017 some 64,400 were active in research and development.

Long-term decisions should be based on the results of detailed analyses. For this reason, Bosch analyses megatrends and identifies fields of business that are a good fit for the company. Figure 6 shows the main drivers of e-mobility, for instance. In this strategically important realm, the aim is to push powertrain electrification forward with electric motors, the required power electronics, regenerative braking systems, and much more.

5 Doing Good and Talking About It

For many companies, coming up with sustainable innovations is nothing new. For the Bosch Group, a central challenge lies in communicating the company's sustainability-related activities as effectively as possible. Given that sustainability has become an important factor with regard to a company's image and reputation, it is now also decisive to the company's success. Bosch published its first

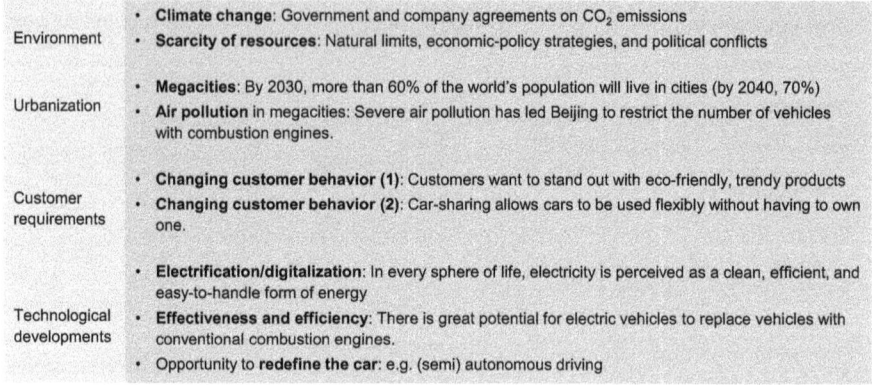

Environment	• **Climate change**: Government and company agreements on CO_2 emissions • **Scarcity of resources**: Natural limits, economic-policy strategies, and political conflicts
Urbanization	• **Megacities**: By 2030, more than 60% of the world's population will live in cities (by 2040, 70%) • **Air pollution** in megacities: Severe air pollution has led Beijing to restrict the number of vehicles with combustion engines.
Customer requirements	• **Changing customer behavior (1)**: Customers want to stand out with eco-friendly, trendy products • **Changing customer behavior (2)**: Car-sharing allows cars to be used flexibly without having to own one.
Technological developments	• **Electrification/digitalization**: In every sphere of life, electricity is perceived as a clean, efficient, and easy-to-handle form of energy • **Effectiveness and efficiency**: There is great potential for electric vehicles to replace vehicles with conventional combustion engines. • Opportunity to **redefine the car**: e.g. (semi) autonomous driving

Fig. 6 The main drivers of powertrain electrification

environmental report in 1998, and this marked the beginning of communication activities related to the company's sustainability activities. Since then, communications have been continuously improved and expanded. Today, the Bosch Group sets the standard for innovative sustainability communication. Environmental reports were also published in 2001/2002 and 2003/2004. In the years that followed, reporting was expanded, and in 2005/2006 and 2007/2008 two comprehensive print reports on corporate responsibility were published, one of which observed the guidelines of the Global Reporting Initiative (GRI). Over time, these reports have become increasingly detailed. While the first environmental report focused exclusively on data from Germany, the second report included European data, and the third comprised data from around the world. Reporting on sustainability-related topics was then expanded when the corporate responsibility reports were published. Moreover, these reports have since been complemented with annual brochures that present sustainability data and targets.

5.1 Sustainability Communication Online

At the same time, the development on an environmental web portal in 2004 marked the beginning of online reporting. It was then replaced in 2006 by the sustainability portal. Until 2008, the ever-more comprehensive Sustainability Report was the cornerstone of the Bosch Group's sustainability-related communications. From 2008 onward, sustainability communications were fundamentally overhauled with the aim of reaching a broader target group. The CSR-Homepage is now shapes the core of sustainability communication at Bosch. It is part of the Bosch Global homepage and was launched in 2008. It comprises information in various forms on sustainability-related sub-topics. This information aims to provide readers with a quick overview and is not very detailed.

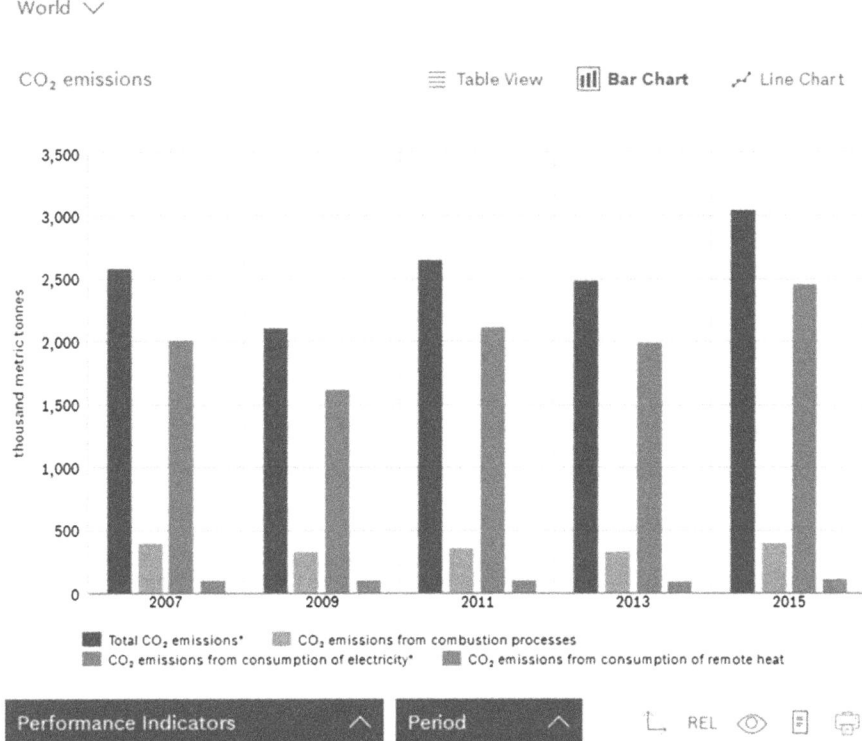

Fig. 7 The Bosch Group's interactive data tool

At the same time, an interactive data tool offers access to the Bosch Group's ecological, social, and financial data. This tool provides a very detailed look at the Bosch Group's key CSR figures. Figure 7 shows the broad range of information on offer. Key figures relating to water consumption, waste, CO_2-emissions, associate numbers, accident rates, or energy needs are listed both at the global and regional levels. In addition, each individual graphic can be displayed in a number of ways. Certain key figures are also presented in a very detailed manner, broken down into sub-parts, and can be tracked back to 2007. In addition to this, all data can be exported to Excel, which makes it possible to examine them offline.

The Bosch Group's CSR homepage also includes dialogs on environmental and sustainability-related topics in which renowned personalities from science, business, politics, and the non-profit sector are invited to express their views on current topics. The subjects discussed include social issues that go beyond the Bosch fields of activity. This exchange has no direct relationship to the company's specific interests, or to meeting the CSR targets that Bosch has set. Rather, the talks aim to demonstrate the company's openness to the concerns of different interest groups. Bosch also considers sustainability a social issue, one that it aims to contribute to.

The CSR Homepage is complemented by the Sustainability blog. Since 2010, Bosch has published weekly articles about CSR topics and innovations, many of them addressing subjects in different regions. The current topics addressed on the blog contribute to increasing transparency and help promote reader loyalty. In addition to this, since 2011 readers have been offered the possibility of registering for the quarterly CSR newsletter, which is published in German and English. At the moment, the offer counts almost 1000 registered subscribers, and the number is rising. This type of sustainability communication is geared toward a very specific group of people. By registering for the newsletter, subscribers consciously select information on sustainability and Bosch Group-related topics they are expressly interested in. This allows the company to provide high-quality, targeted information to potential associates and business partners.

5.2   Annual Report and German Sustainability Code

At Bosch, sustainability is a board of management issue. This is reflected in the Annual Report, which has included a section on environmental protection since 2001. Today, the report contains sections on environment and health, as well as on compliance. It also communicates figures on CO_2 reduction in the form of a graphic. While the Annual Report does not present a detailed account of sustainability-related information, it reaches a target group whose primary interest is not sustainability. From the 2017 reporting year onward, companies with more than 500 employees will have to comply with EU Directive 2014/95/EU, which stipulates that these businesses must disclose information related to non-financial topics and diversity.

The German Sustainability Code (DNK) is a proven standard that companies can draw on to expand their sustainability communication. Today, the DNK already meets EU reporting standards and even exceeds these requirements in some instances. The report, which is about 20 pages in length, describes the company's comprehensive sustainability strategy, which includes the areas of process management, environment, and society, as well as the relevant key figures. This form of reporting already provides sound orientation for the reporting requirements that are entering into force in 2017. Bosch has also committed to complying with these requirements, and thus signed the DNK's declaration of intent, which is updated annually, in September 2015. The declaration contains 12 targeted areas of action, which are described in detail. For instance, under the heading of innovation management, the company discloses how it contributes to resource conservation with the right processes and innovative products and services. This enables Bosch not only to make the resource use of its customers more sustainable, but also its own. The current and future effects of its products and services on the economy and product portfolio are also assessed. Bosch also shows which ecological and social challenges the company will face in the future, and the ways in which it systematically controls and improves its sustainability performance. The Bosch Group responds to challenges such as demographic change, the reduction of greenhouse gas emissions throughout the value-added chain, and the maintenance of healthy ecosystems by finding technical answers to ecological questions. As early as 2000, Bosch launched its Design for Environment initiative, which helps conserve resources by considering sustainability criteria in the early phases of development.

5.3   Communication on Progress

Broad-ranging sustainability communication improves transparency and has a positive effect on an organization's image. Standardized assessments such as the DNK allow stakeholders to compare companies with one another. Another reporting format that Bosch has been using since 2004 is the United Nations Global Compact (UNGC). The initiative was launched in 1999 by Kofi Annan, the former Secretary General of the United Nations. Companies, organizations, and public entities can sign a voluntary agreement committing to observe a list of basic sustainability-related principles. Until now, more than 10,000 organizations have signed the Global Compact, making it the biggest network of its kind. The UNGC comprises 10 principles in the areas of human rights, working conditions, environmental protection, and the fight against corruption. Within the UNGC, an international network with a regional structure has been created, within which companies can engage in discussions with one another about sustainability-related topics. From the perspective of communications, the commitment to publishing an annual progress report entitled the "Communication on Progress—COP" is of particular relevance. The report must contain information on the company's efforts to put the 10 UN Global Compact principles into practice. This requirement is based on the idea that the process of pursuing sustainability is an ongoing one that must continuously be optimized.

At Bosch, this progress report is part of the annual Sustainability Report. Since 2011, the report has been published in a compact, 24-page format that focuses on four main areas:

– Environment
– Products
– Associates
– Society

5.4   Blueprint for Regional Reports

Thanks to its compact format and concise presentation, the Bosch Sustainability Report provides a valuable platform for a broad range of readers who may not wish to read a detailed account of sustainability activities at the company, but who would like to gain a good overview. The Bosch Sustainability Report can also be used as a template for regional sustainability reports (see Fig. 8). The report is adapted to meet

Fig. 8 The Bosch Group's sustainability and annual reports

local needs and is published in the local language. In this way, Bosch can address topics that are of interest for specific regions and with which local readers can identify. For instance, listing the milestones that a region has reached not only helps motivate associates, it can also have a positive impact on Bosch's reputation in individual regions.

5.5   Social Media

Thanks to the opportunities the Internet offers, communication has undergone a major transformation in recent years. Indeed, the Internet opens up entirely new possibilities with regard to reaching a broad range of target groups. Information can be made available regardless of place and time, and many forms of dialog-based communication are possible at a number of levels. Hierarchical structures or social status play a secondary role here. In order to present the company in a modern way, Bosch has also expanded its presence in social media. Since 2012, the company has had comprehensive information and dialog offers on the biggest international platforms, among them Facebook, Youtube, and Twitter. The overall aim is to achieve a full presence on all relevant channels and improve Bosch's reputation in the process. Thanks to an interdisciplinary editorial team, users are invited to engage in dialog about innovation and current corporate topics such as mobility, energy, or sustainability. In this way, new target groups such as bloggers and other online multipliers can be reached. As a result of the high level of anonymity that social media offer, the initial hurdle is very low and active participation is high. Another significant benefit of social media is that the reach of communication can be increased in a short period of time with very little effort and at a low cost. Direct feedback via likes, comments, and discussion rounds provides the company with insights and know-how that could otherwise only be gained through elaborate studies. When it comes to communications, Bosch has a very broad set-up and uses a range of different media to reach stakeholders in all areas and across social classes.

6 Tone from the Top

Dr. Volkmar Denner, chairman of the board of management of Robert Bosch GmbH, explains the Bosch approach to sustainability: "We have a long tradition of combining economic activity with social responsibility. Together with our committed associates, we want to help people gain a foothold in our society and offer them a better future". Sustainability activities at Bosch are organized in the following way:

- **The sustainability office** is the point of contact for internal and external requests. It keeps in touch with associations and identifies problems and areas of improvement within the walls of the company.
- **The sustainability advisory board.** Members include the following corporate departments: purchasing, production, infrastructure, HR, and environment, as well as the presidents of various divisions.
- **The sustainability steering committee** sets the main targets and monitors their achievement. The CEO himself sits on this committee, together with another member of the board of management and representatives of the sustainability advisory board.

These structures aim to send a clear signal to all stakeholders: "At Bosch, sustainability is a management issue—one that must be addressed in all business units." Sustainability is a central topic for the board of management, and this shows once again the importance it is accorded at Bosch.

7 Conclusion

CSR concerns the entire company, across functions and hierarchical levels. It must thus be reflected in the organization's vision and corporate strategy. It must also be implemented in its product portfolio and in value-added processes, in addition to being embodied by the company itself.

The Bosch Group's "Invented for life" strategic leitmotiv is the starting point for the product portfolio. Examples in this article show that protecting the environment and securing the lasting success of the company's business are not mutually exclusive. In fact, environmental protection drives the development of many innovations, new products, and services. For this reason, Bosch spends more than 50% of its research and development budget—more than three billion euros—on developing products that help conserve resources and save energy. The company already generates almost 40% of its sales with such products.

Companies are well advised to come up with innovative approaches to sustainability-related communication. This is especially important in today's fast-paced, fast-changing society. As mentioned already, communication has undergone a transformation that has been beneficial, but also fraught with challenges. "Information overload" is one consequence of modern communication. The sheer quantity of information that users must process is a problem in itself, as they are continuous bombarded from all sides by competing companies and organizations. In light of this, the art of successful communication now lies in presenting different target groups with information that addresses their specific needs. To achieve, this, Bosch uses a broad range of media. The information published in each channel is packaged in a way that makes it attractive and easy to understand for the target groups in question. In the best of cases, this information can contribute to further improving the company's reputation.

Business and society are closely interrelated. Good business often depends on a healthy society. For this reason, companies are well advised to contribute to social well-being. Robert Bosch already promoted this idea as early as 1886, the year he founded his company. Throughout its history, Bosch has shared its founder's views. This values-oriented attitude is reflected in the "We are Bosch" mission statement.

Reference

Bosch Sustainability Report. (2017).

Bernhard Schwager   After 20 years within Siemens, Bernhard Schwager joined the Bosch Group in 2005. In May 2006, he was appointed president of the German Association of Environmental Professionals, and in May 2008 chair of the German Environmental Management Systems and Audits committee, part of the German Standards Institute. At Bosch, Schwager is the head of the sustainability office. In this capacity, he acts as contact person for diverse stakeholder groups and promotes sustainability issues. As part of his work, he represents the company in various national and international organizations and he is the author or co-author of numerous books and articles.

CSR and Innovation: Anchoring Sustainability in Henkel's Innovation Process

Uwe Bergmann

1 CSR: Business Strategy—Innovation

1.1 The Social Responsibility of Business

The term CSR—and expectations for how companies address their social responsibility—is often understood in a superficial and overly-simplified way. In many instances, it is used as a synonym for voluntary "citizenship" projects outside of core business activities. Other interpretations suggest that CSR simply involves adherence to minimum requirements for moral and ethical behavior, such as the 10 principles of the United Nations Global Compact. Similarly, the discussion of CSR is dominated by standards, certifications and reporting requirements, although these only provide a framework for business activities. The full breadth of topics and considerations that the term CSR covers is much more comprehensive, as illustrated by the United Nations Sustainable Development Goals: These 17 goals and 169 targets provide a common understanding and a shared focus on the challenges and opportunities that need to be addressed in order to drive progress toward sustainable development. CSR offers diverse opportunities for first-movers to open up new markets and gain a competitive advantage by leading progress toward sustainable development. For companies aiming to seize this advantage, innovative products, processes and business models play a decisive role.

U. Bergmann (✉)
Corporate Communications, Henkel AG & Co. KGaA, Düsseldorf, Germany
e-mail: uwe.bergmann@henkel.com

© Springer International Publishing AG, part of Springer Nature 2018   195
R. Altenburger (ed.), *Innovation Management and Corporate Social Responsibility*,
CSR, Sustainability, Ethics & Governance,
https://doi.org/10.1007/978-3-319-93629-1_11

1.2   Competitive Advantage and Innovation in the CSR Discussion

In 2001, the EU Commission published a greenpaper with the title, "Promoting a European framework for Corporate Social Responsibility" (Commission of the European Communities 2001). This document addressed questions of social responsibility, environmental impact and the cultivation of natural resources. Alongside these considerations, the greenpaper also explicitly addressed the market opportunities and competitive advantages presented by CSR:

> The European Union is concerned with corporate social responsibility as it can be a positive contribution to the strategic goal decided in Lisbon: "to become the most competitive and dynamic knowledge-based economy in the world, capable of sustainable economic growth with more and better jobs and greater social cohesion.

The rise of environmentally-compatible technologies, business models, and brand identities in the period since the publication of this landmark greenpaper indicates that companies see opportunities to strengthen their competitive position by going beyond simply conforming with social and environmental legislation. Alongside these developments, CSR considerations have begun to carry more and more weight in investment portfolio management strategies. These trends demonstrate that CSR—often used interchangeably with the term "sustainable business practices", despite conceptional differences—has gathered significantly increased business relevance.

Increasingly, companies, investors and politicians use CSR-related topics to position themselves. For example, at the beginning of the 1990s only a few dozen companies published a report outlining their activities and impact related to environmental, CSR and sustainability topics: Today, this reporting is standard practice, particularly among companies listed on the stock market—with some nations even introducing laws making environmental or non-financial reporting compulsory. By March 2017, the website CorporateRegister.com had accessed almost 84,000 reports from almost 14,000 organisations. As interest in the topic has risen, adherence with social and environmental standards—often referred to as "social license to operate"—has also gained increased importance (see Fig. 1). Alongside this, customers and consumers show increased interest in social and environmental performance—which opens up opportunities for companies to differentiate themselves from competitors by taking a leading role in driving progress toward relevant challenges. In some sectors, new or significantly expanded markets have emerged: Examples range from renewable energy and organic food, through to industrial solutions that contribute to making products and processes more environmentally-compatible, or improve health and safety.

This development has also been reflected in political attitudes to CSR. In October 2011, the European Commission published, "A renewed EU strategy 2011–2014 for Corporate Social Responsibility" (Communication from the Commission to the European Parliament, The Council, The European Economic and Social Committee and the Committee of the Regions 2011). This document remains relevant as a milestone in defining how the term CSR should be understood. Directly following the introduction—and its definition of the term "CSR"—the text points out implications regarding innovation and the ability to compete:

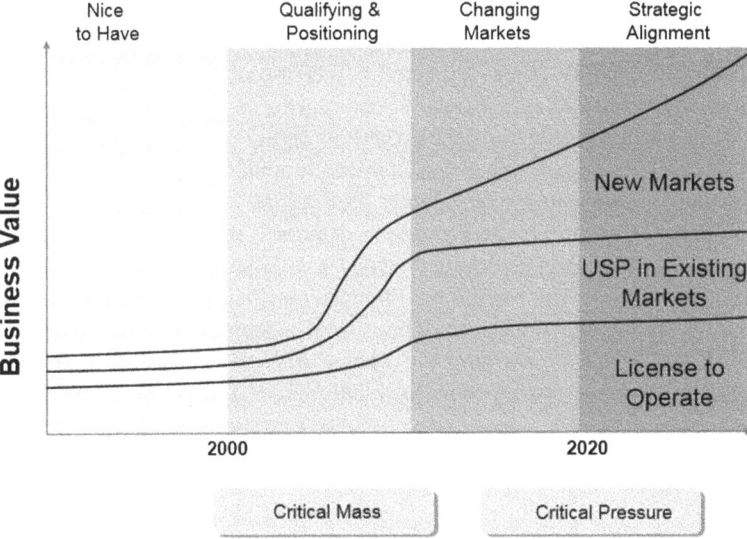

Fig. 1 Business relevance of sustainability/CSR. Source: Henkel

A strategic approach to CSR is increasingly important to the competitiveness of enterprises. It can bring benefits in terms of risk management, cost savings, access to capital, customer relationships, human resource management, and innovation capacity.

Because CSR requires engagement with internal and external stakeholders, it enables enterprises to better anticipate and take advantage of fast changing societal expectations and operating conditions. It can therefore drive the development of new markets and create opportunities for growth.

By addressing their social responsibility enterprises can build long-term employee, consumer and citizen trust as a basis for sustainable business models. Higher levels of trust in turn help to create an environment in which enterprises can innovate and grow.

Companies, politicians and stakeholders increasingly began to ask themselves how this potential could be leveraged. In particular, they wanted to understand how the abstract concept of "social responsibility"—or the equally abstract and often-used synonym "sustainable development"—could be integrated into companies' systems, processes and cultures.

1.3   Increasing Challenges

Four mega trends will shape the competitive landscape for companies in the coming decades:

- *The need to decouple economic growth and resource consumption.* In many regions around the world, this has developed from an academic concept into a tangible challenge. The global environmental footprint of humankind is already greater than the planet's resources can sustain, and the global population is expected to rise to 9 billion by the year 2050. At the same time, rising economic performance is leading to increasing consumption: With demand—and competition—for available resources expected to increase in the coming decades.

- *Unprecedented global transparency and interest in sustainable business activities.* This has been enabled by the growth in digital media and the availability of devices such as smartphones and tablet computers, as well as through the increasing language ability of younger generations.
- *Increasing regulation and new defacto standards introduced by non-governmental organizations (NGOs) are having increasing influence on access to markets.* This includes voluntary certification standards initiated by NGOs, as well as purchasing guidelines from industrial and retail companies.
- *Increasing consumer awareness of social and environmental topics—despite prevailing lack of willingness to compromise on performance, price or convenience.*

These challenges present businesses with opportunities and risks. Against this backdrop, conducting business in a socially responsible and sustainable manner is increasingly recognized as a factor in driving profitable growth, increasing efficiency and minimizing risk—forming a strong foundation for long-term economic success. For this reason, companies see the benefits of integrating CSR considerations into the development of new products, processes and business models. Alongside this, they also recognize the importance of collaborating with partners along the value chain and—in particular—engaging employees by expanding their knowledge of CSR topics and motivating them to consider the implications for their day-to-day work. Adherence to standards, certification programs and mandatory reporting requirements is not enough.

2   Integrating CSR into Henkel's Values and Strategy

2.1   Anchored in the Company's Tradition

Throughout Henkel's long tradition, one key principle has always applied: It is important *how* profit is earned. This fundamental understanding of the need to act responsibly along the entire value chain has been anchored in the company's culture since it was founded.

When the merchant Fritz Henkel founded his laundry detergent company in 1876, he had the vision of making people's lives easier, better and more beautiful. From the beginning, the company assumed responsibility for its employees, the communities it operated in, and society. In 1912, the company established a first aid center and employed an on-site nurse. Since 1927, systematic accident prevention work has successfully improved employee safety. In 1959, Henkel began carrying out regular environmental quality checks for detergents and household cleaners—which formed the basis for environmental quality checks for its products and production sites.

In 1972, Chairman of the Henkel Management Board Dr. Konrad Henkel warned: "Companies that only think in terms of profit will soon have a lot to lose." He was convinced that it was important for the company's economic, environmental and social goals to be aligned. In preparation for the Earth Summit in Rio de Janeiro in 1992, then Chairman of the Henkel Management Board Professor Dr. Helmut Sihler became one of the first business representatives to sign the International Chamber of Commerce (ICC) Business Charter for Sustainable Development. In 1993, Henkel became a member of the World Industry Council for the Environment: One of the organizations that merged in 1996 to form the World Business Council for Sustainable Development, of which Henkel is also a founding member. During his time as Henkel Chairman, Dr. Hans-Dietrich Winkhaus introduced the company's guiding principle of "competitive advantage through environmental leadership". This tradition of commitment to sustainability has shaped Henkel's long-term approach to doing business, and is part of what makes Henkel unique. It gives the company the organizational capacity to confront key challenges and act as a trusted partner for its stakeholders.

2.2 Sustainability Strategy and Targets

One of Henkel's values is: "We are committed to leadership in sustainability." Henkel uses the term sustainability, rather than social responsibility or CSR, although the terms are often used interchangeably. While CSR is often restricted to addressing social topics and adherence to standards, sustainability more effectively addresses both social challenges and the opportunities that they present for business.

As a leader in sustainability, Henkel aims to pioneer new solutions while developing its business responsibly and increasing its economic success. This commitment impacts all of the company's activities along the entire value chain. On this basis, the company defined its sustainability strategy for 2030:

> Our 20-year goal for 2030 is to triple the value we create for the footprint made by our operations, products and services.

Henkel's sustainability strategy is based on the Vision 2050 of the World Business Council for Sustainable Development (WBCSD): "In 2050, 9 billion people live well and within the resource limits of the planet" (WBCSD 2010). For Henkel, this means helping people to live well by generating value while using less resources and causing less emissions. This is the idea at the heart of its sustainability strategy: Creating more value—for its customers and consumers, for the communities it operates in, and for the company itself—while reducing its environmental footprint at the same time. This will be made possible by innovations: Products and technologies that enable better living standards, while at the same time reducing consumption of raw materials.

According to the latest data and prognoses for economic development, standards of living and the "Planetary Footprint" (e.g. WBCSD 2010), the global economic system must become five times more efficient by 2050 in order to enable 9 billion people to have a good standard of living within the limits of the earth's resources. Henkel derived

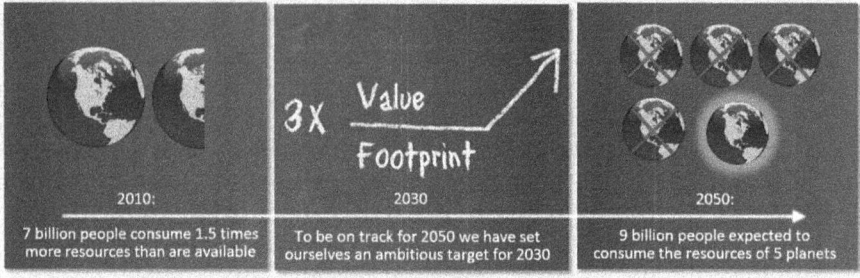

Fig. 2 Necessary response to global challenges. Source: Henkel

its long-term goal from this challenge: The company wants to triple the value it creates for the footprint made by its operations, products and services by 2030 (see Fig. 2).

Henkel refers to this goal of becoming three times more efficient as "Factor 3". One way of improving its efficiency would be by tripling the value the company creates while keeping its environmental footprint the same. Alternatively, Henkel could keep the value the same and reduce its footprint by one third, or improve both value and footprint. To reach its goal by 2030, Henkel will have to improve its efficiency by an average of 5–6% each year.

To implement its Sustainability Strategy, Henkel is concentrating its activities on six focal areas that reflect the key challenges of sustainable development as they relate to its operations (see Fig. 3). Three of the focal areas describe how Henkel wants to create "more value" for its customers and consumers, employees, shareholders and for the communities it operates in. The three other focal areas describe the ways in which Henkel wants to reduce its environmental footprint. In order to drive progress in these focal areas along the entire value chain, the company relies on the commitment of its people, its products, and collaboration with its partners.

Using the energy of its more than 53,000 employees and the unique scope of its business, Henkel aims to further expand its leadership and intensify its contribution to sustainable development. To reflect the growing importance of sustainability for its stakeholders and its long-term economic success, in 2016 Henkel developed key drivers for the coming years: It aims to strengthen its foundation, boost employee engagement and maximize its impact.

Henkel's long tradition in the field of sustainability—which is embedded in its values—together with its clear strategy and performance form a strong foundation from which to further expand its leadership and intensify its contribution to sustainable development. Henkel reached its targets for 2011 to 2015, improving the relationship between the value it created and its environmental footprint by 38% overall. By 2017, the efficiency increase had risen to 43%. On the road to achieving its long-term goal of "Factor 3", the company also wants to improve its performance in these areas still further in the coming years.

When it comes to implementing its sustainability strategy, it is Henkel's people who make the difference—through their dedication, skills and knowledge. They

Fig. 3 Henkel's focal areas.
Source: Henkel

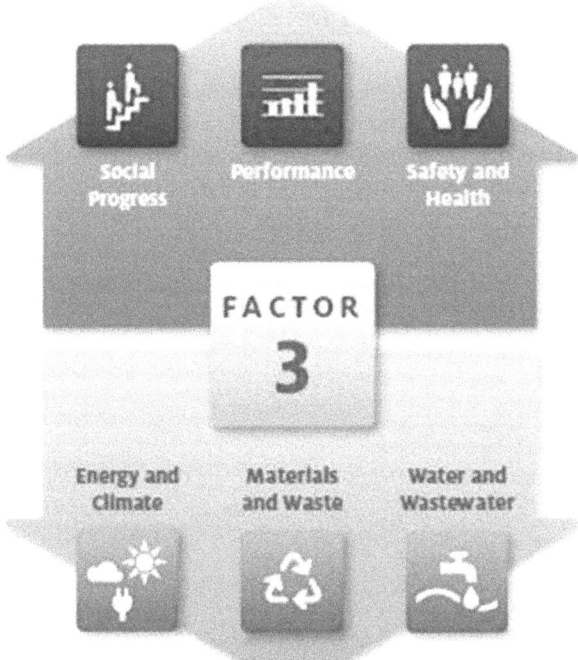

make their own contributions to sustainable development, both in their daily business lives and as members of society. They interface with the company's customers and drive innovation, develop successful strategies, and give Henkel its unique identity. Because of this, Henkel wants to further develop and boost its people's engagement in sustainability. For example, Henkel aims to reach 20 million people through its employees' social engagement activities by 2020.

Henkel wants to strengthen its contributions to addressing major global challenges and maximize the impact it can achieve with its operations, brands and technologies. For this reason, the company has set itself additional ambitious goals that address two of the most pressing global challenges. Henkel is working towards a climate- positive *contribution*: it aims to reduce the carbon footprint of its *operation* by 75% by 2030 and to leverage the potential of its brands and technologies to help its customers and consumers to reduce CO_2 emissions. To do so, Henkel also wants to continually improve its energy efficiency by using more energy from renewable sources. In addition, the company wants to leverage the potential offered by its brands and technologies along the value chain to help its customers and consumers save 50 million metric tons of CO_2 between 2016 and 2020.

3 CSR in Each Phase of the Innovation Process

3.1 Research and Development at Henkel

Henkel's vision is: "Leading with its innovations, brands and technologies." Innovations play a central role in securing the company's future ability to take a proactive approach to markets that are constantly transforming (see Fig. 4). Product innovations will play an essential role in decoupling increased quality of life from resource

Fig. 4 R&D expenditures. Source: Henkel

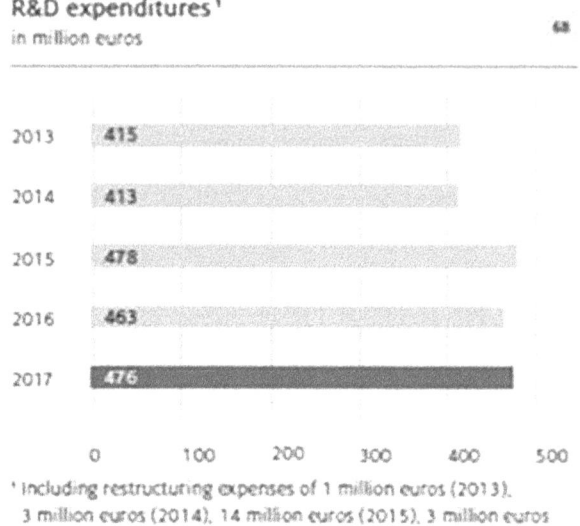

R&D expenditures[1]
in million euros

2013	415
2014	413
2015	478
2016	463
2017	476

0   100   200   300   400   500

[1] Including restructuring expenses of 1 million euros (2013), 3 million euros (2014), 14 million euros (2015), 3 million euros (2016), 7 million euros (2017).

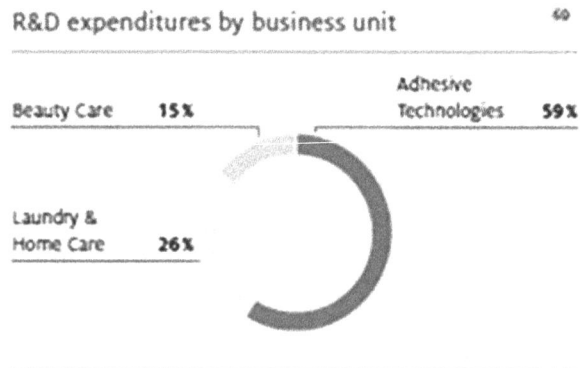

R&D expenditures by business unit

Beauty Care   15%

Adhesive Technologies   59%

Laundry & Home Care   26%

use. This is not a question of developing individual "green" products where only the environmental profile has been improved: Henkel's aim is to continuously improve all products across its entire portfolio, taking every aspect into account. A high degree of innovation is very important in achieving this.

In 2017, Henkel employed an average of around 2700 people in research and development. The scientists work in specific fields of competency, in areas across biology, chemistry and technology to develop innovative basis technologies for new products and production processes. Henkel holds nearly 8100 patents to protect its technologies around the world.

The results open up new opportunities for Henkel to expand its product portfolio—and move the company toward another of its targets: Every new product must make a contribution to sustainability in at least one of its six focal areas.

3.2 Integrating Sustainability into Innovation Processes

Criteria for assessing sustainability have been systematically anchored in the Henkel innovation process since 2008, to steer product development in line with its sustainability strategy from the outset (see Fig. 5). This means that, at a given point, Henkel's researchers must demonstrate the specific advantages of their project in regard to product performance, added value for customers and consumers, and social criteria ("more value"). They also have to show how it contributes to using less resources ("reduced footprint").

Henkel works with various measurement methods to optimize the "Value" and "Footprint" dimensions. These allow the actions to be identified that have the greatest effect on sustainability along the value chain. Henkel uses the results to develop innovations with improved sustainability performance. Only by considering the entire life cycle can the company ensure that the action taken will improve the overall sustainability profile of its products.

The Henkel Sustainability#Master® (see Fig. 6) combines various instruments for measuring sustainability. This evaluation system centers around a matrix based on the individual steps of the value chain and on our six focal areas. The goal is to

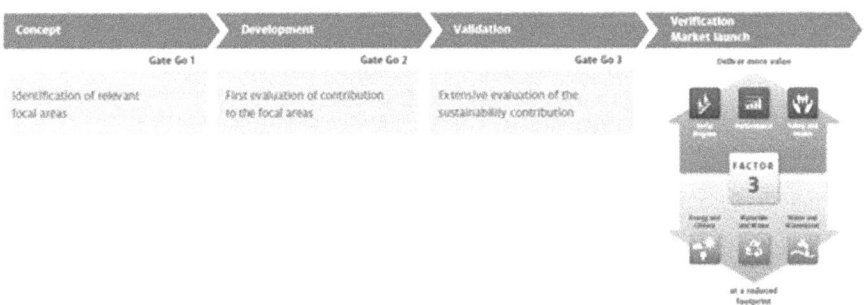

Fig. 5 Henkel innovation process. Source: Henkel

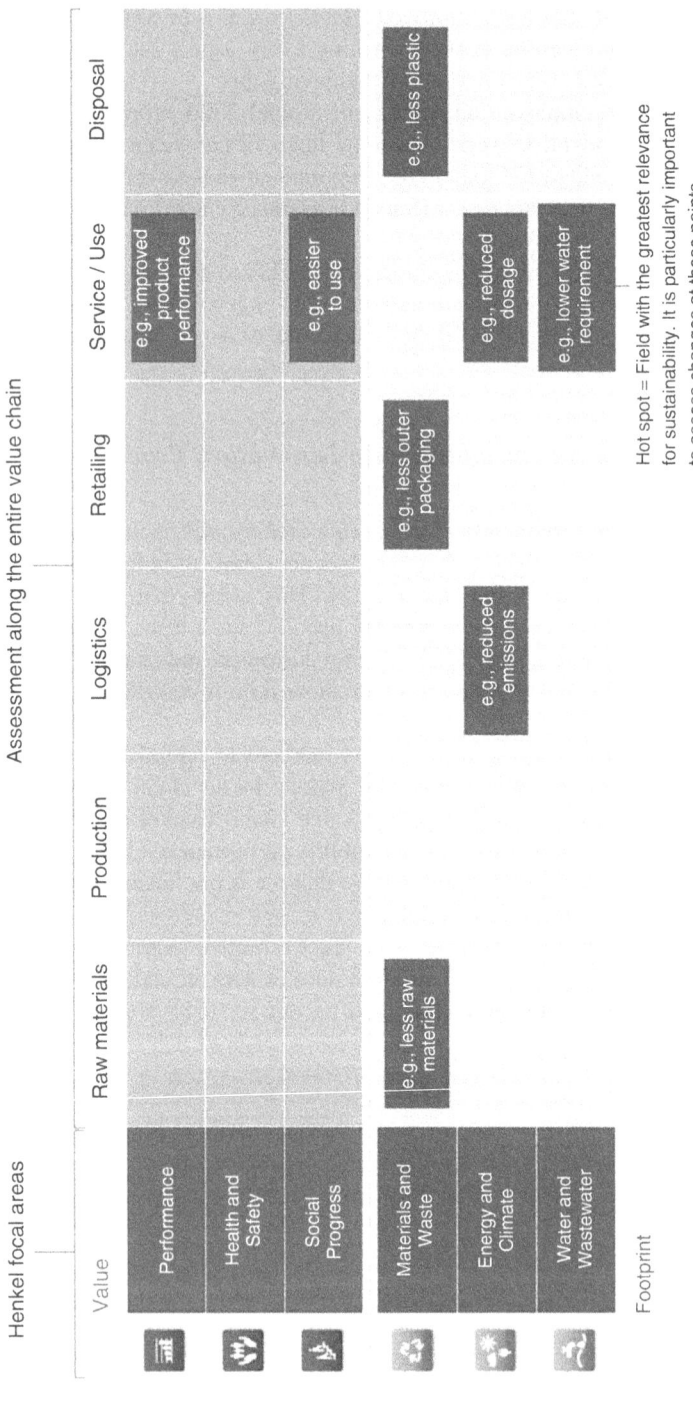

Fig. 6 Henkel Sustainability#Master® . Source: Henkel

increase the value of the product and simultaneously reduce its environmental footprint. Hot spots can be identified for every product category on the basis of scientific measurement methods. These are the fields with the greatest relevance for sustainability—this applies to both the "Value" and the "Footprint" dimension. The specified hot spots can also be used to compare the sustainability profile of two products or processes. This allows sustainability profiles to be prepared for each product category.

Henkel's researchers use these findings for innovation and continuous product improvements.

3.3 Improvement Through Life Cycle Assessments

Sustainable development needs a systematic approach. With the help of life cycle analyses—and knowledge have acquired during many years of work on sustainability—Henkel's experts analyze the complete life cycle of its products. As early as the product development phase, Henkel is able to assess what environmental impacts occur, to what extent, and in which phase of a product's life. Building on these results, improvement measures can then be applied where they are most needed and can be most efficiently implemented.

A review of the life cycle analyses of various product categories shows that the impacts on the environment often occur at very different points during the lifetime of a product. Suitable improvements can therefore often take widely differing forms. For example, the life cycle analyses of a laundry or dishwashing detergent show that energy consumption and hence the associated carbon dioxide emissions are highest during use in the washing machine or dishwasher. In such cases, Henkel focuses on developing products that can be used in a manner that saves energy and water. Other product categories call for an increase in the resource-efficiency of our own processes. Additional approaches for improving the environmental profiles of our products include the use, wherever appropriate, of renewable raw materials, improving the level of biodegradability, and reducing and enhancing packaging materials.

4 CSR and Sustainable Innovations at Henkel

4.1 Innovations That Deliver More Value with a Reduced Footprint

Consumers expect products to satisfy the criteria of quality, environmental compatibility and social responsibility. To go beyond these expectations, Henkel wants to leverage the potential offered by its brands and technologies along the value chain to help its customers and consumers to save 50 million metric tons of CO_2 by 2020. It aims to achieve this by developing products and technologies that enable customers and consumers to consume less energy during the use phase, and therefore have a lower CO_2 footprint. Similarly, the company provides clear information on how to use products responsibly, and offers a range of products and technologies that enable customers to avoid energy consumption and CO_2 emissions—for example, by facilitating lighter vehicles.

4.2   Example 1: Liquid Applied Sound Deadener in Comparison with Bitumen Melt Sheets

Liquid-applied Sound Deadeners (LASD) from Henkel provide a sustainable alternative to the bitumen melt sheets that are commonly used to reduce noise and vibration in vehicles. Bitumen melt sheets need to be cut to the right size and shape before being placed into the car body by hand. The process is not only labor intensive: The cutting leads to a considerable amount of wasted material and the sheets themselves can produce dirt and dust particles, which can lead to the need for repair processes in the paint application phase (see Fig. 7).

Henkel's LASD technologies, like its water-based Teroson AL 7155 acrylic system, can be fully automated spray applied by using application robots. This means it can be applied more precisely and cleanly, and also offers increased flexibility. Its paste-like form makes it more efficient to transport and reduces storage space in a warehouse. LASD also provides a 20% weight saving potential compared to bitumen melt sheets, which supports the trend toward reducing overall vehicle weight in order to reduce fuel consumption and the related CO_2 emissions when the car is being driven.

4.3   Example 2: Igora Royal Highlifts

The Henkel Sustainability#Master® reveals the sustainability profile of the new Igora Royal Highlifts hair coloring line. Fibre Bond Technology integrated into the color cream professionally lightens hair and is particularly gentle. Hair breakage is minimized.[1] Application is particularly simple and works without any additional steps, mixing, or weighing out of color additives. At the same time, the new colors produce the coolest shades of blonde (see Fig. 8).

Integrating Fibre Bond Technology into the color cream reduces its environmental footprint by avoiding the use of a separate product component. This also reduces packaging, resulting in an overall reduction of 11 metric tons of CO_2 every year. In addition, 67 metric tons of water can be saved in production each year. The

---

[1]Compared to lightening with the Igora Royal 10 and 12 lines without Fibre Bond Technology

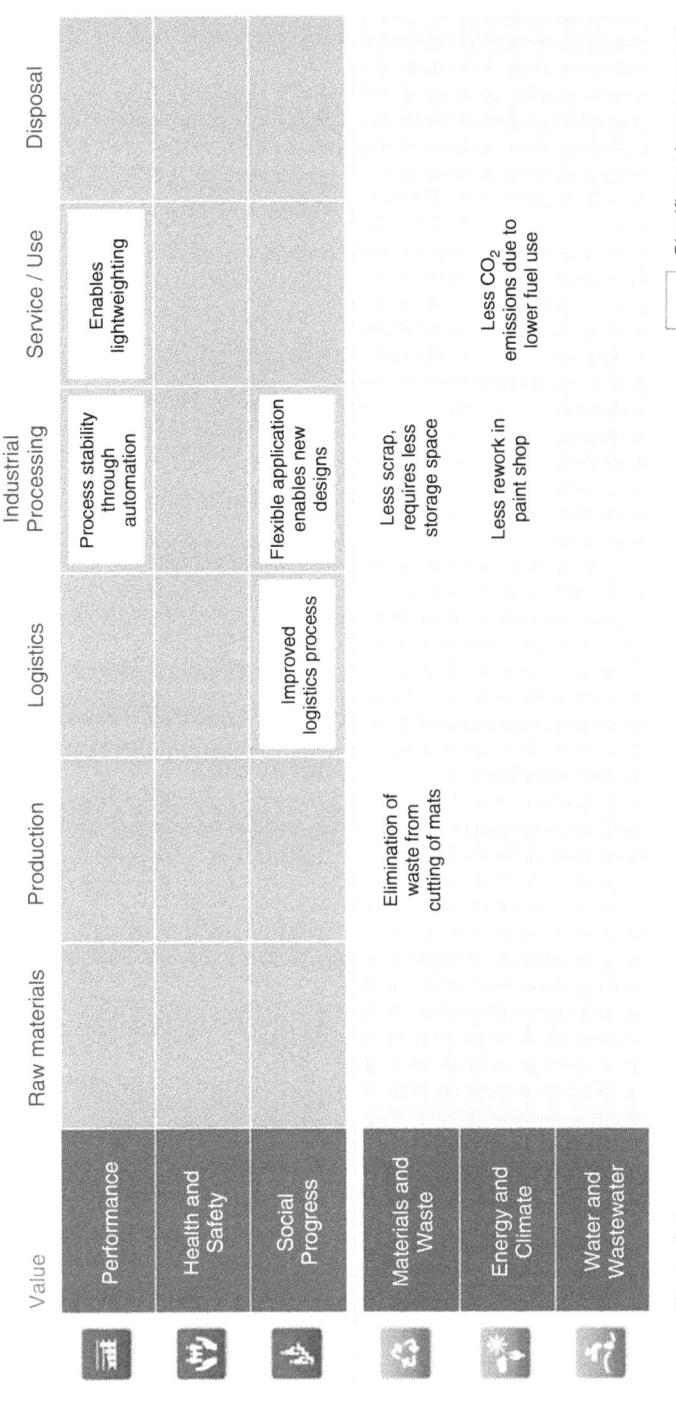

Fig. 7 Henkel Sustainability#Master® Liquid-applied Sound Deadeners (LASD). Source: Henkel

Fig. 8 Henkel Sustainability#Master® Igora Royal. Source: Henkel

elimination of this product component also has a positive effect on storage, both during production and for the customers.

4.4 Example 3: Somat Gold Compared to Somat Gold Phosphate-Free

Henkel has not used phosphates in any European dishwashing product since April 2016, thus helping to reduce water pollution—9 months before the entry into force of a corresponding EU Regulation in 2017. Since Somat products deliver excellent performance even with low-temperature programs and without pre-soaking, they have already made a significant contribution to reducing energy and water consumption in the past. An additional factor in improving the sustainability performance is the narrower and lighter packaging made of recyclable cardboard, which makes logistics more efficient (see Fig. 9).

5 Summary

Sustainable and socially responsible business practices will become increasingly important for companies' long-term economic success. By 2050, the planet's population is expected to reach 9 billion people: At the same time, consumer behaviours are being transformed by a growing middle class in emerging nations, and natural resources such as fossil fuels and water are becoming increasingly scarce.

Reducing standards of living or abstaining from consumption (what academics refer to as "sufficiency") is not a realistic solution for the majority of the world's population. For this reason, society must find a way to increase standards of living while reducing the related resource consumption and emissions.

Companies must develop innovations, products and technologies that enable higher standards of living while reducing consumption of raw materials at the same time. Solutions for the future will be developed through collaboration with partners from along the value chain.

Companies that hold leading positions in sustainable and socially responsible business will have a competitive advantage. Customers, suppliers, non-governmental organizations (NGOs), and governments are placing increasing focus on how sustainably and responsibly their partners are doing business. This makes sustainability a strategic success factor.

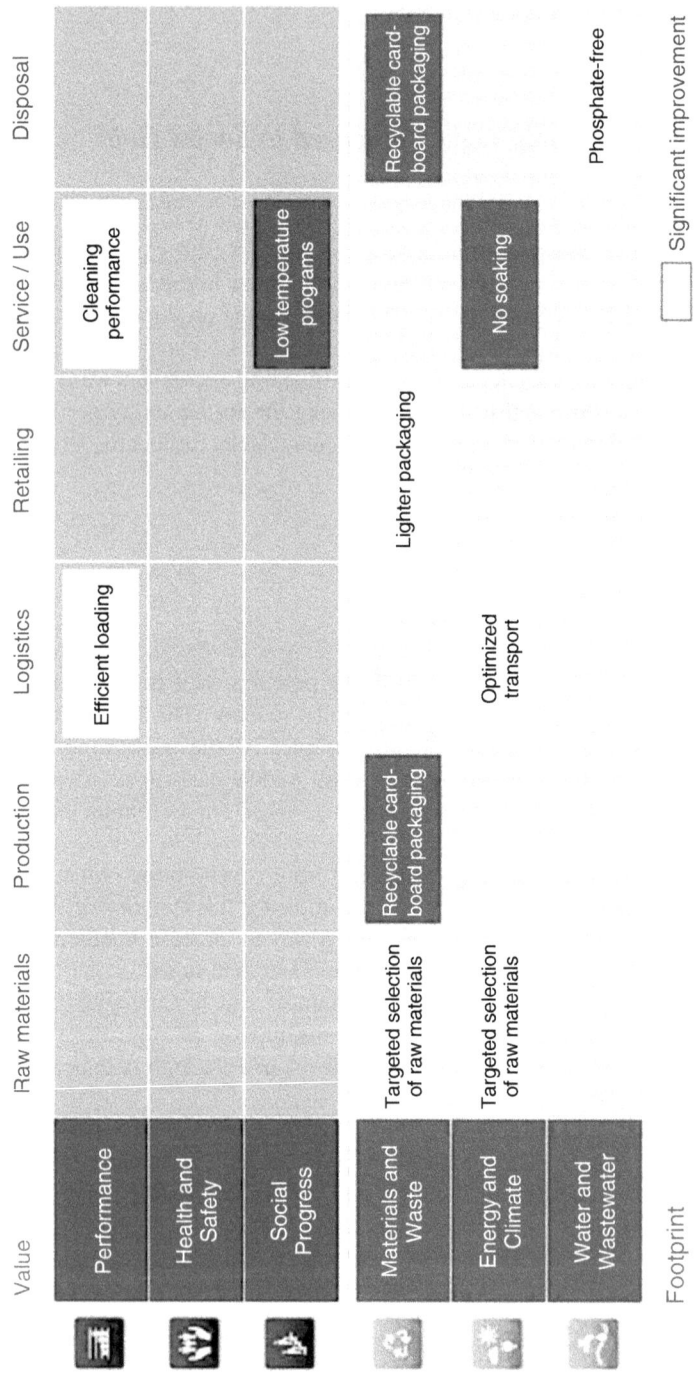

Fig. 9 Henkel Sustainability#Master® Somat Gold phosphate-free. Source: Henkel

References

Commission of the European Communities. (2001). *Promoting a European framework for corporate social responsibility* (Dissertation). Brussels.

Communication from the Commission to the European Parliament, The Council, The European Economic and Social Committee and the Committee of the Regions. (2011). *A renewed EU strategy 2011–14 for corporate social responsibility*. Brussels.

WBCSD. (2010). Vision 2050: The new agenda for business. WBCSD, Conches, Geneva

Uwe Bergmann heads the Sustainability Management department at Henkel and coordinates Sustainability across the company. He holds an M.Sc. in Environmental Technology from Imperial College, London and a B.Sc. in Environmental Sciences from the University of East Anglia, Norwich. Prior to joining Henkel in 2000, he worked as a Researcher and Consultant at the Institute for Environmental Management at the European Business School, in Oestrich-Winkel, Germany.

GoGreen Technologies: Environmental Innovation at Deutsche Post DHL Group

Katharina Tomoff and Achim Jüchter

Although the importance of environmental aspects in management has grown continuously over the past years, the meaning and value assigned to the topic by people and organizations still vary widely. For some organizations, environmental responsibility is about being prepared for regulatory changes, for others it's about anticipating changing customer needs. It also can be about a number of other things, such as cost efficiency, energy security, employer attractiveness or shareholder value. And sometimes it's simply driven by a sense of environmental consciousness.

The views on what constitutes environmentally responsible action are equally diverse. When it comes to improving environmental performance, different focus areas can be observed—while some organizations focus on fostering employee awareness and responsible behavior in daily routines, or driving technological innovation, others focus on establishing policies and guidelines, or creating room for new ideas and innovative solution-finding.

Within Deutsche Post DHL Group, most of the previously mentioned aspects have played a major role in the setup and management of its environmental program. Still, especially over the past years, the path of technological innovation has gained more and more importance in the environmental agenda, leading to solutions such as Deutsche Post DHL Group's very own StreetScooter, an electric vehicle for letter and parcel deliveries developed in-house, aerodynamic teardrop trailers for its Supply Chain operations, and the DHL Cubicycle for inner city express delivery services.

The drivers and strategic rationale for green innovation in Deutsche Post DHL Group as well as the framework for fostering respective approaches and selected outcomes will be discussed in greater detail in the pages that follow.

K. Tomoff (✉) · A. Jüchter
Deutsche Post DHL Group, Bonn, Germany
e-mail: katharina.tomoff@dpdhl.com; achim.juechter@dpdhl.com

© Springer International Publishing AG, part of Springer Nature 2018
R. Altenburger (ed.), *Innovation Management and Corporate Social Responsibility*,
CSR, Sustainability, Ethics & Governance,
https://doi.org/10.1007/978-3-319-93629-1_12

1 The Role of Environmental Management and Innovation at Deutsche Post DHL Group

The drivers for responsible logistics match, in many cases, those of other businesses. For Deutsche Post DHL Group, the initial motivator behind its environmental management activities was its responsibility to society and the environment to reduce the negative impact of its business operations. Additional drivers have since weighed in, including stricter legislative requirements, growing customer demands and increasing investor and employee awareness, underscoring the need to engage (see Fig. 1).

Aside from these more or less generic drivers, still others emerge specific to the logistics business. The transport of goods, for example, brings with it high energy demands. The scarcity of fossil energy resources and the resulting increase and volatility in fuel and energy costs become important factors that call for increased energy efficiency and alternative solutions. Of all the environmental dimensions, greenhouse gas emissions clearly account for the biggest share of the environmental footprint and have therefore taken over as the top item on green agendas across the logistics industry.

Acknowledging the need for change, Deutsche Post DHL Group was the first globally operating logistics company to set itself a carbon efficiency target, aiming to improve its carbon efficiency by 30% by the year 2020 over the baseline year 2007. By the end of 2016 Deutsche Post DHL Group had already reached its 30% efficiency target, well ahead of schedule. This led to the announcement of a new strategy in early 2017.

In line with these target settings, the GoGreen program was founded in 2008 to drive the Group's environmental performance. The program is organized along five

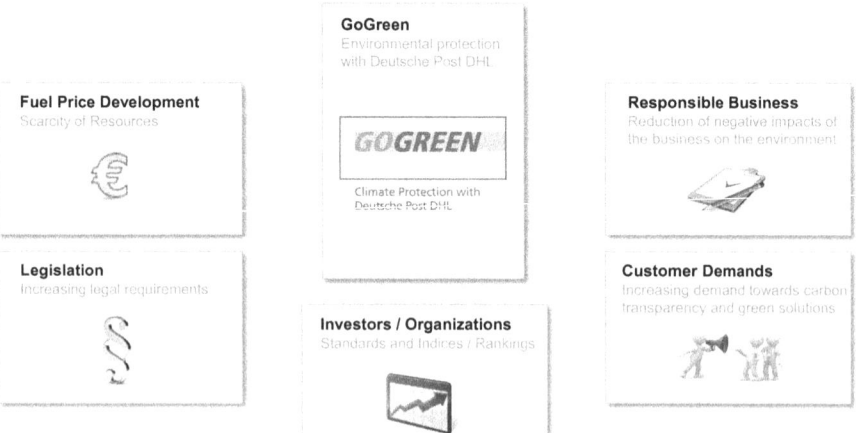

Fig. 1 Drivers for green logistics

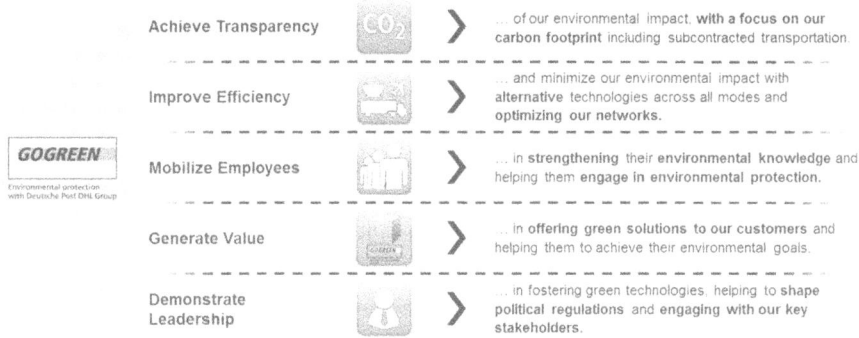

Fig. 2 Deutsche Post DHL Group's GoGreen program 2008

pillars—providing **transparency** over the relevant environmental performance indicators, identifying, planning and implementing **efficiency improvement** levers, **mobilizing employees**, developing and offering **green products and services** and **engaging** in legislative developments and stakeholder relations (see Fig. 2).

The development and introduction of innovative green technologies naturally finds its main application in the program's "Improving Efficiency" pillar. But clearly, efficiency is not achieved through technologies alone. Employee mobilization and making employees more aware of the environmental impact of their daily routines also play a significant in enhancing efficiency. Measures such as driver training and "GoGreen office checks" have delivered positive results at Deutsche Post DHL Group. Still, reducing emissions through technological innovations and solutions offers a variety of advantages that have made the approach particularly applicable and successful within the organization:

- Green technologies are often perceived as an attractive, "premium" option against the conventional status quo. This in turn eases and increases buy-in from employees and managers for environmental engagement.

  - Example: electric vs. conventional diesel van

- Technological solutions offer the opportunity to achieve high levels of carbon efficiency of up to 50% and beyond. Without the right solutions, achieving the targets of the Paris Climate Agreement will be unlikely.

  - Example: advanced biofuels in transport, LED lighting in buildings

- For organizations and employees already facing formidable challenges, targets and complexity, technological solutions can help to improve the environmental footprint while limiting or mitigating additional complexities or managerial efforts.

  - Example: speed limiter and anti-idling systems vs. individual management of driving behavior

Despite these advantages, there are also challenges affiliated with green technologies, most notably:

- Innovation risk
- Operational risks from applying solutions in early stages of maturity
- Operational complexity out of the necessity to

  - match the available solutions with the given use cases in the best suitable way
  - adapt operational processes to enable solutions

- Potentially higher upfront investment costs that only amortize later through subsequent operational savings

Maximizing the benefits and managing the challenges of green technologies is therefore intrinsically linked to successfully managing their development and driving their implementation.

2 How to Foster Green Innovation

Driving the organization to innovate and providing an environment for innovations to develop and grow successfully requires a variety of preconditions and frameworks at different managerial levels. Deutsche Post DHL Group has built its innovation agenda around three fundamental priorities:

- Strategy and Culture
- Policies and Guidelines
- Centers of Excellence and Knowledge Sharing Communities

2.1 Strategy and Culture

Integrating green innovation into corporate strategy and demonstrating its strategic value is a key element—it turns green innovation into a top-of-mind item on a company's everyday agenda and anchors it in the corporate culture. The following criteria have been identified as playing a decisive role in accompanying such a change:

1. *The green strategy must be incorporated into the company's overall strategy.*
   While details need to be elaborated in a separate green strategy, there is a caveat to offering it as a stand-alone roadmap: it will always have the flavor of an "on-the-side-strategy" that gets tended to only once the "real business" is accomplished. To avoid this, Deutsche Post DHL Group integrated its environmental program into its Strategy 2020, giving it explicit mention twice in two of its

Focus.	Connect.	Grow.
We focus on what has made us successful ...	**We connect across the organization ...**	**We expand in new segments ...**
1 Logistics as our core	1 One global team	1 Leader in eCommerce related logistics
2 Committed to the needs of our stakeholders & our planet	2 Certified specialists for everything we do	2 Accelerate footprint shift towards emerging markets
3 A family of divisions	3 Connected approach in operations, commercial, green solutions and shared services	3 Tap new market opportunities for organic expansion
... to achieve industry-leading margins.	**... to achieve quality leadership & service excellence.**	**... to achieve sustainable above-market growth.**

Fig. 3 Environmental responsibility within Deutsche Post DHL Group—Strategy 2020

three strategic pillars, with accompanying materials detailing it further (see Fig. 3):

(a) The strategy places the GoGreen program alongside the Group's three stakeholder groups (customers, investors and employees) as its fourth strategic commitment, with clear targets set for all four bottom lines.

(b) Focus is also placed on the green agenda's role in customer solutions, making a clear connection to its business relevance and value-add.

2. *Realizing a more sustainable future calls for thinking big and providing a clear vision.* This is what Deutsche Post DHL Group did when setting its new climate protection target, which calls for reducing all logistics-related emissions to zero by the year 2050. Committing to a target without necessarily having the entire line-up of measures in place to achieve it is a great driver of innovation.

3. *Managing toward a clear mission and vision requires clear interim goals across the main action areas of the green strategy.* For Deutsche Post DHL Group it was important to set four measurable goals and obtain the commitment to deliver against them from the Group's divisional CEOs and its top management. The target time frame (2025) is long enough to allow for innovation ramp-up but short enough for external stakeholders, especially customers, to see the results.

4. *Key aspects of the business need to be addressed—especially the people aspect.* Deutsche Post DHL Group's GoGreen strategy addresses operations, through its overall carbon emission approach and its specific focus on last-mile delivery solutions, and provides a clear sales target that highlights customers as the most relevant stakeholder group and greatest beneficiary of the Group's environmental performance. Though implicitly involved in the realization of these goals, employees are given their own place in the new strategy, drawing attention to the important role they play in driving the topic in terms of both expertise and motivation (see Fig. 4).

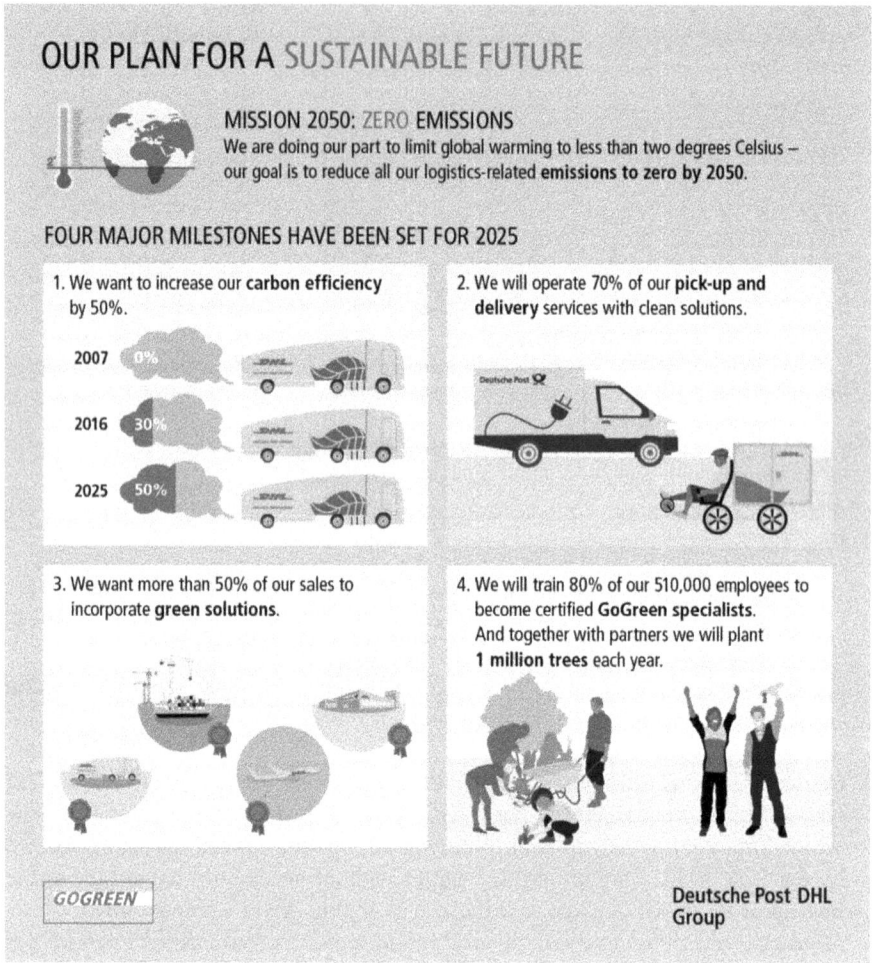

Fig. 4 GoGreen Strategy with mid-term targets (2025) and a long-term mission (2050)

2.2 Policies and Guidelines

Policies and guidelines are often perceived as rather restrictive tools for managing procedures and behaviors in organizations. Still, in the right format, they can also support innovation. The major benefit of integrating environmental aspects into existing policies and guidelines is that it elevates them to the status of a core business topic.

The most relevant regulatory mechanism within Deutsche Post DHL Group is its Corporate Investment Policy. The decision to include environmental considerations in this specific policy was based on the understanding that the financial decision point on investments often hinges on the proffered level of technological development. Since 2010, as a rule, replacement assets using energy or generating carbon must realize carbon savings or other environmental benefits over the course of the investment period. A rating scheme has also been introduced to assess each investment's level of technology enhancement. The top level (A) is reserved for solutions that are both highly innovative in character and of strategic relevance for the organization's environmental performance. In line with this regulation, a database of best practice solutions has been created that gives applicants an overview of internally used and successfully proven efficiency solutions.

This construct serves to foster innovation in different ways:

- It gives strategic direction to the organization by demonstrating that innovation is generally higher rated than "business as usual" solutions.
- The ability to display the strategic value of an investment beyond just financial figures gives middle management in particular more confidence to bring forward proposals for innovations that might still be in the early development stage and cannot yet benefit from scale effects and therefore a fully matured business case.
- By integrating best practice platforms into the process and thereby establishing a regular educational touch point between operational managers and green innovations, the growth of solutions and synergies across different organizational entities can be accelerated and through that, support solutions in reaching a critical mass.

The approach has proven to be a successful driver for green innovation at Deutsche Post DHL Group. One factor for its success is that the mechanism does not prescribe specific solutions. In a diversified operational environment like global logistics, unified solutions can lead to very different results depending on location and use cases, causing positive effects in one operation and negative effects in another. By leaving a level of flexibility to managers and establishing instead a more basic driver for education and action, a balanced approach between operational stability and innovative progress is provided.

2.3   Centers of Excellence and Knowledge Sharing Communities

Innovation requires frontrunners, i.e. groups within the organization that have the expertise, mindset and financial and human resources to develop and introduce new solutions that other organizational units can later benefit from. Ideally, these centers of excellence are close to but not directly in charge of daily line management. This approach guarantees a deep understanding of the business needs and requirements while providing the necessary distance and space to focus on mid- to long-term innovation rather than short-term challenges.

A number of Centers of Excellence have been established within Deutsche Post DHL Group over the years, and all of them have contributed significantly to the growth of green innovation within the Group. A crucial role has fallen particularly to the business service functions, such as fleet and real estate management, that are close to large operational entities. These departments combine technological expertise with a deep understanding of daily operational procedures. Additionally, by serving larger operational entities with usually standardized operations, innovations no longer face the conflict between being closely tailored to the user's individual needs and still having enough volume to allow for scalability and economic success. Examples of green innovations at Deutsche Post DHL Group that have been developed or introduced in such a way:

- The StreetScooter electric delivery vehicle by StreetScooter GmbH and Deutsche Post Fleet GmbH, development and service functions for Deutsche Post DHL's largest van fleet in Germany
- The aerodynamically optimized Teardrop Trailer by DHL Supply Chain Fleet Engineering UK, a service function for DHL's largest truck fleet, in cooperation with the trailer manufacturer Don Bur

In addition to individual innovation spots such as these, creating a broader network for knowledge sharing across the group also plays an important role in driving innovation in an efficient and sustainable manner. Knowledge exchange in that sense does not just cover technological questions around solution-specific performance or challenges, but also managerial questions around how to organize the successful growth of a solution across all phases, from first idea to final roll out.

To enable Deutsche Post DHL Group managers to profit from the expertise that is continuously being made in this field, and to provide them with a network of support for their respective undertakings, the company has established topic-related GoGreen expert groups for road and real estate. Organized centrally in the Group's Corporate Center, these groups convene regularly, bringing together cross-divisional and cross-functional experts to discuss the planning, implementation and outcomes of current technology developments with the aim to:

- Create a consistent knowledge level across the company
- Create synergies and foster technology transfers
- Efficiently use resources by avoiding redundancies in undertakings
- Ease solution finding for recurring challenges

Set up in 2009, these expert groups have sparked what has become a sustained spirit of innovation across the Group, establishing a support network for managers, giving focus to the overall agenda and spreading the results across the organization.

3   From Burn Less to Burn Clean: Green Innovations in Deutsche Post DHL Group Operations

The efficiency approach at Deutsche Post DHL Group is guided by two basic principles: "burn less" and "burn clean". "Burn less" measures help the company reduce the energy and fuel consumption in its operations, while measures under

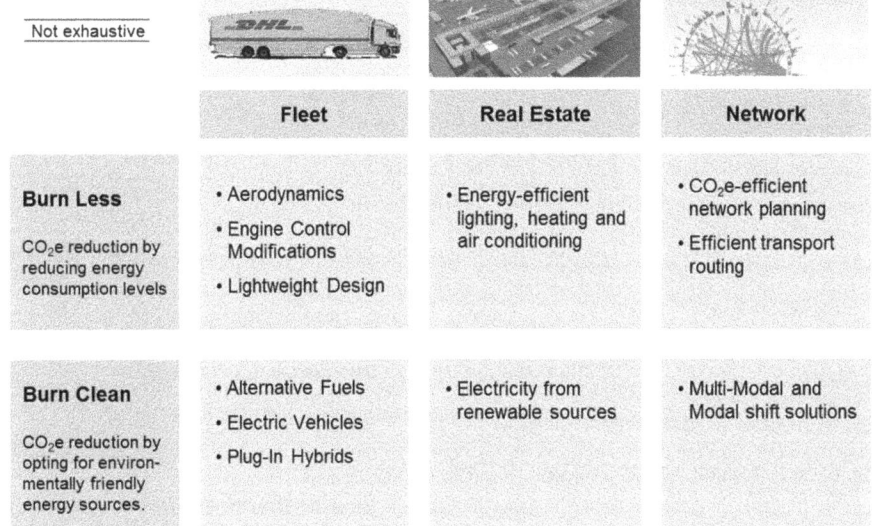

Fig. 5   The Deutsche Post DHL Group efficiency framework

"Burn clean" promote the use alternative non-fossil sources of energy and fuel to capture additional emissions savings (see Fig. 5).

To be successfully applied in operations, technologies must perform against a variety of requirements. These are determined by a range of aspects, such as the field of application, geography and use case, and can most accurately be split into four overarching categories: cost, operational feasibility, environmental benefits and local or legal factors (see Fig. 6).

In this respect, there is no single efficiency technology that can be applied to all areas of operation in a way that is suitable, practicable and cost-effective. For this reason, Deutsche Post DHL Group continually develops and tests a variety of measures to increase its portfolio of solutions and improve its emissions on a broad scale. Some of these measures, primarily from the area of ground transport, are treated in greater detail below.

3.1   Burn Less: The Aerodynamically Optimized Teardrop Trailer

The teardrop shape on vehicles and trailers is not new—such aerodynamic shapes have been on the road for decades. Still, for logistics, it has found its renaissance thanks to a joint project between the British trailer manufacturer Don Bur and DHL Supply Chain Fleet Engineering UK. The end result of the project—a commercial vehicle shape that mimics the natural aerodynamic properties of a

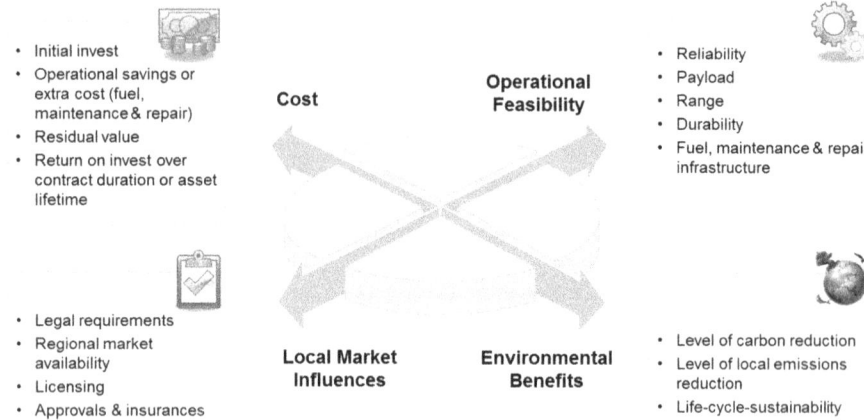

Fig. 6 Basic technology requirements—Road fleet

liquid teardrop—was introduced to the market in 2009. The "Teardrop Trailer" reduces air drag and cuts fuel consumption by up to 10%.

The solution was developed with the UK market in mind. Unlike in mainland Europe, there is no height restriction for trucks in the UK. Trailers and boxes can be built higher than the usual four meters—a potential that had remained unexploited until then.

The Teardrop Trailer makes use of this open-end allowance, using the extra height to improve aerodynamics and lower both fuel consumption and transport cost (see Fig. 7). Since the principle can be applied to all kinds of vehicles, the solution has since established itself as a standard in the Group's UK operations. Today the technology has been applied to more than 1800 vehicles in more than 120 different versions, ranging from box trucks to semi- and drawbar trailers.

In 2014, Deutsche Post DHL Group brought its solution to mainland Europe for the first time and today several of these trailers are being operated on lanes across Germany, Benelux and France (see Fig. 8). This was made possible due to the fact that in some transport cases in mainland Europe, the full legally permitted maximum height of assets is not always required. The company is now able to use the aerodynamic teardrop shape to realize efficiency gains in countries other than the UK.

From a general perspective, the Teardrop solution exemplifies one of the fundamental approaches to improving efficiency, namely focusing and reducing assets to the level that is truly required for the respective job. Following this principle, this therefore means that for high volume, low weight freight shipments, for example, mega- or double-decker trailers would be the more suitable solution for driving efficiency.

Fig. 7 Teardrop Drawbar Combination

Fig. 8 Mainland Europe Teardrop semi-trailer

The trend to standardize assets according to the "one size fits all" principle, where assets are designed to fit a broad range of use cases found on the market, is a result of the desire to reduce asset cost through scale and provide flexibility to the user. Despite the benefits of this approach, it still has its downside for each particular use case—these assets represent a compromise as they are over engineered for some aspects of the task and under engineered for others. Especially for the trucking segment, where assets cost only accounts for 25% and fuel cost for 75% of the total cost of operation, the benefits of the Teardrop shape in many situations outweighs the cost benefits of purchasing standardized trailers. With the Teardrop solution, tailoring does require greater managerial effort in terms of planning and implementation. There is also less flexibility in terms of transfer across operations. For this reason, long term agreements with customers or standardized networks offering suitable defined loads are a basic prerequisite for their application and a limiting factor towards their up-scaling.

3.2 Burn Clean: The StreetScooter Electric Vehicle for Last-Mile Mail and Parcel Delivery

Another example of a Deutsche Post DHL Group green innovation is StreetScooter, its very own electric vehicle for letter and parcel delivery. The decision to develop and build an electric delivery vehicle in conjunction with a small start-up company, which later became an in-house solution, was a result of two things: one, the need for a zero-emission last-mile solution, and two, the lack of economically reasonable offerings on the market.

Driven by its agenda to reduce greenhouse gas emissions as well local air pollutants, in 2009 Deutsche Post DHL Group began screening and testing solutions that could promise fully zero-emission delivery service for its customers. Due to high capacities and payloads, vehicles are still the tool of choice for these operations. The readily available electric powertrain also offers a viable zero-emission. In fact, the daily distances driven in postal and parcel delivery operations do not require the high ranges offered by conventional vehicles and perfectly suited for battery-powered vehicles.

Despite the basic availability, the core challenge standing in the way of large-scale application has been the high cost of these vehicles, ranging anywhere from 50 to 300% more than the cost of a conventional vehicle.

Again, Deutsche Post DHL Group's approach to overcome this challenge was to build a solution tailored to the use case. Moreover, in this case, the particular success factor was to take a broader perspective on the cost structure.

The conventional view on the cost of electric vehicles has been to calculate the cost of the conventional vehicle minus obsolete components, such as the combustion engine or transmission, plus the cost for new components, such as batteries, electric engine and converter. In the end, this view reflects the idea of retrofitting an existing conventional vehicle to electric.

The approach taken for the StreetScooter (see Fig. 9), on the other hand, was to go for a purpose design vehicle that allows for the cost structure to be influenced and optimized in a broader way—by analyzing each single cost component, from development and production cost to full operational costs, such as fuel, maintenance, repair and second life use.

The project was initiated in 2010 by StreetScooter GmbH, a spin-off of RWTH Aachen University in Germany, which aimed to develop a small passenger car called the StreetScooter Compact. In 2011, the cooperation with Deutsche Post DHL Group was founded, which shifted the focus to a commercial vehicle for letter and parcel delivery, the StreetScooter Work. StreetScooter GmbH was acquired by Deutsche Post DHL Group in 2014 and now functions as its own E-Mobility business unit within the Group. The small series production of the vehicle was launched in 2015 and was expanded in 2016. To date, Deutsche Post DHL Group has more than 2000 StreetScooters on the road, and the number keeps growing as the existing conventional diesel fleet gets is gradually replaced.

Fig. 9 StreetScooter fleet in Passau, Germany

The following provides a selection of different measures and concepts applied for cost optimization in the vehicle development and rollout of the StreetScooter.

3.3 Development

The classical view on development is that the faster a project needs to go or the more innovative the solution is, the bigger the effort and investment required for its realization (see Fig. 10). StreetScooter (see Figs. 11 and 12) turned this idea around—its take was that by going forward with a particularly small development team, the development time could be cut by half and the invest by even 90%. Respectively, when the company was acquired by Deutsche Post DHL Group in 2014, the team only consisted of 30 employees.

To still enable the design of a fully functional vehicle, StreetScooter established a new development methodology, called "disruptive network approach". The methodology is based on building a collaborative network of suppliers that are connected via a joint online development platform. Every supplier is responsible for the development of a particular component or module in the vehicle. To still ensure a proper integration of the various components into one joint vehicle concept, the know-how is bundled and coordinated in lead engineering groups for the different segments, i.e. body, electronics, interior, etc., and coordinated by a small central team.

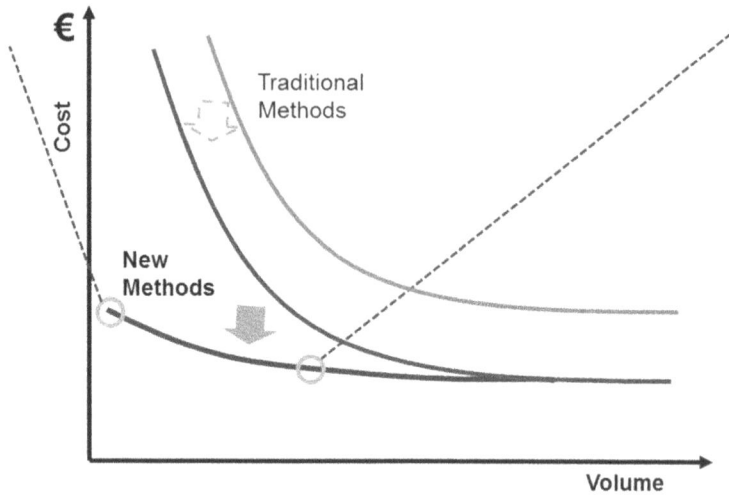

Fig. 10 "Return on Engineering" approach by A. Kampker, Deutsche Post DHL Group

Fig. 11 StreetScooter model work

3.4 Vehicle Design

Vehicle design followed the same basic principle as the Teardrop, i.e. analyzing the use case in detail and building the asset tailor-made around it, avoiding unnecessary functionalities on the one hand and improving and optimizing the important characteristics on the other.

Fig. 12 Charging StreetScooter fleet

To give the engineering team the best possible insight into the use case, a group of 150 couriers accompanied the development process, testing and reviewing the vehicle concept throughout the various development stages. Additionally, in cooperation with Deutsche Post fleet management, the cost drivers of vehicles in postal and letter operations were analyzed in detail to identify further saving opportunities in areas such as maintenance and repair.

One look is enough to reveal that the resulting vehicle differs from what's seen on the market:

- The plastics body reduces not only the production cost in a small series production but repair costs as well, especially after low-speed accidents. It also generates weight reductions that transfer into a higher payload for the daily logistics services.
- The frame is designed in a highly durable manner to last for longer than the usual vehicle lifetime. This aims at enabling a refurbishment and second-life life use of the vehicle in operations.
- The vehicle is designed for a dedicated range of 80 km in real driving to fully meet the needs of daily operations but still avoid additional weight and cost from an oversized battery.
- The short start-stop driving cycles related to delivery operations and the frequent opening and closing of doors makes it difficult for the driver cabin to heat up properly. For this reason, the conventional heating is replaced by a seat and front window heating. This reduces the energy consumption of the vehicle in winter time and reduces the total battery capacity needed.
- The cargo box in the back was designed in a fully cubic way to maximize the effectively usable space. Additionally, the sliding doors are equipped with electronic locks to enable an easier opening and closing for the couriers. The higher

floor height provides improved ergonomics for the couriers when lifting heavy packages in or out of the vehicle. The usual downside of the higher floor, namely the limited viewing angle to the side and back, is compensated via two wide-angle cameras at the side and back that send images to a screen on the driver dashboard.

- An in-house developed charging solution designed especially for large commercial fleets is also being used. All Deutsche Post DHL Group vehicles are charged at their facilities. To reduce the cost per charging point, a cloud solution has been applied in which the intelligence lies neither with the single charging point nor with the control units in the hubs. Instead, all vehicles as well as the hubs are connected with a central platform that gathers the vehicle data and operational frame conditions and determines and steers the charging strategy against a variety of parameters. These parameters include battery lifetime optimization and reduction of peak energy demands, which leads to additional cost savings over the entire operations and vehicle lifetime, beyond just the pure infrastructure cost.

The Teardrop and the StreetScooter are just two examples of innovations that have been developed and introduced over the past years at Deutsche Post DHL Group as a way to improve its environmental performance as well as that of its customers. The entire portfolio spans a broad field of activities, from new network designs and shifting deliveries from vehicles to bikes (see Fig. 13), to developing intermodal services between Asia and Europe and improving building and facility efficiency with thermal reflective paint.

Overall, what drives the Group's environmental engagement is its basic understanding that green innovation—whether by lowering operational cost, preparing for regulatory changes or helping customers to achieve their strategic targets—contributes as much to improving its environmental performance as to its bottom-line economic success. Deutsche Post DHL Group remains steadfast in its commitment

Fig. 13 DHL Cubicycle

to green innovation and its GoGreen program and will continue to explore new opportunities in the future.

Katharina Tomoff has been head of Deutsche Post DHL Group's Shared Value department since 2013. The department focusses upon achieving economic success through social responsibility. Her department also covers the Group-wide environmental protection programme Go Green, for which she has been responsible since 2010. The new environmental protection objective announced in March to reduce all logistics-related emissions to zero by 2050 is being promoted by various measures such as fleet modernisation, using vehicles with alternative sources of power and intelligent warehouses. Before joining Deutsche Post DHL Group she worked for SICK AG, a medium-sized company producing sensors for automating logistics and production processes.

Achim Juechter received a Diploma in Automotive Engineering and a Master in Business Administration from the University of Technology Aachen (RWTH) as well as a Master in Automotive Engineering from the Tsinghua University of Technology in Beijing, China. Having joined Deutsche Post DHL Group in 2009, he has been responsible for the GoGreen technology development in the Group's ground fleet until 2016. In his current role as Head of the GoGreen Efficiencies & Solutions Team, Achim leads the global planning and coordination of DPDHL Group´s environmental efficiency levers in operations as well as GoGreen customer services.

CSR as a Driver of Innovation at Swiss Post

Anne Wolf, Michael Heim, and Lorenz Wyss

1 Introduction

In Zurich's Seebach district, a yellow Swiss Post electric scooter cruises quietly around a corner; a little while later, letters and magazines fall into a private letter box with a gentle push. Farmer Hügli in the Bernese Jura produces biogas on his small farm—as part of Swiss Post's first CO_2 offsetting project on Swiss soil. And he's about to hand over two parcels to the mail carrier, to save himself the trip to the post office: this service is called pick@home.

What do these things have to do with each other? They are examples of how innovation and corporate social responsibility (CSR) interact with each other at Swiss Post. They show that Swiss Post makes life easier for people with modern services, and that it takes its responsibility towards society and the environment seriously.

At first glance, connections like these may not be so easy to discern. This article therefore aims to show how innovation at Swiss Post comes about as a result of addressing issues of social responsibility and sustainability. How are CSR and innovation strategically anchored at Swiss Post, how do they work together and how do they inspire each other? Can Swiss Post integrate social and environmental issues into the company's innovation process in order to develop solutions for current problems, as Altenburger (2013: V) calls for? How do product, service and business model innovations designed to open up new markets and increase competitiveness come about? This article looks into these issues. It will use practical examples to show that CSR at Swiss Post is not the much-cited cherry on the cake but rather, like the spirit of innovation, a corporate culture that is practised with a

A. Wolf (✉) · M. Heim · L. Wyss
Kommunikation/Corporate Responsibility, Post CH AG, Bern, Switzerland
e-mail: anne.wolf@post.ch

© Springer International Publishing AG, part of Springer Nature 2018
R. Altenburger (ed.), *Innovation Management and Corporate Social Responsibility*,
CSR, Sustainability, Ethics & Governance,
https://doi.org/10.1007/978-3-319-93629-1_13

clear ambition to achieve sustainability 3.0 as set out by Dyllick and Muff (2016) or CSR 3.0 as set out by Schneider (2015) as part of a responsible approach to corporate management.

> **Swiss Post**
> Swiss Post is a diversified company that operates in the communication, logistics, passenger transport and financial services markets and employs 61,000 people from more than 140 countries in over 100 different occupations. For the changing consumption patterns, lifestyles and working habits of many people, driven by the ongoing process of digitization, it develops both integrated solutions and modular products and services, building bridges between the physical and digital worlds. Swiss Post develops and maintains a high-quality universal service, while at the same time meeting the goals of its owner, the Swiss Confederation.

2 CSR and Innovation at Swiss Post

2.1 Strategically Anchored Sustainability

Companies frequently see social and environmental challenges such as climate change as risks, but also increasingly as an opportunity for innovation—this also applies to Swiss Post. To ensure that CSR can be used effectively to leverage innovation, Swiss Post defines its own normative requirements as follows.

Vision "Simple yet systematic": This is the motto that encapsulates Swiss Post's corporate vision. Digitization and new technological possibilities are rapidly changing customers' requirements and habits with respect to mobility, consumption and communication, and are therefore directly shaping Swiss Post's markets. With innovative services, products and integrated solutions that connect the physical and digital worlds, Swiss Post wants to make it easier for customers to operate in today's complex environment, giving them greater scope to succeed. The vision also sets out Swiss Post's ambition to create benefits for society and make a significant contribution to a modern infrastructure in Switzerland.[1]

Core Values In the core values, the term "sustainable" is defined as the basis for how we think and act: "We act in a sustainable manner and gear our activities to long-term business success. We strive to achieve an appropriate balance between acting in an environmentally friendly manner, embracing our social responsibilities and achieving business success."[2]

---

[1]https://www.post.ch/en/about-us/company/our-principles

[2]https://www.post.ch/en/about-us/company/our-principles/core-values

CSR Strategy[3] The 2017–2020 CSR strategy—or CR in the nomenclature of Swiss Post[4]—states that for Swiss Post, CSR means taking responsibility through sustainable corporate management: promoting economic growth and boosting competitiveness while acting in an environmentally sustainable and socially responsible manner. This holistic approach to responsibility is based on the principle of sustainability (environment—economy—society) with its inter- and intra-generation equity. In the strategy, Swiss Post defines climate and energy, responsible procurement, corporate citizenship and the circular economy as the key areas of action, with goals, strategic thrusts and measures.

2.2   Interlinking with Innovation Management

Innovative processes have shaped Swiss Post, which throughout its history has reinvented itself again and again, as illustrated by the evolution in transport from the mail coach through diesel to hybrid and electric Postbuses. Identifying, testing and introducing technological innovations is as much a part of Swiss Post's DNA as the ongoing development of its services and products, adapted to a changing society and thus to the requirements of its customers. Digitization, which is rapidly transforming the daily lives of many people and their consumption patterns, lifestyles and working habits, is today one of the biggest challenges for Swiss Post, which must continually develop new products, services and integrated solutions for its markets.

Innovation in this context is a deliberate and systematic change process towards something as yet unknown, "the new" (Kaschny et al. 2015: 20). At Swiss Post that could mean developing new products, services or processes. To be considered truly innovative—and therefore sustainable—these must meet a need and be recognized as useful in their interpretation and application process; they must co-produce their own prerequisites for recognition by finding acceptance in a social interaction and "sensemaking" process (Svetlova 2008: 166). Moreover, they are not completed with their launch, and can instead be further developed by the interaction process with their users. However, a company like Swiss Post cannot innovate only in relation to its markets: it can also do so by establishing contact with new stakeholders, formulating or demanding adherence to codes of ethics as an industry pioneer, entering into partnerships with or becoming a founding member of NGOs.

The Development and Innovation Unit Swiss Post established the Development and Innovation unit to drive innovation systematically and develop new business opportunities. The unit launches cross-unit projects and promotes market-oriented programmes, supporting Swiss Post for example in transforming its core business in

---

[3]https://www.post.ch/en/about-us/company/responsibility

[4]Swiss Post refers to CR, while the term CSR is used in this text. On the mostly analogous use of the two terms, see also Loew and Rohde (2013: 19).

the digital world and building bridges between the physical and digital worlds. Its innovation specialists are always looking for new ideas, bring knowledge, expertise and new partners to Swiss Post; connect and inspire. To ensure that social and environmental issues are integrated into the company's innovation process, CSR is an integral part of innovation management at Swiss Post. Cooperation between innovation specialists and CSR specialists is defined at the organizational level at Swiss Post; innovation projects with a bearing on CSR are supported from the outset by the CSR department, which helps to shape and drive them forward with its expertise and experience. In all other promising projects, the CSR specialists can formulate standards and requirements and contribute their ideas and objections.[5]

The Open Innovation Team Because organizations in an increasingly diverse world of widely distributed knowledge cannot rely solely on their own innovative strength, they are increasingly dependent on the expertise of external partners. With its start-up screening and trend scouting programmes, Swiss Post's Open Innovation team keeps a close eye on market and technological developments in order to identify interesting business opportunities for Swiss Post at an early stage. This is partly in response to growing demand from consumers for sustainable solutions, which create lucrative business areas for specialized start-ups, for example in the circular economy. Social start-ups, which are experiencing a real boom, are another potentially interesting area of collaboration for the Open Innovation team. Because their business ideas are not primarily aimed at maximizing profits, but at finding solutions to social problems, social entrepreneurs manage to carry off the balancing act between entrepreneurial thinking and social added value. Swiss Post makes allowance for the trend towards growing numbers of social start-ups at the organizational level: since 2016, it has a project manager dedicated exclusively to these projects, forming the hinge between Development and Innovation and CSR.

Innovation Management Tools "Our staff's dedication and innovativeness are key to our success," as the Swiss Post vision has it. To tap into this potential systematically, Swiss Post has developed a range of different tools for innovation and ideas management.

The so-called INN process is effectively the company's internal "innovation pipeline" for larger projects and new business ideas: between 2009 and 2017, more than 100 projects passed through the INN process successfully. Here, the notion of sustainability is particularly important: after all, the abbreviation INN in the original German stands for innovation, new business and sustainability. Swiss Post allocates around CHF 10 million to the INN process each year, its contribution to the start-up financing of innovative and sustainable projects. CSR issues are well anchored in the INN process: projects designed to achieve social or environmental improvements are also passed on to the CSR team, which provides support for the projects and advice to the people behind their ideas, increasing the chance of successful implementation. The CSR department is integrated into the INN jury,

---

[5]https://www.post.ch/en/about-us/company/innovation/development-and-innovation-unit

which assesses these projects and carries out an initial selection on a monthly basis, and contributes to making its decisions. Member of Executive Management and the jury discuss innovation issues and trends at regular events, and a meeting of innovation managers from all Group units is held four times a year and attended by CSR management. This ensures a continuous dialogue between innovation and CSR.

In 2016, Swiss Post's Ideas Management, which works closely with Innovation Management, received an international award. One of its tools is the company suggestions scheme Postidea, which allows employees to post and structure ideas on a web platform. Projects with a bearing on Swiss Post's environmental or social responsibility are forwarded to the CSR department, which helps to implement the suggestions for improvement. Among the factors motivating staff are awards such as the "Innovator of the Month" and events such as the Boost Camps, where project ideas proposed by employees are subjected to focused development over just a few days so that a quick decision on their potential can be reached.

3   Mutual Inspiration: Together on the Road to CSR 3.0

Innovation and CSR have many faces—particularly within a diversified company like Swiss Post, which operates in such different markets. This makes it all the more important to have a modern understanding of CSR as an integral part of corporate core competency that is closely aligned to the strategic goals of the company and includes an approach to corporate responsibility across the entire value chain. Swiss Post is therefore trying to contribute to a holistic solution to pressing social problems with a forward-looking attitude. Its ambitions are clearly in the direction of sustainability 3.0 as set out by Dyllick or CSR 3.0 as set out by Schneider, who notes: "With CSR 3.0, the company is the driver—not the driven—and the one who does the strategic and guiding agenda setting"—and focuses on innovation (Schneider 2015: 39).[6] Swiss Post therefore also sees social responsibility as a source of innovation, and concurs with Schneider when he notes that CSR must be in a state of constant development that is never complete "if the idea of a self-inventing and fertilizing CSR is to be retained" (Schneider 2015: 21 ff.).

One prerequisite for any company seeking to evolve towards CSR 3.0 into a proactive framer with a strategic CSR that thinks globally and acts locally is outside-in thinking from the perspective of society and its issues. Swiss Post therefore also poses the following questions: What are the burning issues, today and in the future? Where are the points of contact with our core competencies? What can we do to

---

[6]In his CSR model of maturity level, Schneider (2015) describes how our understanding of CSR has changed: from philanthropic CSR 1.0 with social sponsoring activities outside the core business through CSR 2.0 with entrepreneurial and societal value added to CSR 3.0, in which the company becomes a proactive political framer with a reflected and strategic CSR.

contribute to the resolution of social sustainability challenges and what opportunities arise from them? Who are our stakeholders—and what do they expect from us? For the preparation of the areas of action in the 2017–2020 CSR strategy, for example, stakeholders were asked about their expectations in terms of a credible CSR with respect to the 2020 time horizon. As the resulting relevance matrix shows: in addition to the "perennial" issues of climate and energy, external stakeholders also attach a great deal of importance to working conditions at suppliers—an area to which Swiss Post, with its policy of responsible procurement, for example of work clothing, has been successfully committed for years.

This outside-in thinking is also deeply anchored in the Development and Innovation business unit, where the Open Innovation team systematically monitors and assesses social trends and conducts focused analysis of projects of a CSR nature.

4 Practical Examples

Where social interests coincide with company interests, we reach the so-called "sustainability sweet spot" (Grieshuber 2012: 585; Altenburger 2013: 9), the area where the most fruitful innovations can emerge—whether they be new products, processes, sustainable business models or management innovations, or the development of new markets and customer groups. The following examples are designed to give an initial idea of the diversity of the areas at Swiss Post in which CSR and innovation are mutually inspirational—although this is not the place for a more detailed, in-depth presentation.

4.1 How Do People Want to Use Postal Services?

As varied as Swiss Post's markets are, they do have one thing in common: customers expect personalized services, "anytime and anywhere". The systematic change process for the optimization of access points—with Swiss Post coming to customers and not the other way round—should also be seen in the light of CSR 3.0: Swiss Post is looking for innovative solutions, and making allowance for the requirements of all generations. It develops services for city dwellers and country dwellers, for mobile commuters, and for elderly people who value personal contact. For example, because they provide a second source of income for shopkeepers, the postal agencies in village shops contribute to the preservation of rural infrastructure. And customers benefit from longer opening hours and can take care of their postal transactions while doing their shopping.

In addition, Swiss Post is expanding its nationwide physical presence with additional access points such as acceptance and collection points and parcel terminals which customers can access 24 hours a day. There is also new thinking with respect to the role of delivery staff and parcel carriers, whose role has been redefined:

parcel carriers can now collect consignments from the home or office, accept payments or bring money, a home delivery service valued by the older generation. Customers benefit from services that are more flexible in terms of both time and place. And the environment benefits from these innovative new solutions because repeated delivery journeys are no longer required if the recipient has not been reached. Journeys, shipments and paper are also saved by online financial services and apps that allow financial transactions to be carried out and accounts managed on a smartphone or PC. Likewise, the PostBus unit continues to evolve and offer more and more integrated solutions for continuous mobility chains, from car sharing to bike rental and excursion tips.

4.2 How Can Greenhouse Gas Emissions Be Reduced?

Swiss Post is the largest logistics company in Switzerland and operates an energy-intensive business, which is why one of its main concerns is to increase its greenhouse gas efficiency. The Group target is to increase CO_2 efficiency by at least 25% by 2020 (base year 2010): this means reducing CO_2 emissions for every consignment transported, every passenger carried, every transaction and every heated square metre in Swiss Post buildings. This climate goal is based on the recommendations of the climate science community and the Paris climate conference agreement of 2015. From 2010 to 2016, Swiss Post improved its CO_2 efficiency by more than 16%. Numerous measures involving the interaction of CSR and innovation in terms of technologies, processes or behaviour contribute to this: to reduce the use of resources and the level of pollutant emissions, Swiss Post focuses on a wide range of eco-innovations, exploits potential efficiencies and promotes renewable energies.

Alternative Drive Technologies In its efforts to achieve sustainable, innovative mobility, Swiss Post promotes alternative drive technologies, launches pilot projects on a regular basis and collaborates with external partners. With a great deal of success: the vehicle fleet is increasingly powered by green electricity, biodiesel and biogas. For example, there are 37 Swiss Post hybrid buses on Swiss roads in 2017. In a pioneering move at the end of 2016, Swiss Post took its last petrol-driven scooter out of service. This means that all of the roughly 6300 scooters used by Swiss Post's mail carriers are now quiet, environmentally friendly and electric. With a trailer attached, they can also transport up to three times more than a two-wheeled vehicle. They are powered exclusively by "naturemade star" certified green power[7] from Switzerland and require around six times less energy than a petrol scooter. This reduces Swiss Post's CO_2 emissions by around 4600 tonnes per year, or 733 kg of

---

[7]Swiss Post sources 100% of its electricity from renewable, "naturemade basic" certified energies. At least 10% of this electricity is high-quality "naturemade star" certified green power. https://www.post.ch/en/about-us/company/responsibility/climate-protection-energy-efficiency-and-renewable-energy?wt_shortcut=klima&wt.mc_id=shortcut_klima@query=naturemade

CO_2 per vehicle. The vehicle fleet now largely consists of the three-wheeled Kyburz DXP model, which was developed and optimized to meet Swiss Post's requirements in a long-term collaboration with a Swiss manufacturer as part of its strategy of sustainable supplier development. The requirements of the employees were also taken into account: the scooters are ergonomic, put significantly less strain on the hips and back and are safer to handle.

Second Lease of Life for Scooter Batteries as Energy Storage Don't throw it away—think instead in terms of the circular economy and how one innovation can give rise to the next: After their initial use, batteries from electric scooters have not yet outlived their usefulness. After 7 years of use, the batteries have a storage capacity of around 80%, which can still be made use of. In one project, Swiss Post is currently trialling discarded batteries as stationary energy storage units in buildings with solar systems. Twofold innovation: the batteries from electric vehicles are reused and solar energy is stored.

Electricity from Renewable Energy Swiss Post has been increasing its share of renewable energy for years now. It covers 100% of its electricity requirements with renewable sources from Switzerland, of which 10% is "naturemade star" certified green power. Its ten photovoltaic systems on the roofs of Swiss Post buildings feed around 5 gigawatt hours of solar electricity into the grid every year, an amount that covers the average annual requirement of over 1000 households. Ten additional plants are also planned to generate solar power for the company's own use.

Swiss Post's CO_2 Offsetting Project in Switzerland Swiss Post funds its climate protection project "Eco-electricity from biogas plants on Swiss farms", which was developed in collaboration with Genossenschaft Ökostrom-Schweiz and South Pole Group.[8] The greenhouse gas methane is collected in biogas plants and used to generate eco-electricity and heat. This use of renewable energy sources saves the same amount of CO_2 as is generated by the mailing of letters, parcels and goods. This means that the Swiss Post's so-called "pro clima"—Shipment consignments are carbon neutral. The project also includes a corporate citizenship component: Swiss Post has provided the farmers' families interested in the climate protection project with an e-cargo bike each for the transport of loads and children. The bikes are powered by self-produced eco-electricity and give farming families access to environmentally-friendly, sustainable mobility by making transportation for the biogas plant, the farm, the market or of children more pleasant and comfortable. Thanks to their labelling, the cargo e-bikes also promote the message behind the climate protection project.

Transport Supplier Development When Swiss Post realized in 2005 that its transport partner was not using the entire freight capacity in its trucks, it conducted a joint assessment of new articulated trailers with the forwarder, and only a year

---

[8]https://www.post.ch/en/about-us/company/responsibility/carbon-neutral-shipping?query=pro +clima

later, the first double-decker trucks were purchased. While these trucks with a total weight of 40 tonnes require 10% more fuel, they can transport 36 wheeled containers full of parcels instead of 24. Around 35 tonnes of CO_2 are saved per truck and year. Given this successful development, Swiss Post again ordered 16 of the latest generation double-decker trucks in 2016. Today, there are a total of 38 of these vehicles on the road for Swiss Post, showing that Swiss Post's strategy of long-term supplier development with innovative solutions and a positive sustainable impact has proved successful—and pays off.

4.3   How Do People Want to Travel in Switzerland?

Complex innovations at the interface between social needs and technological possibilities can only be successful with strategic partnerships. As a logistics and mobility company, Swiss Post therefore works together with research institutions, cities and municipalities to develop forward-looking mobility solutions. For example, as a partner of an interdisciplinary mobility laboratory, it is involved in pilot projects testing small electrically powered self-driving buses and remote-controlled SmartShuttles. These will be used jointly in the future in an effort to reduce the number of motorized vehicles in city traffic, prevent traffic congestion and reduce CO_2 and particulate pollution. Swiss Post also supports future mobility experts with its sponsorship of the mobility academy "College for Collaborative Mobility", or "cocomo".[9]

Because even a very well-developed public transport system cannot cover all of the population's requirements, Swiss Post is developing transport services and projects that can be combined intelligently: For example, it is currently developing a comprehensive mobility platform that offers access to both public transport and private sharing offers with an app and encourages carpooling. Swiss Post also offers a popular self-service bike rental scheme in many cities. And it designs car-sharing models for companies and municipalities with which they can save money and reduce environmental pollution. In terms of employee mobility, Swiss Post focuses on an energy-efficient vehicle fleet, preferential offers and awareness-raising measures for business trips and journeys to and from work. In addition, Swiss Post offers its employees eCargo bikes at discount rates, which can be used to carry loads of up to 100 kg, and with their built-in bench seats and safety belts, can also transport two children.

---

[9]http://www.wocomoco.org

4.4   How Can Supply Chains Be Designed to Be Socially and Environmentally Sound?

Globalized value chains with many subcontractors and different regional labour laws are a growing challenge for Swiss Post's procurement unit, a challenge it has met since 2006 with its Code of Ethics and Social Responsibility for suppliers with minimum requirements for the protection of people and the environment. In 2012, Swiss Post became the first Swiss company to work with the Fair Wear Foundation (FWF), undertaking to implement high social standards at the producers of the roughly 300,000 items of Swiss Post clothing procured each year.[10] To ensure continued progress, Swiss Post has stipulated that from 2017, all public service tenders must include sustainability criteria. Taking an innovative approach towards fair and safe working conditions at suppliers: this, too, is another step towards CSR 3.0.

Solutions for the Circular Economy Given the scarcity of finite resources, the circular economy is gaining in importance. With its products and services, Swiss Post aims to support a resource-friendly and circular economy. In the field of logistics in particular, huge economic potential is emerging, and Swiss Post is exploring the potential for innovation in various pilot projects and cooperation opportunities.

With its infrastructure, processes and market position, Swiss Post can make a decisive contribution to the development of a circular economy in Switzerland and occupy a key position between consumers, producers, industry and regulators in the development of solutions for the circular economy. Implemented solutions currently include cases like these: working with Hewlett Packard, a system was launched for more efficient use of Swiss Post's printers, with HP collecting and recycling the printer cartridges. Swiss Post collects Nespresso coffee capsules from private letter boxes throughout Switzerland, significantly increasing the rate at which they are recycled. The same applies to old clothes, which are collected from private households by the mail carriers. Recycling projects are in place for Swiss Post employees' work clothing and shoes: these are made more difficult by the fact that the Swiss Post logos must be removed before any second use. Here, Swiss Post is working with producers to make them easier to remove. Textiles that can no longer be worn can be recycled into courier bags and other such items. Another example is the collaboration with the start-up SafeRec, a company that works with Swiss Post to safely destroy and recycle computers, tablets and mobile phones, including sensitive data. The Swiss Post project "Tauschprofi" improves the returns process for electrical devices. Customers can trade in a used smartphone or tablet at the post office counter as a deposit for a new device. Swiss Post is also increasingly involved in the area of product-as-a-service, where instead of products being sold, their use is charged as a service. The first implementations were in the area of "light". This relates not simply to lamp sales or installation, but also to the provision of light.

---

[10]https://www.post.ch/en/about-us/company/responsibility/responsible-procurement

Open Innovation is also working on many other sustainable projects: one project for example is looking into smart emptying of waste containers using computer-controlled sensors. Another is investigating the use of smart technology to identify the need for maintenance of buildings and technical installations in order to avoid unnecessary journeys for inspections.

4.5   How Do Digital Transformation and the Internet of Things Contribute to Sustainability?

Swiss Post is developing services for its customers at the interface between the physical and digital worlds that have a reduced environmental impact: for example, digitized letter mail can be sent to an electronic mailbox, which avoids journeys; for people with reduced mobility, Swiss Post's e-voting solution enables barrier-free access to a virtual ballot box; its e-health solutions with electronic patient records simplifies communication and helps save costs in the healthcare sector. In the area of robotics and the Internet of things (IoT), Swiss Post is also driving the expansion and development of new technologies forward: as part of a pilot project, it is building a long range wide area network (LoRaWAN). The IoT can be used to deliver high-quality data for sustainable traffic planning: for example through the use of real-time monitoring data to help synchronize Postbus journeys with passenger volumes and improve the way routes or the number of buses are adapted to meet actual requirements.

In terms of the IoT, Swiss Post can offer a unique infrastructure of stationary and mobile objects throughout Switzerland: When sensors, Internet connectivity and other technologies are added to buses, bus stops, postal agencies, letter boxes, parcel terminals or Swiss Post employees' mobile devices, they become ideal for data collection with minimal energy requirements: in the current "mobile sensing" trial, for example, 1300 Postbuses are collecting information on the weather, road conditions and traffic situation on their journeys. They can measure particulate and noise levels or the pollen count for municipalities, cantons or federal offices.

The reduction in sensor sizes and the ability to operate them cost-effectively are giving rise to practical, time-saving solutions for customers and new business opportunities for Swiss Post. For example, an intelligent sensor connected to the Internet can trigger an automatic order when stocks on a shelf run low. This allows Swiss Post to replenish the supplies of its customers as and when needed and when there is sufficient space in the warehouse. This eliminates unnecessary journeys, which benefits the environment. This can also be replicated in the private sphere: smart buttons could trigger a particular service at the touch of a button, for example to notify the mail carrier that a parcel is waiting for collection from a home. LoRa technology does not require a SIM card or a mains connection. This means that remote or mobile areas of application can also be opened up and served flexibly and appropriately, so that the rural population benefits from the same services as people in cities.

5 Conclusion

These examples from many different areas—from the customer focus of postal services and energy efficiency measures to responsible procurement and smart solutions for modern, urban companies—illustrate the wide range of issues confronting Swiss Post as a logistics and mobility company. They show clearly that it can only face global challenges such as climate change, the digital transformation of society, growing mobility or urbanization if innovation and sustainability interact successfully. To develop forward-looking solutions in line with the needs of the Swiss people, economy and environment, Swiss Post is focusing on new technologies, engaging in active stakeholder management and entering into partnerships with research institutions, cities and municipalities. To ensure that innovation and ideas management are broadly based and dovetail strategically and organizationally with CSR, the development of internal and external innovation proposals is continuing. With a greater focus on outside-in thinking, Swiss Post has set out on the road to CSR 3.0: it is creating new access points driven by the requirements of customers, who want to see increasingly mobile and flexible postal services. It is analyzing new market opportunities based on contributions to the resolution of social challenges and is receptive to promising ideas—such as the circular economy and reverse logistics, and services such as PubliBike offered as part of a sharing economy. In addition, numerous measures are also helping to improve energy efficiency and reduce greenhouse gas emissions. Swiss Post is proving itself to be a socially responsible economic player with its pioneering work in the area of responsible procurement. Its environmental and social responsibilities inspire Swiss Post to generate ideas and solutions, and CSR is evolving into a driver for innovation.

Swiss Post sees an active, opportunity-based sustainability management in the sense of CSR 3.0 as its greatest opportunity for innovation: and this is not only a matter of ethics, it is also an economic decision. Swiss Post believes that the environmental efficiency measures it implements and a good reputation in the social sphere help to increase employee and customer loyalty, open up new market segments and, last but not least, improve its own economic efficiency—to secure the success of the company in the long term.

References

Altenburger, R. (2013). *CSR und Innovationsmanagement: Gesellschaftliche Verantwortung als Innovationstreiber und Wettbewerbsvorteil*. Berlin: Springer.

Dyllick, T., & Muff, K. (2016). Clarifying the meaning of sustainable business: introducing a typology from business-as-usual to true business sustainability. *Organization & Environment, 29*(2), 156–174.

Grieshuber, D. E. (2012). CSR als Hebel für ganzheitliche Innovation. In A. Schneider & R. Schmidpeter (Eds.), *Corporate social responsibility* (pp. 371–384). Berlin: Springer.

Kaschny, M., Nolden, M., & Schreuder, S. (2015). *Innovationsmanagement im Mittelstand: Strategien, Implementierung, Praxisbeispiele*. Berlin: Springer.

Loew, T., & Rohde, F. (2013). CSR und Nachhaltigkeitsmanagement. Definitionen, Ansätze und organisatorische Umsetzung im Unternehmen. www.4sustainability.de/fileadmin/redakteur/bilder/Publikationen/Loew_Rohde_2013_CSR-und-Nachhaltigkeitsmanagement.pdf

Schneider, A. (2015). Reifegradmodell CSR – eine Begriffsklärung und -abgrenzung. In A. Schneider & R. Schmidpeter (Eds.), *Corporate social responsibility* (pp. 21–42). Berlin: Springer.

Svetlova, E. (2008). Innovation als soziale Sinnstiftung. In P. Seele (Ed.), *Philosophie des Neuen* (pp. 166–179). Darmstadt: WBG.

Anne Wolf has been Head of Corporate Responsibility at Swiss Post since 2011. After studying geography, German studies and teaching, her career has included roles at the German Development Service in South America, the consultancy firm Roland Berger Strategy Consultants, as an IT service provider for energy provider E.ON and at Munich Re. There she helped establish the Munich Re Foundation and its administrative office, where she was responsible for strategy. At the same time, she completed an MBA with a thesis focusing on stakeholder management in reinsurance. Before joining Swiss Post, Ms Wolf was an advisor on corporate responsibility/sustainability for Swiss Federal Railways.

Michael Heim has been Head of Corporate Social Responsibility at Swiss Post since 2013, focusing primarily on responsible procurement, the circular economy and corporate citizenship. Before that, he gained 15 years of international and multicultural experience in various roles in supply chain management at Swatch SA and PepsiCo International. He was in charge of responsible supply chain management, management of materials supply, supply chain and systems implementation projects and integration projects. Today, as in his previous roles, he is regularly involved in innovation projects. He holds an MBA from the University of Bern and a Diploma in Sustainable Business (University of St. Gallen, Business School Lausanne).

Lorenz Wyss has worked for Swiss Post since 2011. Via the University of Berne (geography), his career has spanned various sectors and industries before leading him to an innovation management role at Swiss Post Group, where he has been Head of Ideation and Innovation Spaces since 2011. As Head of Ideation at Swiss Post, Lorenz Wyss is an experienced go-getter with a deep interest in idea management and corporate entre/intrapreneurship. He is particularly interested in the way that these concepts can be used to help transform large organizations, and has many years of experience in implementing innovation architecture in a corporate environment.

The Responsible Business Model: Perspectives from the Tata Group

Shankar Venkateswaran and Sourav Roy

1 Introduction

The Tata group of companies, which completes 150 years in 2018, traces its beginnings to 1868 when Jamsetji Nusserwanji Tata, the founder, setup India's first large scale textile mill. It has since emerged to be a global conglomerate with aggregate revenues of more than $100 billion and over 660,000 employees[1] today. The longevity of the group provides an effective counter perspective to the declining lifespan of businesses worldwide—for context, nearly 9 of every 10 Fortune 500 companies from 1955 were either gone, merged, or contracted in 2016[2] while Tata Steel, Tata Power and the iconic Taj hotel in Mumbai have been around for over a 100 years. This journey has also seen the group navigate multiple evolutions in the operating context for its businesses: setting up industries which contributed to a push for economic self-reliance in independent India, adapting to the post liberalization Indian economy in the nineties, successfully riding out the information technology maelstrom at the turn of this century, aligning to new market paradigms after financial crises of 1997 and 2008 and so on.

The Tata story has also been one of sustained growth and value creation. The group has over 100 operating companies across multiple business sectors today (of which over 30 are publicly listed) and has an aggregate market capitalization in

---

[1]http://www.tata.com/aboutus/sub_index/Leadership-with-trust

[2]http://www.aei.org/publication/fortune-500-firms-1955-v-2016-only-12-remain-thanks-to-the-creative-destruction-that-fuels-economic-prosperity/

S. Venkateswaran
Tata Sons Limited (Ex), Mumbai, India

S. Roy (✉)
Tata Steel Limited, Mumbai, India
e-mail: souravroy@tatasteel.com

© Springer International Publishing AG, part of Springer Nature 2018
R. Altenburger (ed.), *Innovation Management and Corporate Social Responsibility*,
CSR, Sustainability, Ethics & Governance,
https://doi.org/10.1007/978-3-319-93629-1_14

excess of US$130 billion[3] making it the most valuable business group in India. It has also established itself as a global conglomerate in the truest sense with operations in over 100 countries worldwide and more than two thirds of its revenue coming from countries outside India. The Tata brand is also among the most respected brands globally, and valued at over $13 billion[4] in 2017.

The growth and longevity of the Tata group, however, cannot be adequately understood without placing them in the context of its relationship with society. While J.N. Tata is warmly remembered as the "father of Indian industry," it was his belief that "the community is not just another stakeholder in business, but is in fact the very purpose of its existence". This provides the unique genetic code of the group,[5] and resonates in a rich legacy of Tata businesses emphasizing social and employee welfare much in advance of any regulatory requirement (provisions like 8-hour working day, leave with pay and profit-sharing bonus were introduced at Tata Steel 2–3 decades before it became law in India and Tata companies have been doing CSR projects for over a century while it became law only in 2013), aligning business and philanthropy with the national development agenda (Tata Steel and Tata Power were as much nation-building projects as businesses) and putting in place sustainable business models, some of which are outlined here.

Today, when environmental and social drivers have an increasingly pivotal influence on their success, businesses are faced with the need to balance their social and environmental (collectively referred to here are societal) impacts with a value creation imperative. This is referred to variously as Corporate/Business Responsibility or Sustainability and these terms are used interchangeably in this chapter. This chapter postulates that the fundamental pathways for enabling businesses to be responsible or sustainable are (a) embedding corporate responsibility into its DNA (b) innovations in products (goods and services) and business that meet customer needs while delivering societal value (c) constantly innovating business processes that deliver such products responsibly (d) making available intellectual capital and other resources embedded in business organizations to solve larger societal problems and (e) obtaining and continuously renewing a social license to operate the business. It then considers how each of these components can be leveraged for responsible value creation and looks back at select experiences from the Tata journey in illustration of each.

---

[3]http://www.tata.com/htm/Market-capitalisation-of-Tata-companies.htm

[4]http://www.livemint.com/Companies/Vc9830oUIQPqSVBYeNsNEN/Tata-Group-is-Indias-most-valuable-brand-Brand-Finance.html

[5]http://www.greenleapreview.com/transformational-sustainability-at-the-tata-group-stuart-hart-aarti-sharma-and-christian-sarkar/

2 Embedding Corporate Responsibility into DNA

Given the long term and widespread nature of business impacts on society and the planet and their impact on business success, it is very critical that companies consistently stay the course on responsibility—over time and across geographies. The challenge therefore is to embed this into the organizational culture or DNA of the business. Experience suggests that this requires a combination of policies, guidelines and codes on the one hand and structures that promote this on the other hand, with the leadership demonstrating and reinforcing the commitment at every opportunity.

The role of the Tata leadership has been visionary and consistent throughout its 150-year old history. Mention has already been made of its founder, J N Tata who believed in the centrality of communities to the purpose of a business. His successors have reiterated this commitment in different ways. J R D Tata, the longest serving chairman of the group, spoke about managing corporate enterprises not merely in the interests of the owners but equally in the interests of employees, consumers, local communities and the nation. Ratan Tata who retired as Chairman in 2012 spoke about how ethical behaviour in business has been the bedrock on which the Tata group has built and operates its enterprises.

These fundamental building blocks have been articulated in various policies and codes that the Tata group companies and employees have adopted. The Tata Code of Conduct,[6] first crafted in 1998, lays down the ethical standards that all Tata employees need to observe in their professional lives vis-à-vis all stakeholders. The Climate Change Policy of 2010 and the Tata Sustainability Policy of 2015, both of which provide the framework for how Tata companies approach these issues, formalize what group companies and employees have been intuitively practicing for a number of years.

The structures to oversee the group's sustainability and corporate responsibility commitments have also evolved. The Tata Sustainability Group[7] (TSG), under the aegis of the Tata Global Sustainability Council comprising CEOs of the largest Tata businesses, provides oversight, guidance, support and thought leadership to all Tata group companies in embedding sustainability in their business strategies and demonstrating responsibility towards society and the environment. TSG, in its integrated mandate on sustainability, was preceded by institutions like the Tata Council for Community Initiatives and the Environment Sustainability team at the group level. In addition, several individual companies in the group have committees and sub-committees at the board and management level that oversee and review the sustainability strategies and performances of their respective companies.

Over time, the Tata group has developed frameworks to enable group companies assess where they are on the sustainability journey and identify opportunities for improvement. In 2012, the Tata Index for Sustainable Human Development was

---

[6]http://www.tata.com/ebook/tcoc/index.html

[7]www.tatasustainability.com

designed to standardize reporting across companies, NGOs etc. More recently, TSG has developed a Sustainability Assessment Framework ("SAF") which is a diagnostic tool to help Tata companies recognize and understand key elements of sustainability, monitor their sustainability progress over time, communicate and share information on best practices and get ideas for improvement to achieve leadership positions on sustainability. This tool is aligned with business performance through a structure based on the Tata Business Excellence Model.[8] A companion framework is the CSR Assessment Framework which enables companies to assess where they are in terms of their community investments and feeds into the SAF. Both these frameworks are currently in the pilot stage and the plan is to integrate them with the Tata Business Excellence Model which all Tata companies use to improve their business performance.

3   Delivering Total Value Through Products and Businesses Innovation

Products, both goods manufactured and services rendered, lie at the core of a business model. Businesses therefore focus on ensuring that the value proposition of their products is always relevant for customers and that these products are made at optimal cost structures.

Responsible businesses are no different in their quest to create value for their customers. What they do additionally is to also find ways to create a larger societal value while trying to tread lightly on the planet—in other words, they focus on total value. This is not a philanthropic project but one that responsible businesses recognize can provide an enduring value proposition for consumers and "future-proof" them. Many companies produce products for consumers at the base of the pyramid but as the examples from the Tata companies below show, there are other ways too of providing total value, both through product and business innovations.

The two automotive companies in the group, the India-based car, truck and bus manufacturer Tata Motors and UK-based luxury car manufacturer Jaguar Land Rover (JLR), have both committed themselves to building products that provide total value. Tata Motors recently developed the Tata Starbus Diesel Hybrid Bus[9] which uses 25% less fuel than regular buses and provides a cleaner alternative for public transport in Indian cities where increasing pollution is a significant concern. The company is developing a fleet of Fuel Cell Buses,[10] based on the series hybrid platform, which can provide clean public transportation in cities where hydrogen infrastructure will be available.

---

[8]http://www.tata.com/aboutus/articlesinside/Business-excellence

[9]http://www.financialexpress.com/auto/car-news/tata-motors-launches-starbus-hybrid-at-rs-2-2-crore/521459/

[10]http://www.buses.tatamotors.com/products/starbus-fuel-cell.aspx

JLR's REALCAR[11] (REcycled Aluminum Car) project is a tangible initiative towards providing a 'climate friendly' product to the consumer at no incremental cost. Launched almost a decade ago and predating the recent step up on eco-friendly automobiles, the project aims to boost the amount of aluminium in their vehicles to lower their body mass, improve fuel efficiency and thereby reduce environmental footprint and running costs for customers. The energy and other costs of using primary (virgin) aluminium, however, are quite high but JLR offset these vide collaborations with key suppliers (like Novelis) to develop a closed loop value chain and use recycled aluminium (which lightens its environmental footprint by using up to 95% lesser energy). The company reclaims over 50,000 tonnes (the weight of 200,000 Jaguar XE body shells) of aluminium in 1 year, has created a new automotive aluminium alloy that accepts increasing amounts of recycled aluminium scrap and deploys this across its key models like the XE and XJ.

BIG: Beautiful is Green,[12] is an initiative by Tata Housing, a company engaged in the development of residential, commercial and retail properties that demonstrates its commitment to environment excellence across its value chain. Tata Housing projects committed itself to develop buildings that are certified by the Indian Green Building Council/LEED and optimize utilization of resources including cement, water, steel and energy by adopting cutting edge technologies. Its projects account for nearly a third of green development in the Indian real estate business, with 55 million sq. ft. of the total 155 million sq. ft. of eco-friendly construction in India.[13]

Tata Steel has been innovating and delivering products that go beyond meeting certification and legislative requirements to improving the sustainability performance of the operations and products of its customers across the globe. For instance, the company has launched over 30 new products including new products for food and paint packaging, new types of tube capable of withstanding extreme temperatures and new products for the car and construction markets. Nest-in[14] is a light gauge prefabricated steel construction solution conceptualized, designed and delivered by Tata Steel keeping in mind the growing need for affordable yet reliable infrastructure options for India. The product is offered to consumers in a wide range of modular options which are compliant with building codes, offer high seismic resistance and are thermally insulated. The structures, being prefabricated, can be constructed in a matter of days and also minimizes resource wastage compared to conventional building methods.

India has over 75 million[15] people, around 5% of its total population, living without access to clean drinkable water and hence, exposed to water borne diseases or spending most of their earnings on addressing this problem. Tata Chemicals, in

---

[11]http://www.cisl.cam.ac.uk/publications/publication-pdfs/cisl-closed-loop-case-study-web.pdf

[12]http://www.tatahousing.in/big/about.php

[13]http://www.tata.com/article/inside/For-a-greener-world

[14]http://www.nestin.co.in/About-Us

[15]https://blogs.wsj.com/indiarealtime/2016/03/22/indians-have-the-worst-access-to-safe-drinking-water-in-the-world/

collaboration with the Tata Consultancy Services (TCS) and Titan, developed a nanotechnology-based water purifier named Swach (meaning clean in Hindi), and brought it to market in 2009. This water purifier uses replaceable cartridges made of natural ingredients like rice husk ash and does not depend on electricity—attributes which make the product an affordable and convenient way to meet a basic need of drinking water for potential consumers and also contributed to addressing one of India's biggest health problems. Swach was voted the consumer product of the year[16] for 2012 while winning several awards on innovation and design. It has emerged as a mainstream business line for the company with the product touching over 8 million lives[17] since inception and is now looking to expand its footprint through alternate marketing channels including NGOs. Tata Chemicals is only building on its legacy of delivering products that have addressed societal needs like the first iodised salt (iodisation minimises the risk of the disease goitre) and its "Sampann" range of pulses, spices and salt that deliver health and nutrition.

From the domain of financial services, Tata AIA Life Insurance Company (a joint venture between the Tata group and AIA) provides an effective instance of business designing a product to address a societal need while setting up value propositions for users and distributors of the product. The rural and low income segments of India's population, largely, depend on erratic income from informal employment or weather dependent agriculture leading to low levels of asset creation—factors which reinforce the need for insurance coverage but paradoxically discourage providers of insurance from reaching them. Tata AIA ventured into the micro insurance space in 2001 to provide life insurance services to these underserved segments for affordable premia and hence, are adapted to the wallets of rural, low income and landless Indians. An equally complex problem was to take these products to deep rural markets, for which the company developed a network of "micro agents" who are trained and licensed to sell Tata AIA products. These "micro agents" are members of the community, preferably women, who have credibility among the villagers and earn their livelihoods from their association with Tata AIA. The company, over the years, has set out several such products and now counts micro insurance as a mainstream business line.

Over the years, the Tata group has set up several new businesses to deliver products that meet customer needs. In more recent years, these have included those that are responding to the new and growing opportunities that sustainability has provided, where total value is integrated. One such is Tata Cleantech Capital Limited[18] (TCCL), a joint venture between Tata Capital and International Finance Corporation (IFC), set up to finance technology options which have a lower environmental footprint (or "cleantech") and sustainable infrastructure development. Offering a full suite of services to help businesses identify, evaluate and access

---

[16]Nielsen survey across 30,000 respondents in 36 markets (https://www.tataswach.com/pages/milestone)

[17]http://www.tatachemicals.com/upload/content_pdf/tata-chemicals_AR_15_16.pdf

[18]http://www.tatacleantechcapital.in/

funds for renewable energy, energy efficiency and water treatment projects, it has funded over 2.6 GW of renewable energy projects and also completed eco-city studies for 5 major cities in India to recognized sustainable infrastructure options. Another is mJunction, which started off as a providing an e-transaction platform for buyers and sellers of steel and other material; it has since diversified its portfolio to include idle assets, stressed assets and retail surplus, thus become a leading player in promoting what is increasingly being considered a major solutions to the global sustainability challenge—the Circular Economy! Similarly, Tata Power Renewable Energy Ltd. that focuses on building and operating solar, wind and waste-gas based power plants is a response to recognition that the energy opportunities of the future lie in the renewables space. It has set up the largest rooftop solar plant in the world which will produce more than 15 million units of power annually and offset over 19,000 tonnes of carbon emissions.

4 Integrating Sustainability into Business Processes

Manufacturing operations are inherently disruptive, impacting both communities and the environment around the plant sites and in the supply chains. Responsible businesses recognize that at the first instance, they must avoid or at least mitigate the negative impacts and then focus on find ways to add value to their surroundings. Going further, they then extend this thinking to their business partners in the value chain.

Many responsible businesses approach this challenge by embedding sustainability principles into their core businesses processes, which are the lifeblood of a business model and go hand in hand with products in creating value. They design and constantly innovate their intrinsic processes to recognize, consider and address social and environmental perspectives and eventually use this as a criteria to evaluate how well its processes are working. The business case for doing this is obvious—it "future-proofs' their businesses from emerging regulations and strengthens their license to operate by building social capital.

As the examples below show, Tata companies have been quite innovative across their value chains in terms of avoiding and mitigating disruption to creating value.

An interesting example of avoiding disruption is the case of Tata Chemicals. The company was evaluating a potential investment in a soda ash extraction and processing plant near Lake Natron in Tanzania in partnership with the Tanzanian Government in 2007 and had initiated appropriate environmental assessments recognizing that the proposed site was an ecologically sensitive wetland region. However, these assessments as well as feedback from a structured stakeholder engagement process indicated potential threats to the lake's chemical composition and the 'sensitive' ecosystem, especially the lesser flamingo, from operations of the proposed soda ash facility. Tata Chemicals, recognizing this potential threat to the fragile ecosystem of the region, exited the project.

Another Tata company, Amalgamated Plantations, which owns tea plantations in north-east India found a different way to avoid disruptions. Its Hathikuli plantation,[19] which is adjacent to the Kaziranga National Park, home to the one-horned Indian rhinoceros, became completely organic a few years ago, which has encouraged newer species of birds, animals and insects to thrive in the region. This despite the fact that this move is adversely impacting the financial viability of the plantation as the higher price of tea is not compensating for the lower yields.

On upstream value chains, the Tata experience in its jewelry business provides useful context. When the Tata group entered the gold and diamond jewelry business under the banner of Titan, it made significant headway in formalizing what was essentially an unorganized and informal sector. The gold supply chain, in particular, was dominated by individual craftsmen from marginalized communities across India who operated from dingy and hazardous workspaces with limited tools and technology at their disposal. As a consequence, the inherent talent pool was drying since the craftsmen often found these conditions unviable. Titan, since 2011, has setup a network of Karigar Parks[20] (Karigar, in Hindi, means craftsman) close to its manufacturing locations which provides opportunities to craftsmen to work in safe, healthy and modern facilities. They are also given financial aid, insurance and medical check-ups while working at these parks. These talented craftsmen who would usually retire by the age of 40 when working in their small workshops, now hope to continue working till 60 at the Karigar Parks. For the company, productivity gains have been huge simply by setting up assembly lines and streamlining processes at the Karigar Parks while ensuring a deeper relationship with key supply chain partners. The significance of these parks lay not only in efficient supply chains for the company but also in raising the average standards of the Indian jewelry industry.

Staying with the upstream, Tata Projects provides an interesting approach to "greening" its supply chains. The company is in the business of executing large and complex industrial and urban infrastructure projects. Its business model makes it highly dependent on its supply chain for goods and services required to deliver these projects and it saw this as a significant opportunity to embed sustainability in its business. Though this project is in its initial stages of execution, what Tata Projects has done is elegantly simple—it has introduced environmental criteria in its vendor evaluation process, recognizes vendors who meet pre-defined criteria as "green vendors" and provides them with incentives such as preferential payment terms, recognition in various forums etc.

All Tata companies are continuously innovating on their technologies and processes within the "fence" to improve material use, energy and operational efficiencies and reduce emissions, effluents and waste. There are several excellent examples across the group but only a few are mentioned here to illustrate the range of

---

[19]http://amalgamatedplantations.co.in/plantations/assam-estates/hathikuli-tea-estate/

[20]http://www.thehindu.com/business/Industry/titan-pitches-karigar-centres-as-industry-model/arti cle5710418.ece

initiatives. Tata Steel is playing a leading role in ULCOS,[21] a Europe-wide initiative to reduce carbon emissions in steelmaking. In 2010, it built a €20 m HIsarna pilot plant[22] at its Ijmuiden site in the Netherlands. HIsarna's revolutionary cyclone converter-based ironmaking process directly converts iron ore and coal into iron, without any pre-treatment of the ore and coal, which has the potential to reduce CO_2 emissions by 20% compared to conventional ironmaking and, used in combination with carbon capture and storage techniques, it should be possible to achieve CO_2 reductions of up to 80%. The Tata Chemicals' soda ash plant in India, which is located in a water-stressed area, completely stopped using fresh water for its operations in 2007; its Make-up Water (MUW) plant uses seawater and generates 4.5 KL of condensate per ton of salt, thus generating on an average 2 million KL of water. Mention has already been made of JLR's REALCAR project which, apart from light-weighting the car for better fuel efficiency also incorporates the principle of circularity. But JLR is not the only company in the group that has embraced circularity. Croma, which is an electronics retailer, has made provisions in its outlets for its customers to deposit their used devices which are then sent to registered e-waste processors. And Tata Chemicals' cement plant is essentially a recycling plant—it chief raw material being the solid waste from the soda ash manufacturing process.

Looking ahead, the Tata group is working towards mainstreaming several new ideas into its business processes. The group was a member of the Natural Capital Coalition which came up with the Natural Capital Protocol in mid-2016; several Tata companies piloted this protocol before it was finalised and released. The group is addressing the issue of social capital valuation and is contributing to the efforts of the World Business Council for Sustainable Development to develop a Social Capital Protocol. In order to go down a low-carbon path, the group explored the use of an internal carbon price to assess capital expenditures and a few of the companies in the group have begun to experiment with this.

Tata Global Beverages (TGB) provides an interesting example of social innovation in downstream value chains. Responding to the challenge of low market shares in rural India, the company, in 2006, set up a rural distribution system called *Gaon Chalo* (loosely translated in the Hindi as "let's go to the village") with 3 stakeholder groups—(a) rural entrepreneurs who helped the products travel the 'last mile' in rural areas while making a living of it for themselves (b) NGOs to help identify potential rural entrepreneurs and (c) existing stockist and dealer networks to help reach TGB brands to the rural entrepreneurs at prices which enable attractive margins on the listed price. *Gaon Chalo* is present in 18 Indian states with direct reach across 70,000 villages[23] and has contributed to TGB's consolidated rural market share growing from 18% to over 26%.

---

[21]http://www.ulcos.org/en/index.php

[22]https://www.tatasteeleurope.com/en/innovation/case-studies-innovation/hisarna%E2%80%93pilot%E2%80%93plant

[23]http://www.tataglobalbeverages.com/company/innovations/innovation-in-communication

5   Contributing Intellectual Capital for Societal Good

Businesses, by virtue of the products and processes they have fine-tuned in their years of operations, have inherent knowhow in terms of skills, knowledge, experience and expertise—in other words, intellectual capital—to contribute to solving many intractable problems and challenges that society and the planet faces. This capital is largely tangible and obvious (familiarity with specific technologies, processes, systems etc.) but can occasionally be latent (volunteering spirit, for example).

The Tata group, like many others, has been acutely aware of the potential its employees and businesses have found ways to make this contribution. Many of this has been through pure philanthropy but there are some initiatives that also have the potential to serve a long-term business interest, though that has rarely been the principal driver. Some examples of both these are outlined in the paragraphs that follow.

TGB, the company behind the Tata Tea and Tetley brands, knows how to create campaigns which resonate with consumers and impart a unique identity to its brand. It chose to leverage this capital to address a societal issue—under representation of the voice of women in the Indian electoral process despite comprising 49% of the country's population. The 'Power of 49' campaign, launched ahead of the 2014 general election in India, leveraged all media platforms and was delivered in two phases—in the first phase it focused on creating awareness among women about the power of their informed vote and in the second, it encouraged women to voice the issues that were critical to them in a 10-point manifesto. The campaign saw participation from 1.2 million women, guided the eventual manifestos of the three largest political parties and inspired the highest female voter turnout in Indian history. It also had a positive brand impact based on independent assessments and also 13.4% increase in TGB branded tea sales during and immediately after the campaign. The company demonstrated how a much stronger connection can be forged with consumers if brand building processes can be designed to spot and reflect societal issues which are of concern to them.

India is on its way to become the youngest country in the world with over 60% of its population in the working age group.[24] This, when contrasted with the fact that a very low proportion of the workforce is formally equipped with vocational skills, provides context for the national agenda of skilling India's youth for employment, entrepreneurship and community enterprise. Tata STRIVE,[25] a Group CSR programme of the Tatas, works towards developing vocational skills of young people from financially challenged backgrounds in alignment with this national agenda and aims to create trained manpower across the entire industrial spectrum as well as develop entrepreneurial talent. It was set up less than a year ago but has already reached out to more than 40,000 young men and women across 13 states of India. In delivering content that is industry relevant and contemporary, Tata STRIVE

---

[24]http://www.esocialsciences.org/general/a201341118517_19.pdf

[25]www.tatastrive.com

significantly leverages the inherent knowhow of Tata companies in specific sectors when designing courses relevant to those sectors. For instance, Voltas and Tata Motors have contributed to creating content relevant to consumer durable and automotive repair trainings. This ensures that the content is practical and usable for the candidate, and also enables sector experts within Tata to make a contribution to society. In the long term, it increases the availability of skilled people in the job market which Tata companies also tap.

Volunteering has always been a part of the Tata ethos with several companies providing opportunities for their employees to serve. To complement these efforts, a specially curated group volunteering programme called Tata Engage[26] was launched in 2014 which currently has two formats. The first is the Tata Volunteering Week organized biannually as a series of volunteering opportunities of about half a day during working hours, meant to introduce employees to volunteering as well as for them to do this with colleagues from various Tata companies who they may not even know. The second and deeper initiative is ProEngage which pairs volunteers for extended periods and on their personal time with non-profits, and provides teams of employees to use their skills and knowledge to solve a problem identified by an NGO. There is already a rich track record of skills across writing, financial modelling, human resource management, marketing, MIS building and strategy formulation having been applied through ProEngage projects. The ambition is to further deepen this through a volunteering programme which allows Tata employees to use their skills to create greater societal impact by going off on a paid sabbatical and work with NGOs. Tata group employees contributed over 1.3 million volunteering hours last year and Tata Engage is one of the top 10 volunteering programmes globally.

m-Krishi is another example of how the Tata group is deploying inherent knowhow for societal impact. This is a multi-lingual, mobile-based, integrated and personalized agro advisory service that the Tata Consultancy Service (TCS) launched for Indian farmers in 2009. By subscribing to mKrishi, even via a low-end mobile phone, Indian farmers can get vital agricultural information, including weather conditions, current and future costs, and selling prices for grains in local and global markets. Drawing on mKrishi's recommendations on amount of pesticide and fertilizer usage for a particular soil condition, and when to harvest and how to maximize their agricultural productivity, farmers have been able to more smartly operate their agricultural businesses. If the farmers are facing any agricultural problems, they can also send their queries directly to experts or policy makers in their local languages, both through written or voice activated SMS and images, and receive personal advice and information on those issues. TCS is now offering this service to over 200,000 under-served farmers in scores of villages across 9 states of India including Punjab, Uttar Pradesh, Tamil Nadu, West Bengal and Maharashtra.

Club Enerji, an initiative started by Tata Power in 2007, works with school children across India to become energy conservation champions. Employees of Tata Power, in collaboration with over 500 schools currently and NGO partners,

---

[26]http://www.tataengage.com/

leverage their technical understanding of electricity to help create instructive content and innovative means to reduce wasteful consumption of electricity. In the last 9 years, this initiative has saved around 3.06 million units and sensitized 3.5 million citizens. While this initiative of an energy company to restrict demand may appear counterintuitive to many observers, it is driven by an inherent conviction within the company to use its technical knowhow towards societal responsibility and brings into play the previously made postulate on responsibility stemming from the DNA of a business.

The group has also attempted to break the influence of exclusionary forces of society within its organizational framework by a group Affirmative Action Policy[27] in 2007. Focused on the four "E"s of Employment, Employability, Entrepreneurship and Education, the policy encourages Tata companies to provide equal opportunities to India's historically excluded communities. Now the group actively exercises positive discrimination in its hiring and business practices—if everything else is equal, group companies prefer to employ more Affirmative Action candidates and engage more business associates from these communities. In doing so, the group is also helping Indian national and state governments meet their goals of providing these communities greater social and economic assimilation opportunities.

Tata companies have also made significant contributions to biodiversity and species preservation. Three examples of the latter are—the Mahseer Conservation Project (the Mahseer are a group of species of freshwater fish most of which face the threat of extinction in the wild) of Tata Power at its reservoirs in Maharashtra, India; Save the Whale Shark Campaign of Tata Chemicals to prevent the world's largest fish, declared endangered in 2001, from being slaughtered near its soda ash plant in Gujarat, India; and protecting the Sage Grouse at Tata Chemicals' plant at Wyoming, USA.

6   Sustaining the Social License to Operate

Increasingly, the license for businesses to operate is derived less from regulations and more from the intangible and implicit social contract they have, especially with their neighbouring communities. Responsible businesses respect this social contract, make efforts to engage constructively with the needs of local communities and recognize that this may need them to go beyond the products, processes and intellectual capital at their disposal.

So what can businesses do to first obtain and then sustain this social license to operate? The first is creating the necessary institutional structures within companies to ensure relevant advice and strong governance for their philanthropic efforts. Secondly, businesses need to be strategic in their approach to philanthropy by (a) addressing real and felt development needs of the target communities, national

---

[27]http://www.tatabex.com/assessment/taap-assessment

and local priorities, (b) identifying clear outcomes they hope to achieve, (c) aligning implementation plans to the inherent long term nature and local context of the social development process and (d) measuring success in terms of impacts and outcomes. And finally, businesses need to foster a strong spirit of partnership and collaboration, recognizing that some of the expertise is not available in-house and both know-how and resources lie with other actors such as grass-root NGOs, philanthropic foundations and governments at the national, state and local levels. Examples of how Tata companies have addressed these essentials is outlined below.

Company law in India, for instance, asks all companies above specified revenue and profit thresholds to setup a CSR Committee of their Boards[28] to oversee their CSR policy and commitments. Tata Global Beverages, Tata Chemicals, Tata Steel and Tata Power, to name just four Tata companies that have operations in regions with sizeable population who are socially and economically marginalized, have also established advisory councils comprising experts from academia, development and sustainability domains who provide direction and insights to the company boards on their approach towards CSR.

The Tata group, in keeping with its philosophy of keeping the community at the centre of its business, has an ownership structure which is driven by a spirit of trusteeship and enables a systemic approach to philanthropy. Two-thirds of the shares of Tata Sons,[29] the holding company of the group, belong to the Tata Trusts.[30] The wealth that accrues from this asset is used to support an assortment of causes, institutions and individuals feeding into a wide array of social development imperatives.[31] In this manner, the profits that the Tata companies earn go back many times over to the communities they operate in. The Tata Trusts, one of the largest and oldest Indian philanthropic foundations, is among the largest private philanthropic agencies and is also responsible for the setting up of pioneering institutions aimed at the greater good of India, including the Indian Institute of Science,[32] the Tata Institute of Social Sciences,[33] and the Tata Memorial Hospital.[34]

The Tata Sustainability Group (TSG), in an effort to foster a strategic approach to CSR, has developed a CSR Assessment Framework (referred to earlier) to help Tata companies recognize key CSR elements, monitor their CSR journey and review gaps in their current approach. It has also articulated a CSR approach based on 10 principles to enable companies to be strategic and impactful in their CSR.

Tata companies have begun emphasizing an impacts based approach and voluntarily adopted a Social Return on Investment (SRoI)[35] framework to assess the

---

[28]https://www.mca.gov.in/SearchableActs/Section135.htm

[29]http://www.tata.com/company/profileinside/Tata-Sons

[30]www.tatatrusts.org

[31]http://www.tatatrusts.org/article/inside/About-Tata-Trusts

[32]http://www.iisc.ac.in/

[33]http://www.tiss.edu/

[34]https://tmc.gov.in/

[35]http://www.tatamotors.com/investors/financials/70-ar-html/csr.html

impact of its CSR initiatives. Tata Motors, for instance, ran a pilot SRoI assessment to study impact of employability projects on training tribal youth in several trades over a 10 year period in Maharashtra. The company is working on institutionalizing the SRoI framework across other locations in India. Tata Chemicals has also piloted a social impact exercise.

On aligning local contexts with addressing real development needs, the Okhai[36] initiative of Tata Chemicals provides an effective illustration. The company has a soda ash and salt manufacturing plants in Mithapur, in the Okhamandal region of Gujarat, India where it generates livelihoods, directly and indirectly, for more than 100,000 people through its core operations. However, it also noted the absence of any other significant income earning opportunity for the rural women of Okhamandal which led to the genesis of Okhai, a self-help group based enterprise supported by the Tata Chemicals Rural Development Society (TCSRD). It recognized the rich tribal design heritage of the region and enabled the local women to leverage their design and embroidery skills to earn a livelihood, move towards economic empowerment and eventually towards social empowerment. Periodic progress evaluations emphasize impact based indicators which allowed the company to get a real sense of success, identify gaps in implementation, learn lessons and set out forward plans—all factors which have helped significantly in the current push on digital marketing for Okhai and also, the ambition of the company to replicate similar programmes at its other locations.

Tata Consultancy Services (TCS), in 2014, committed Rs 1 billion towards building dedicated sanitation facilities for girls in schools.[37] This commitment was part of the flagship Swachh Bharat Campaign of the Indian government and addressing the complex interplay of sanitation and education for girls in India. By constructing usable toilets in schools, TCS aimed to boost enrolment rates, reduce dropout rates and, in turn, improve learning outcomes for girl students. Given the scale of the project, the company structured their approach on the lines of a mainstream business endeavour—a special taskforce was constituted for the end-to-end implementation, structured process put in place to identify partners and technological solutions and pilot exercises completed to understand feasibility of various solutions. TCS also rolled out a Management Information System (MIS) that served a threefold purpose: it enhanced programme monitoring through projections of planned vs actual progress; provided evidence of work completion through the feature of 'uploading site pictures'; and enabled a real-time control over programme activities through its mechanism for online sign-off of checklists defined for implementation partners. The overall and state specific progress dashboards were shared with the Ministry of Human Resource Development (MHRD) and State Authorities on a weekly basis. The company had constructed over 1500 toilets by 2016[38] and

---

[36]http://www.okhai.org/

[37]https://www.tcs.com/tcs-completes-building-sanitation-facilities

[38]http://indianexpress.com/article/business/companies/swachh-vidyalaya-abhiyan-tcs-mahindra-account-for-80-of-completed-toilets-by-india-inc/

laid out a wonderful blueprint for the application of management principles and platforms to achieve a philanthropic goal.

Tata Steel, since 2009, has been working on reducing neonatal and infant mortality rates in over 150 villages[39] of the Seraikela block in Jharkhand, India. This project aims to address inadequate public health services, low levels of health awareness in the community and the absence of a sustainable scalable delivery model, all of which were the root causes of high mortality levels in Seraikela. However, a project of this nature required collaboration which led the company to adopt a public private partnership model and bring on board multiple partners—the Tata Steel Rural Development Society (TSRDS) implements the project along with the American India Foundation (AIF) who had operational experience on similar projects, Department of Health and Family Welfare (Government of Jharkhand) who anchored the public health systems in the region and SEARCH which is a NGO with the requisite technical expertise on neonatal and infant health. The project has been running successfully for over 7 years now with neonatal and infant mortality rates coming down by 46% and 39%, respectively, in Seraikela. It is also a useful illustration of multiple partners bringing different skills to the table, working towards a common goal and now looking to take this partnership to other parts of the country.

In a world with growing dependence on advanced technology platforms, delivery models for development are also being reconfigured. Tata Communications recognized the role of technology in enhancing access to information and improving the quality of education for children from underserved communities. The company partnered with the Parikrama Humanity Foundation (PHF) to initiate an integrated development programme for underprivileged children in Bangalore, India which creates high caliber education infrastructure and improved sports infrastructure powered by strong internet connectivity. Over 1700 children across 5 schools have already benefited from the initiative in the last few years.

The collaborative approach to philanthropy is also reflected in group level initiatives such as disaster response programme anchored by the Tata Sustainability Group (TSG). TSG has reached out to over 400,000 beneficiaries through post disaster relief and rehabilitation projects within and outside India in the last 3 years. In doing this, it has worked with over 30 NGOs and recognizes the importance of these partnerships in ensuring that the project takes local contexts in to account and also extends the impact of rehabilitation beyond the timeline of the project. TSG, for the Tata group, has also codified guidelines for disaster response[40] which identifies 10 values that define a Tata response—one such value being *"We will seek to work with other groups and entities so as to best serve the interests of the affected communities."*

---

[39]http://www.tatasteel.com/investors/annual-report-2016-17/Integrated%20Report%20and%20Annual%20Accounts%202016-17.pdf

[40]http://www.tatasustainability.com/disasterResponse.aspx

7   Conclusion

It is very clear that challenges affecting humanity and the planet are complex, their origins are in great part anthropogenic and their impacts increasingly global. Thus, "Spaceship Earth" requires all its inhabitants to stop being a part of the problem and become a part of the solution.

There is little doubt that that businesses have the great potential to be a significant part of the solution. The above examples show how just one group of companies have over the years found ways to be profitable, grow and continue to do so in some cases over a century while significantly providing value to all its stakeholders.

More and more business leaders today recognize that they have to, pro-actively, use the great capability and power they have for a greater good. The active role that business leaders played in the formulation of the Sustainable Development Goals and the Paris Agreement on Climate Change, both in 2015, is evidence enough that business is serious about its responsibility. The time has now come for action and it is responsible businesses that have to show the way.

CONTRIBUTING INTELLECTUAL CAPITAL FOR SOCIAL GOOD

Leveraging inherent knowhow to address societal challenges

Tata Global Beverages: Rural distribution model

Tata STRIVE: Creating skills for jobs and entrepreneurship

Tata Engage: Societal impact through employee volunteering

mKrishi: Technological solutions for farmers

Tata Power: Reducing wasteful energy footprint of consumers

Tata Affirmative Action: Solving for workplace inclusion

Tata Power & Tata Chemciasl: Biodiversity conservation

RESPONSIBLE
BUSINESS
MODEL:
PERSPECTIVES
FROM TATA

INTEGRATING SUSTAINABILITY INTO BUSINESS PROCESSES

Designing processes to recognize and address societal issues

Tata Chemicals: Exiting Lake Natron investment

APPL: Organic tea plantations and wildlife conservation

Titan: Reconfiguring hazards in the gold supply chain

Tata Projects: Incentivizing environment positive suppliers

Tata Global Beverages: Innovating rural distribution models

Croma, m-Junction: Addressing the E-waste challenge

TSG: Exploring natural and social capital valuations

SUSTAINING THE SOCIAL
LICENSE TO OPERATE

Recognizing and
respecting social
contracts through
strategic
philanthropy.

Tata Trusts: Among
India's largest
philanthropic
contributors

DELIVERING TOTAL VALUE THROUGH PRODUCTS

Creating societal value while treading lightly on the planet

Tata Motors: Reducing eco-impact of public transport

JLR: Applying circular economy to vehicular emissions

Tata Housing: Embedding green principles in construction

Tata Steel: Enabling B2B customers to reduce eco footprint

Tata Chemicals: Enabling access to affordable drinking water

Tata AIA: Reaching insurance coverage to rural households

Tata Cleantech Capital, mJunction: Sustainability led business

Tata Steel: Neo Natal
care

TCS: Promoting
sanitation in India

Tata Chemicals:
Livelihoods based on
local skills

TSG: Responding to
disasters & CSR
Assessment Framework.

EMBEDDING CORPORATE RESPONSIBILITY INTO DNA

Combining policy, institutional structures and leadership commitment

Tata Code of Conduct | Climate Change Policy | Tata Sustainability Policy

Tata Global Sustainability Council | Tata Sustainability Group |

Sustainability Assessment Framework | Tata Business Excellence Model

Exhibit 1 Responsible Business Model

Shankar Venkateswaran Till recently, Shankar was Chief—Tata Sustainability Group which drives all Tata group level initiatives in the sustainability space and guides Tata Sons and the group companies on sustainability and CSR initiatives. A mechanical engineer from IIT Madras and an MBA graduate from IIM Calcutta, he has over 35 years of work experience in the corporate and development sectors.

Before joining Tata Sons, Shankar was with PricewaterhouseCoopers (PwC) India as Director, Sustainability. He started his career at AF Ferguson & Co and moved on to the corporate sustainability and CSR sector as chief executive of Partners in Change. He has also served as Executive Director, India, of the American India Foundation.

One of the leading thinkers and speakers in the areas of sustainability and corporate responsibility, Shankar has held board positions at Mobile Creche, Srijan, ActionAid India and Ghana, Tata Chemicals Society for Rural Development, Tata Power Community Development Trust and Aatapi

Trust. He was also a member of the guidelines drafting committee for the National Voluntary Guidelines for responsible business.

He enjoys tennis, theatre and reading in his leisure time.

Sourav Roy Sourav presently leads the CSR effects of Tata Steel Limited, which is among the oldest and deepest engagements with society across India Inc. Previously, he anchored the sustainability strategy work of the Tata Sustainability Group (TSG) and worked with Tata companies to integrate social and environmental aspects within their approaches to business. His parallel engagement within TSG was on setting up disaster response frameworks for the Tata group while managing post disaster relief and rehabilitation programmes within and outside India.

He joined the Tata group through the Tata Administrative Services, flagship leadership programme of the group, in 2007. His experience with Tata companies range across operations strategy, M&A/corporate restructuring and project finance in diverse sectors like metals & mining, power, logistics, infrastructure and hospitality. He has also worked with the Grameen Bank (Bangladesh) and other organizations in the Indian development sector before joining the Tata group and retains a keen interest in development finance.

Sourav holds a graduate degree in economics from the Delhi University with post graduate diplomas in management and business law from IIM (Ahmedabad) and National Law School (Bangalore), respectively. He loves playing any sport, understanding the history of cinema, trekking and reading a good book in his spare time.

Impact of Corporate Social Responsibility on Innovation Activities: The Case of Xerox

Jutta Willamowski, Yves Hoppenot, Stefania Castellani, and M. Antonietta Grasso

1 Introduction

In this chapter we show how Corporate Social Responsibility (CSR) has influenced the research of a large manufacturing company, namely Xerox, during its transition to increase its services portfolio. We explain how a deep and long lasting CSR culture has influenced and effectively nurtured two (successful) research projects in the area of sustainability and the instruments that made this possible. We present each of these projects in detail and show how they facilitate more sustainable behaviour. The first case is to reduce paper waste in the workplace and the second to lower single car occupancy in commuting. Finally, we discuss how these projects enabled us to develop a more general approach to behaviour change in the workplace.

2 Xerox Engagement in CSR and Innovation Processes

Xerox is an $11 billion technology leader that, starting with xerography, has continuously innovated the way in which the world communicates, connects and works. Some of its most well known product innovations include a breadth of printing equipment and digital services that bridge the analog with the digital world. In this context CSR in the third millennium is a legacy of strong corporate citizenship (Abbott and Monsen 1979) first laid down by company president Joseph Wilson in the 1960s as key to combining good business with good citizenship. This

J. Willamowski (✉) · Y. Hoppenot · S. Castellani · M. A. Grasso
Xerox Research Centre Europe, Meylan, France
e-mail: jutta.willamowski@xrce.xerox.com; antonietta.grasso@xrce.xerox.com

© Springer International Publishing AG, part of Springer Nature 2018   263
R. Altenburger (ed.), *Innovation Management and Corporate Social Responsibility*,
CSR, Sustainability, Ethics & Governance,
https://doi.org/10.1007/978-3-319-93629-1_15

original statement still impacts directly or indirectly many facets of the company today. As an international market leader, Xerox recognises the opportunity and responsibility it has to positively impact the world.

Over the past 40 years, Xerox has demonstrated its leadership in this direction by supporting educational and community projects around the world, designing "waste-free" products built in "waste-free" plants, investing in innovation that delivers measurable benefits to the environment and many other integrated initiatives that touch Xerox communities, employees and stakeholders. This commitment aligns with Xerox's early adherence to the international ENERGY STAR® program, which ensures that companies create more energy-efficient products, helping to reduce global emissions. In addition, Xerox is a pioneer in remanufacturing its products and supplies: its remanufacturing and recycling practices diverted over 140 million pounds of waste from landfills in 2005. Xerox also mandates sustainable designs for its products, supplies and paper. For example, Xerox proprietary solid ink printing technology generates 95% less consumable waste than comparable laser printers.

Xerox's expertise in document management, especially workflow and digital imaging, is focussed on improving worker productivity and reducing dependency on paper documents. When paper is necessary, Xerox continues to develop high-quality recycled-paper products along with default features in its equipment for automatic two-sided printing and energy-saving "power down" features. Strict policies ensure that Xerox paper is sourced from sustainably managed forests. Suppliers must show that they have taken action to safeguard forest areas of significant ecological and cultural importance and that all fiber comes from legiti-mate sources.

Sustainable growth extends to nurturing the next generation of company leaders investing in communities. Xerox has a long-standing commitment to maintaining an inclusive, progressive workplace environment that values diverse ways of thinking, cultural differences and new perspectives. The company views diversity in the workplace as "a moral imperative and a business and competitive advantage," and diversity practices are deployed throughout all core Xerox operations, from staffing to succession planning to supplier diversity. These initiatives and others have resulted in a number of accolades and awards over the years from a variety of external organisations. Xerox was ranked No. 1 in its industry for "social responsi-bility" in Fortune magazine's 2006 "America's Most Admired Companies" survey. In 2005, Xerox was No. 10 on BusinessEthics' list of the nation's "100 Best Corporate Citizens" and the company earned the 2005 U.S. Community Service Award from the U.S. Chamber of Commerce Center for Corporate Citizenship for its Social Service Leave Program and its Community Involvement Program, which provides seed money to help Xerox volunteer teams work on projects ranging from raising guide dogs to teaching disabled children skating skills or maintaining parks.

The innovation tradition is an equally long lasting one since it was founded on xerography. R&D in the 1970s contributed to key innovations in computing like the Ethernet, the Personal Computer and Grapical User Interfaces. Xerox innovation has contributed to the development of more than two-dozen technology platforms central to the core printing and document technology business. Corporate R&D

has ensured over the years a balanced portfolio of research and technology projects with an impact on short, mid and long term time horizons. Approximately one third of research investments directly support business partners, where Xerox researchers collaborate closely with development engineers to accelerate the uptake of new technologies and services. Another third is directed at creating the next generation technologies and services or to incubate new offering concepts. Research on next generation technologies includes both improvements to current technologies as well as disruptive technologies. The remaining third explores future opportunities and new insights into the potential of new technologies. Here we focus on pioneering areas that can create new business opportunities for Xerox in the future. One of these exploratory efforts has focussed on a key topic of CSR which is sustainability. In the remainder of this chapter we explain both the instruments and the results of the effort to focus on innovation contributing to sustainability.

3 How Internal CSR Culture Has Influenced Topics

3.1 Green Services Initiative

The ENERGY STAR® technology, which reduces the energy consumption and environmental impact of Xerox machines is widely recognised. Being a leader in office and professional printing equipment, the company turned its attention at the beginning of 2000 to expand its offerings in services such as document outsourcing and workflow automation. As the company was refining its strategy, the head of R&D at the time, Sophie Vanderbroek, felt that CSR values, and especially sustainability, sat so squarely at the heart of the company, that innovation activities should go beyond the hardware programs to include services. To reach this objective, and to understand how to more deeply serve Xerox customer needs in this area, the global R&D organisation was given a charter to put together a program inspired by a quote from then CEO, Anne Mulcahy:

> We have always been keenly aware that by helping our customers run their businesses more productively we incur a major responsibility to foster sustainable development ... now we are beginning to see corporate responsibility as a new frontier for market development and revenue growth.

An internal "Green Program", was created to explicitly nurture this type of innovation. On the hardware side contributions had to feed into two main initiatives: an "eco-friendly concept printer" which brought together advances in energy consumption, ink, recycling and spare parts and "reusable paper" technology, focussing on alternatives to traditional paper. On the services side, a part of this exercise was to understand what "green services" could be and develop a call for project proposals in this space. To feed this process a number of brainstorming sessions were organised and a scope of topics defined, to which more specific proposals could then be submitted. More precisely, the following list was produced:

- Process improvements: Xerox could investigate how to improve its own processes then, based on what it had learned internally, improve its external customer offerings. Innovation in this space was called to contribute with product development process re-engineering to assure *cradle to cradle* product lines. This approach would have been beneficial both internally and externally, offering the same approach to our customers.
- Services for multifunction device fleet management: this investigation would address ways to lower power consumption through statistical models of fleet use, while assuring that the Quality of Service be preserved. These models could address optimized device wake up and adjusting the quality of the document to the task at hand (e.g. verifying that for draft use, minimal required quality be produced).
- Services and technology to support the paperless office: Services in this area were called to address the transition of paper workflows to their digital counterpart, exploiting virtual worlds, electronic books, etc. These services, requiring new work practices, would have also required user studies to help define user interfaces, novel use practices, acceptance levels, etc.
- Organisational and personal green *thermometers*. On the organisational side, services that would assess how a company is performing and how it is perceived in environmental terms were called for. These services would for example enable understanding customer concerns by mining public discussion forums with information extraction tools and allow a company to match their communication to address them. Services could represent how individual actions impact organisational processes and resource use by visualizing personal parameters contextualized with global parameters (e.g. contribution of personal printing to global environmental cost).

The creation of the Green program reinforced the message to researchers that the company was not only seriously engaged in CSR topics like sustainability, but wanted to go further and increase its engagement. The main research lines generated a number of project proposals across the various research centres, in hardware and software. The fact that the call included services promoted the participation of the Xerox Research Centre Europe (XRCE), a centre that dedicated to services related research. In particular, some of the authors of this chapter, who are researchers in the area of Computer Supported Cooperative Work of XRCE and experts of office work, contributed to the definition of the Green Program call. This put them in an optimal position to define and feed appropriate project proposals. Considering our background at the time and the topics listed in the call, we proposed an initial project aiming at providing a service supporting more sustainable paper usage in the work place. We were already knowledgeable of the large, existing body of studies describing "affordances of paper" (Sellen and Harper 2003), i.e. the variety of reasons for using paper documents in an office environment and in support of daily work. We were also aware of the fact that shared resources are typically subject to a so called "tragedy of the commons" paradox (Hardin 1968): they are used extensively without paying attention to the damage such use does to the common good (i.e. in this case the environment). Concerning the call, the topics listed included a high level vision of "green thermometer" services within which we proposed the Print Awareness Tool, which has since been integrated in a Xerox offering. A second related project proposal took shape later that addressed the topic of promoting more sustainable commuting practices. The call ended up initiating a full line of research about how to support individuals to change their behaviour within the work context, e.g. helping them to act and/or work in a more environmentally friendly way.

In the remainder of the paper we describe the team of researchers who fostered this activity, the Work Practice Technology area, the two projects carried out and how the line of activities stemming from the overall CSR company engagement generated a broader research effort around Behaviour Change support in the workplace.

3.2   The Work Practice Technology Approach to Innovation: The Case of Behaviour Change Support

Achieving successful design and innovation is a process where attention has to be paid not only to technical possibilities, but also to the social and work environment within which the resulting system or innovations will sit. It is therefore useful for designers to understand how well any technical system will fit with the activities and needs of the users in a proposed setting (the application domain). In the Work Practice Technology (WPT) area we employ ethnographic observational studies to produce a socio-technical understanding of the human practices and put those at the service of our innovation processes through domain and user knowledge. Ethnography is a method of data capture that works through the immersion of the researcher into the environment being studied, collecting detailed material (notes, documentation, recordings) on the 'real-time' and 'real-world' activities of those involved. Periods of immersion can range from intensive periods of a few days to weeks and months (more common in system design studies), and even years. A common product of most ethnography is the development of a 'rich' description—a detailed narrative—of the work or activity in question, often in the form of sets of perspicuous examples, which may then be further analysed or modelled in various ways and for various means. In summary, ethnographic studies for design serve as a resource to qualitatively evaluate current socio-technical system operations and as a resource to understand what sort of problems design could address, how people doing particular activities with particular needs might be supported, or how an innovation concept might mesh with or disrupt particular work or activities. The studies can identify problem areas where people are making mistakes, where the system 'design' or outputs are causing problems for those using them, or where human ingenuity is compensating for problematic technical systems.

3.2.1 The Print Awareness Tool

The Print Awareness Tool (PAT) (Willamowski et al. 2016) makes users aware of their print behaviour and the impact it has on the environment. It aims at motivating users to recognise and hence reduce unnecessary printing resulting in less paper waste. Studies have shown that around 15% of the paper printed in an office environment is never collected from the printer or even if collected is not really used (this figure varies slightly across industries). It goes straight to the waste basket. PAT targets such waste supporting work organisations in their general efforts to reduce their impact on the environment.

In the following, we will first illustrate the context in which the PAT project started, how the Green Program was a key element in the initial phase and the hurdles it had to face to move from research to commercialization. We then describe the system that resulted from the project and how it works. We finish by reporting on the first pilot deployments of the system, highlighting the key mechanisms that make the tool successful with users. We illustrate how PAT not only motivated its users to reduce paper waste, but how it worked beyond that as an 'organisational probe' providing users the opportunity to question the paper-based workflows they deal with every day. We discuss how organisations may use this effect to optimize their processes and thus further reduce their impact on the environment.

Context

When the Green Program was created, outside Xerox, paper reduction and paper-to-digital efforts were becoming more frequent in the business world including Xerox customers. There was a clear demand for services and tools that could help reduce paper usage. Inside Xerox, the topic of paper reduction was perceived as a business threat by the print services business groups because the revenue model was based on the consumption of paper. They feared that reducing the client's paper usage would challenge this model. This made it very difficult for a project like PAT to follow a linear trajectory from research to transfer.

In this situation the Green Services initiative was the key enabler that allowed us to pursue the PAT project. Indeed, we could start the research line even without the explicit support and engagement of the operational business groups and carry out the foundations of the research work independently. This early work then gave promising results which helped to sustain the motivation on the project. This proved effective because, as times went on, more evidence emerged from the market that the project was addressing a growing demand from Xerox customer needs, and could actually become the enabler to accompany customers toward a better use of paper resources and process digitalization.

To motivate and clarify our approach we will now shortly discuss previously existing approaches to print reduction and highlight why these approaches were unsuccessful and not working as expected. Similar to other organisational change management programs, the typical approach to paper reduction was a top-down program where management decides rules and changes to put in place that then impact each individual employee's way of working. One typical approach in the paper reduction domain was to limit employee access to print services by restricting the number of pages an individual could print and "consume" within each working day, week or month. Corresponding quotas were defined by an assigned administrator for each person without a detailed understanding of their work or tasks but simply according to their generic role. To be effective, the quotas were strict and thus often counterproductive, hindering employees in doing their work properly: Each time an employee reached their quota they could not longer print. As the documents were required to do the work properly, the employee had to spend time finding a workaround, asking the administrator for an exception with respect to their quota or a colleague for a favour and printing the document instead. Needless to say such an approach was extremely unpopular.

Our aim with PAT was to find a working alternative to traditional top down change management, an alternative that optimized print resource usage whilst adressing the different needs across the organisation and a global organisational objective. Instead of fixing a strict quota the PAT approach sensitizes and engages the individual employees in a mixed top-down, bottom-up fashion in the organisational paper waste reduction effort. The PAT concept started from the following observation: *while noticing waste in their workplace, individual employees do not relate that waste to their own behaviour. They rather attribute it to others and are not aware of their personal paper consumption and impact.* This is where the PAT approach comes in: it provides individuals with real time precise information on their print consumption. By doing so, it makes them feel responsible and motivated to reduce printing whenever possible, i.e. in ways compatible with their work. The explicit initial aim of PAT is to provoke reflection on personal transient printing, i.e. whether it's 'waste' printing or reflects one's personal organisational work habits that could be reconsidered.

The PAT System: How It Works

PAT materializes as a desktop widget sitting that intercepts print actions in the background. It analyzes these print actions and provides the user with intuitive high level feedback about their printing behaviour, how it has evolved over time and how it compares to the behaviour of their work team and other colleagues. When a user registers with PAT they go through a self-assessment step with questions on printing habits that cause them to reflection upon their personal print behaviour. The self-estimate is then compared to their real, observed behaviour and there's typically a gap. This comparison thus produces an initial "Aha! Effect", which is, in our observations, a first step in motivating the user to improve and reduce unnecessary printing. The PAT desktop widget is permanently visible. It displays a flower that loses its petals over time materializing in a very intuitive way the user's impact on the environment (see Fig. 1).

Fig. 1 The PAT widget, a
daisy losing petals over time
and materializing the user's
impact on the environment

To determine the user's impact on the environment PAT computes a cost for each print action in terms of a virtual currency called Green Points (GPs). An organisation can tune this currency to reflect the printing habits it wants to focus on, reduce and penalise. For instance, to reduce the printing of emails, the organisation would charge printing a page of email with more GPs than printing a page of a project report. Based on each individual user's observed past behaviour, and at the beginning of each month, PAT allocates a personalised GP credit to each user. With each print action the user then consumes some of their GPs. The credit does not however constitutes a strict quota or limit, i.e. even once all GPs are consumed the user may still print. The resulting debit will simply be reported through the interface and highlighted to the user. In contrast, whenever, at the end of the month, the user has not consumed their credit, PAT adds the remaining GPs to their savings which represent the overall reduction they have achieved over time. Following a user's print action the flower widget is updated in real time providing constant high-level and up-to-date feedback to the user. To further help the employee to understand and improve their behaviour, the user can expand the PAT widget to examine the print behaviour in more detail (see Fig. 2). This expanded view focusses on particular, costly print habits, provides the user with a comparison over time, against peers and with tips for improvements.

As mentioned, PAT has a mixed top-down and bottom up approach, materialized by the use of GPs. The personal GP savings contribute to the overall organisational savings pot, helping the user to appreciate thier contribution to the global organisational effort. Indeed, our studies showed that when a company wants to associate its print reduction effort to an environmental message inside a CSR campaign, it is important that the work organisation concretely prove and illustrate its engagement and environmental motivations. Employees might be reluctant to participate in corporate paper reduction efforts if they feel they are disguised as greenwashing, but in reality motivated by financial savings. To address this issue, PAT lets the work organisation explicitly propose a number of corporate actions for the environment (planting trees for instance) in which the organisation may regularly choose to reinvest entirely or in part, the GP savings achieved.

Pilot Studies

To provide the business groups with a proof of concept of the PAT system, we developed a first prototype which was successfully deployed and tested internally in our research centre. This prototype was also demonstrated to Xerox customers who

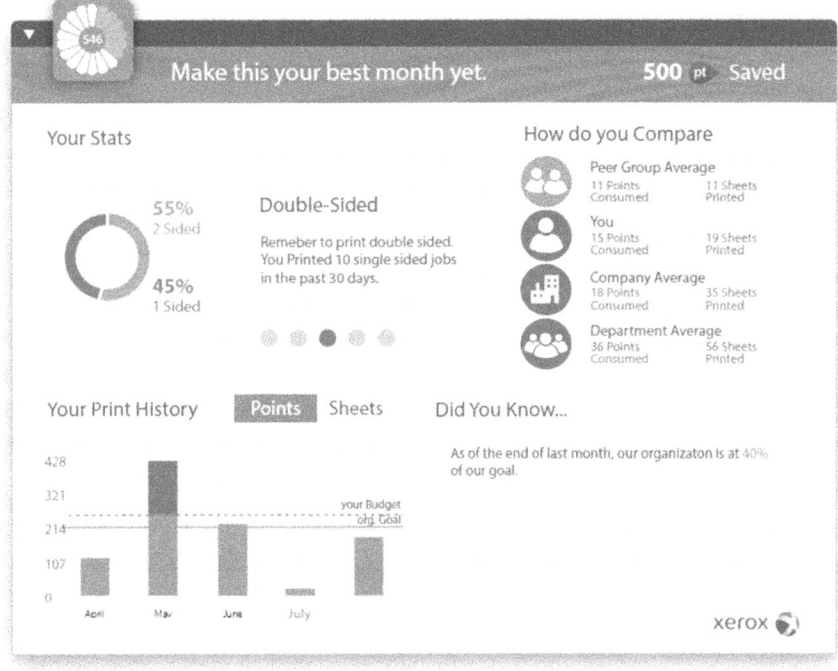

Fig. 2 The detailed view: feedback about key behaviours, evolution over time and comparison against Peers

visited the local technology showroom where it generated a lot of interest. Following this traction, the business groups finally reacted and PAT was transferred and commercialized. When it became available we were invited to accompany the first customer pilot.

We briefly present the most striking observations and findings from the internal and customer pilots. The former essentially allowed us to validate the effectiveness of PAT and its components with respect to reducing transient printing, whilst the second illustrated how PAT functions more broadly as an organisational probe, helping to question more generally the paper usage and processes within an organisation.

PAT to Reduce Paper Waste

In the interal pilot, the *self-assessment step* proved to be very effective. Many users reported that, when going through the self-assessment, they had no idea of their typical print consumption and habits and had difficulties estimating them. At the end of the self-assessment, when confronted with their real observed past print con-sumption (which was often much higher than expected) users started to consider paper waste as a *personal* issue. This effect was maintained and reinforced over time by the *constant presence and visibility of the PAT widget*: Indeed, with the PAT flower widget sitting permanently on their desktop and updating in real time with each print job issued, users reported that they felt a strong pressure inciting them to think twice before hitting the print button. As a consequence PAT users adopted various changes in behaviour to reduce their printing and paper consumption. For instance, instead of printing *all the documents* they considered reading *any time in the future*, they now only printed the documents they needed to read quickly and only the pages really relevant for them. Some started to read and even annotate on screen instead of paper. Others refrained from printing intermediate document versions, a behaviour they previously had, re-printing a document each time they had made a slight change. Finally, users also optimized their personal processes to reduce unnecessary printing.

All these behaviour changes address transient printing, which was the initial target of the PAT system. However, already during this first pilot, we had an intuition of a larger possible impact: by making printing a *personal* issue PAT motivated its users to question the necessity of each single print job they issued (or had to issue). Users thus spontaneously started to pinpoint paper intensive *organisational* work processes they were involved in, processes that they could not change and that hindered them from printing less and "doing better".

Before dicussing the above mentioned effect in more detail we want to insist on one final observation, the very positive reaction of the users to the re-investment proposition. At the end of the 6 month pilot our research centre proposed to all participating users to re-invest the savings they had achieved into one of a set of corporate-proposed actions for the environment; all users keenly participated. Interestingly, this same re-investment functionality was also perceived as very convincing and compelling by many customers during their visits.

PAT as an Organisational Probe

Having proved effective in reducing paper waste, the PAT prototype was transferred into a commercial product with essentially the same functionalities but in a simplified and more polished version. The first customer was a French retail chain widely known for its strong inclination towards sustainability and CSR. They really liked the bottom-up change management approach and the environmental angle we had taken with PAT. From the start the customer invited us to advise, accompany and follow the pilot deployment at their headquarters and to study the impact and effectiveness of the tool in an external customer setting.

The first important upcoming issue with the customer was to decide how to promote and manage the deployment of the tool on its site. Logically the customer's IT department seemed an appropriate candidate to become the tool champion (as deploying the tool requires supporting the installation on every user's personal computer). However, the IT department had been previously responsible for managing another paper reduction effort based on imposing a strict print quota on every user. This attempt had not been well received by the employees. Therefore, and also to insist on the organisational motivation for sustainability, the customer finally explicitly chose to assign the responsibility to promote and manage the tool to their sustainable development department.

During the deployment of the tool we carried out three on-site visits, and conducted three workshops and two series of interviews with PAT users and the customer's local team managing the pilot on site. This study confirmed that the PAT users significantly reduced their transient printing. Furthermore it confirmed also our previous intuition that, even if PAT is introduced as a personal tool to its users, it can have a larger impact and serve as an organisational probe creating the opportunity to discuss, question and reconsider more widely organisational paper-based workflows. Indeed, during our study on the customer site, most of the employees explained that they had never had the opportunity, time and space to discuss their daily work and their related printing habits and duties. Once PAT had been installed for all employees, discussing paper usage and the corresponding processes became possible and even natural: Employees quickly identified the various organisational processes they were involved in and that generated heavy printing. They explained that all these processes with their related printing tasks were deeply embedded in the daily working routine. Before, such printing tasks had always, somehow historically and traditionally been *taken for granted*, that is to say never directly questioned. Additionally, employees often attributed them *by default* to supposed organisational or legal reasons. Over time however, when digging deeper, the discussions brought up many interesting observations, critical assessments and even suggestions of improvements for the various workflows in place.

Nevertheless, as we were external to the organisation, it was not our designated role to gather all the precise pain points raised with respect to particular organisational workflows, let alone reconsider and redesign them to alleviate problems. To systematically capitalize on the additional opportunity PAT provides as an *organisational probe*, i.e. to enable an organisation to collect all emerging ideas, we therefore suggested accompanying any future deployment of PAT on a customer site with the appointment of a corporate print champion in their organisation. A print champion could gather all the feedback from the PAT users and animate the corresponding follow up dicussions and reflections on problematic processes. They could elaborate possible improvements with the work force itself. Not only would this allow the identification of very practical solutions in the sense that the people doing the job and intimately aware of the context be involved in elaborating them; these solutions would also be much easier to put in place as they are not imposed top-down but originate from the employees themselves. This would make organisational processes more effective, to transform them into digital processes where suitable, and to reduce paper consumption within the organisation even further and beyond a portion of the transient paper directly addressed with PAT.

The PAT research project not only proved to be of commercial interest but effectively started a full line of research in our team, around the topic of Behaviour Change (BC) and more specifically BC in the context of the workplace. Shortly after the positive results from the pilots, we started an investigation to map what we had learned up to then to other company activities, in this case to the topic of promoting sustainable commuting. By this time the Green Program no longer existed, but, as it had effectively made legitimate a stronger innovation focus on sustainability topics, we still received internal support, which would have probably not been the case a few years earlier.

3.2.2 Sustainable Commuting

The main objective of the Sustainable Commuting project was to understand how to help commuters reduce the use of the personal car, especially in Single Occupancy Vehicle (SOV) mode where they are alone as a driver, and which technology could be provided in support of this behaviour change. The project therefore focusses on a different dimension of the issue of sustainability in the workplace which is the daily commute between home and the workplace. Our approach was based on the idea of "accompanying" people in BC rather than "persuading" them to change, taking into account the complex social structure in which the behaviour occurs. Since the beginning of the project, and based on the PAT results, we have acknowledged the importance of the roles that the various stakeholders play in the organisation and governance of commuting activities, and in particular the central role that work organisations can have in supporting BC processes (Willamowski et al. 2014; Castellani et al. 2014).

While for paper use practices we could start from a large body of literature, we estimated that around commuting the user practices and company role in accompanying a change deserved more preliminary investigation. The first step of the project was therefore to conduct targeted studies to understand which initiatives already existed to promote the reduction of SOV usage encouraging the adoption of more sustainable transportation means such as public transit or car-sharing. We therefore studied the Workplace Travel Plan (WTP) initiatives from local governments in different countries, and more particularly in Europe and France, to learn about their degree of success and the barriers to their successful adoption. We became aware of the importance of the role of the different stakeholders in urban mobility and commuting, including the public administration, the commuters of course, but more particularly the work organisations which somewhat generate the commuting activities together with their employees (Willamowski et al. 2014; Castellani et al. 2014).

The study of the initiatives related to WTPs revealed a number of issues with their deployment. A WTP typically consists of a set of policies and incentives to reduce SOV usage and adopt greener commuting practices, such as, targeted subsidies for buses, trains or car-sharing. WTPs are used by local governments to engage work organisations as active players promoting sustainable commuting among their employees. Up to a certain extent, WTPs have been successfully used in some countries, but it is still difficult to have work organisations and commuters actively engaged in such efforts. Our analysis of our own and previously existing studies of these kinds of initiatives pointed to several factors that constitute a barrier to the success of WTPs. A first factor is the poor synergy amongst the initiatives that the three stakeholders put in place to organise commuting in a more sustainable way thus reducing their global impact. A second factor is that WTPs' benefits, costs, and impact are often difficult to estimate for companies as they affect more soft values like customer perception and employee well being. This can prevent the engagement of work organisations which typically want to be able financially assess the expected benefits before allocating resources. The third factor is that WTPs are typically not adapted to the specific situation of each work organisation and commuter in a dynamic and personalised manner. Finally, WTPs are often put in place more as a punctual effort rather than a long term one, yet organisations and individuals are not always equally open for change—and BC has been shown to be more propitious at specific life time moments (Willamowski et al. 2014; Castellani et al. 2014).

Possible WTPs enablers for success should include the capability of intercepting the most favourable moments where mobility changes can happen, while intersecting them with the business needs of the work organisation (Roby 2010; Castellani et al. 2016). Also, WTPs tend to work best in organisations that remain engaged and that adopt a proactive role, i.e. organisations that can map the benefits of WTPs to their own objectives. These objectives include engaging in CSR actions, becoming an "employer of choice", and reducing real estate costs (Willamowski et al. 2014; Castellani et al. 2014). Our approach has therefore been to work on foundational requirements to support commuting stakeholders (commuters, work organisations and public administrators) in a permanent program of measures towards more sustainable commuting.

From these studies we created a vision for Sustainable Commuting illustrated in Fig. 3. In this vision the work organisation is a central stakeholder acting as a social context where information on commuting habits and ways to make them more sustainable is made available to commuters as well as personalised services and incentives that support the change.

Another important element in our work has been to understand commuting from the point of view of the commuters, the main actors in the wished for behaviour change process. In order to also understand from the commuters point of view the potential obstacles, pitfalls, and enablers when trying to motivate them to change their behaviour we conducted further studies to understand how commuters organise their mobility activities, how they make decisions on transportation means, etc. (Castellani et al. 2016). These studies were meant in particular to inform the design of technology interventions (tools and services) to promote sustainable commuting habits among commuters and within their work organisations. From the analysis of our studies and existing literature, several aspects emerged that informed our technology design.

First of all, commuting is a complex activity to organise because commuting behaviours are determined by many factors including constraints and preferences that people have as well mobility habits within their private networks. Taking these into account, the choice of transportation means is typically based on an *informal calculus* that people do based on perceived costs and benefits of the possible options (Castellani et al. 2014). Several elements may play a role in the transportation mode decision-making:

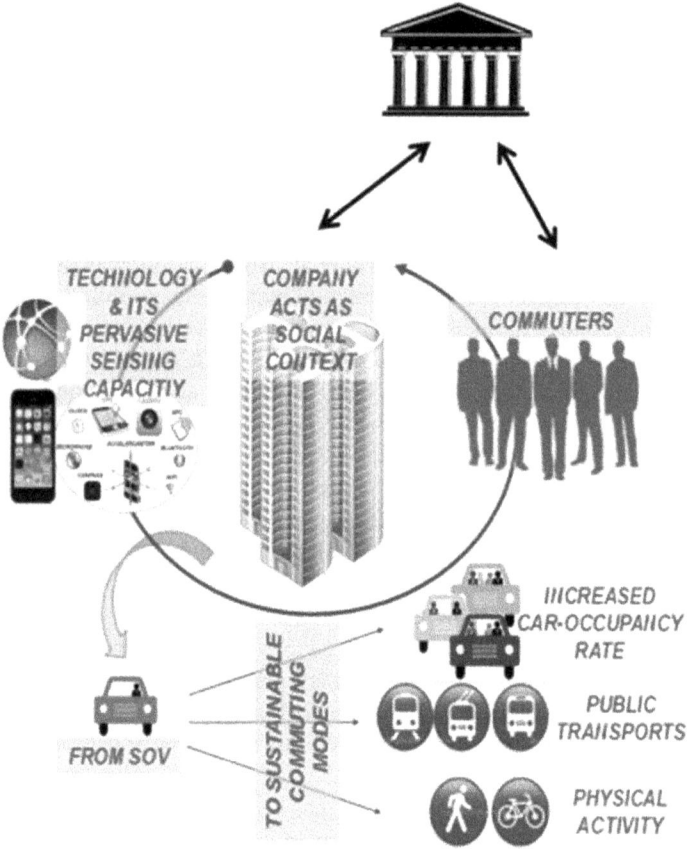

Fig. 3 Sustainable commuting vision of technology intervention

- **Reliability** and **flexibility** are fundamental factors when personal and work constraints are strong;
- **Financial cost** may be a factor of varying relevance, but avoiding "financial waste" is important for everybody;
- **Comfort** and **enjoyment** can have an influence when personal and work constraints are not too strong;
- **Sustainability** impact is often a secondary issue when compared to work related constraints or family duties that people may have.

Moreover, commuting is structured in several parts (for example it can be structured in three parts: journeys from home to workplace, during lunch breaks, and from workplace to home) with potential different needs and preferences for each part but with interdependencies that have to be dealt with. We know that there is not one mobility solution that fits all situations (He et al. 2010), given the large spectrum of infrastructural, personal, and work constraints. Also, during our studies we identified a number of categories of behaviours with different combinations of personal, work, and infrastructural constraints for which there are potentially different BCs paths (Castellani et al. 2016).

Studies also clearly showed that mobility is made up of complex routines that are not prone to be continuously reconsidered, even when levers for change exist. Once habits are settled it is very difficult for commuters to change them. However, there are moments when these changes in the organisation of commuting may be needed, for example when changes in personal life occur, e.g. family situation changes, new job starting, etc. Also, exceptional events, e.g. strikes, car breakdows, etc. and colleagues' suggestions can lead people to consider and use new ways of commuting. Different times of the year, e.g. school holidays, seasonal changes, etc. can also trigger changes in behaviour. So, commuters may be interested in tools that help them to (re-)organise commuting when they need it and that provide personalised help contextualised to their specific situation. These moments are more often propitious to behaviour change and should be intercepted to promote more sustainable commuting.

In support of the vision we devised an architectural counterpart with the services required to support it. The findings of the study suggest that a technology platform designed to gather data about employee commuting patterns and preferences, and to present data on commuting options and associated costs, trajectories, carbon footprints, etc. to employees, may indeed help organisations with implementing WTPs more successfully.

Conceptually the platform consists of three building blocks:

- A sensing layer, in charge of collecting the mobility information required by the other services.
- Personal mobility services finetuned to promote sustainable commuting. The first is a current mobility and simulated assessment support, which contains a travel diary to help commuters to visualise their commuting habits, compare actual behaviour with perceived behaviour, and reflect on them in an informed way to improve them. This layer should also provide the possibility of simulating different commuting scenarios and measuring their costs and benefits. The simulation of itineraries is done with the support of a personalised multimodal trip planner. In particular, the personalised trip planner helps commuters examine the different parts of their commuting journeys computing itineraries for users on the base of their set of preferences and constraints according to reliability, flexibility, cost, environmental impact and fitness. This layer is also in charge of spotting moments that are more propitious to change and suggest corresponding new set-ups. Such trip planning is meant to be used both during the initial commuting set-up and at propitious moments of change e.g. the car at the garage, relocation etc.
- An organisational layer, through which the company can continuously monitor the commuting practices and level of satisfaction, and then in concertation with the public authorities and by taking into account the specific needs of its

employees, provide them with information, incentives, and interventions to address the specific set-up of their work organisation.

3.3   Discussion

In this chapter we described the genesis of two research projects that originated from a research vision strongly influenced by the company culture of CSR. Without this explicit CSR context grounding and entrusting a corresponding research program and the call for research projects in the CSR domain, the same research topics would certainly not have been addressed and these projects would never have seen the light. The areas of research addressed in these projects were new and rather far from what the company's research centres had previously addressed. However, the CSR project call was sufficiently motivating for the authors to start investigating them and the general company CSR culture sufficiently strong to support them over a long enough period of time to achieve significant results.

Both projects presented aimed at facilitating BC in a workplace. Thoughout these projects we developed a deep understanding of the various *barriers* that may exist and *enablers* that may be exploited when fostering BC programs in a work context. Changing behaviours at work is complex: it happens at the frontier between work and private life and the behaviour itself is often secondary compared to work or private constraints, imperatives and objectives. In addition, top down approaches to behaviour change which are often applied in work organisations often fail. Our approach is based on the idea of enabling behaviour change rather than imposing it, and on the consideration that changing human behaviour is not just a simple matter of "persuading" people, but requires first a deeper understanding of the context in which the behaviours occurs with their related barriers, enablers, and motivations. This understanding can then inform the technical design to concretely support behaviour change in practice.

For the first project, we started by studying paper use and paper waste in the office. This in turn informed the development of PAT, a system motivating its users to reduce paper waste at work. PAT achieves this objective by providing employees with ambient awareness on their good and bad printing habits. It explicitly involves both the individual employee and the work organisation in a common effort to reduce unnecessary printing. In the second project, we applied a similar approach to the domain of mobility where the objective has been to provide the stakeholders of commuting—that is commuters, work organisations, and public administrators—with tools that facilitate the move from Single Occupancy Vehicles to more sustainable means of transportation (Castellani et al. 2014).

These two projects allowed us to acquire empirical and domain-specific knowledge on BC support at a workplace that we confronted and enriched in parallel with existing BC prior art and literature. Our approach is particularly well aligned with what Michie et al. (2011), and Patterson and Grenny (2007) propose: similar to us, they recognise that behaviour is dependent on a combination of capability, opportunity, and motivation. They recognise that a person's behaviour is not only dependent on the individual itself but also on the context, and in particular on the social environment and the existing infrastructure. As a consequence, when defining BC interventions they also consider structural enablers and barriers that determine and impact people's behaviour. Their main strength is to go beyond considering behaviour and BC as a direct result only of the person's *individual willpower*. In this sense they are well aligned with and they further detail the latest psychological models such as the model of Planned Behavior (Ajzen 1991) stating that an individual's behavioural intention cannot be the exclusive determinant of behaviour. To sum up, our approach has many parallels with theirs, but focusses in contrast on the particular context of BC in a workplace.

References

Abbott, W. F., & Monsen, R. J. (1979). On the measurement of corporate social responsibility: Self-reported disclosures as a method of measuring corporate social involvement. *Academy of Management Journal, 22*(3), 501–515.

Ajzen, I. (1991). The theory of planned behavior. *Organizational Behavior and Human Decision Processes, 50*(2), 179–211. https://doi.org/10.1016/0749-5978(91)90020-T

Castellani, S., Grasso, A., Willamowski, J., & Martin, D. (2014). Sustainable Commuting @Work. In *Proceedings of workshop on USCIAMO: Urban sustainable, collaborative and adaptive mobility.* EAI Endorsed Transactions on Ambient Systems, vol. 1, October 2014.

Castellani, S., Colombino, T., Grasso, A., & Mazzega, M. (2016). Understanding commuting to accompany work organisations and employees behaviour change. In *Proceedings of IEEE second international smart cities conference*, Trento, Italy, 12–15 September 2016.

Hardin, G. (1968). The tragedy of the commons. *Science, 162*(3859), 1243–1248.

He, H. A., Greenberg, S., & Huang, E. M. (2010). One size does not fit all: applying the transtheoretical model to energy feedback technology design. In *Proceedings of CHI'10* (pp. 927–936). New York: ACM.

Michie, S., van Stralen, M. M., & West, R. (2011). The behaviour change wheel: A new method for characterising and designing behaviour change interventions. *Implementation Science Journal, 6*, 42.

Patterson, K., & Grenny, J. (2007). *Influencer: The power to change anything.* New York: Tata McGraw-Hill Education.

Roby, H. (2010). Workplace travel plans: past, present and future. *Journal of Transport Geography, 18*, 23–30.

Sellen, A. J., & Harper, R. H. R. (2003). *The myth of the paperless office.* Cambridge, MA: MIT Press.

Willamowski, J., Convertino, G., & Grasso, A. (2014). Leveraging organizations for sustainable commuting: A field study. In: CHI'14 Workshop on "What have we learned? A SIGCHI HCI & Sustainability".

Willamowski, J., Mazzega, M., Hoppenot, Y., & Grasso, A. (2016). From eco-feedback to an organizational probe, highlighting paper affordances in administrative work. In Proceedings of COOP 2016.

Jutta Willamowski holds a Ph.D in Computer Science from the Université Joseph Fourier, Grenoble, France. She carried out her Ph.D in the Sherpa project at INRIA Rhone-Alpes After her Ph.D she held an industrial post-doctoral fellowship from INRIA and the company ILOG in Paris. In 1996 she joined the Xerox Research Centre Europe where she was first involved in projects around knowledge management, middleware, and contextual computing. Then she moved to the Computer Vision research area working on topics like image categorization and automatic image enhancement. In 2007 she joined the Work Practice Technology area where she first focussed on the facilitation of colour workflows and troubleshooting. More recently she contributed to the Print Awareness Tool, a system providing its users with feedback on their printing habits and motivating them make it more sustainable. Besides she was also involved in projects around crowd sourcing and sustainable mobility.

Yves Hoppenot is Innovation Portfolio Manager for the Xerox Research, Product Development and Engineering group where he is responsible for the technology transfer between research and the Document Outsourcing Group. He previously worked for the Xerox Innovation Studio in Grenoble, France where he enjoyed the everyday challenge of taking a research prototype and developing a demo for customers that shows what the technology is capable of—even if at an early stage of research. He joined the Xerox Research Centre Europe in 2001. Hoppenot graduated from the École Nationale Supérieure des Télécommunications, Brest in France with a degree in engineering and received a Master of Science degree in image and signal processing from Tampere University of Technology, Finland.

Stefania Castellani is Senior Scientist in the Work Practice Technology Area at the Xerox Research Centre Europe. Stefania's research interests include the creation of interaction mechanisms, semi-automatic and flexible work processes support, and collaborative and distributed systems. She has been involved in projects on adaptive workflow, negotiation infrastructures, advanced search mechanisms, legal case reasoning, remote device troubleshooting, and call centre work performance and motivation management. More recently, she has worked in the research area of Urban Mobility and the design of sustainable commuting technology to encourage behaviour change. Stefania is co-author of more than 40 publications in international conferences and journals and has more than 35 patents issued or pending.

M. Antonietta Grasso is Senior Scientist and head of the Work Practice Technology Area at the Xerox Research Centre Europe for about 10 years. In this role she has lead a few projects from research to transfer to the business groups in areas including document management, call centre agent support, urban mobility. Her research is in the area of Computer Supported Cooperative Work and Human Computer Interaction, and more specifically around ICT in support of community learning and change. Recent years focus has been around the areas of change behaviour support of paper to digital and sustainable commuting transitions. She has published more than 50 papers at international conferences and journals and she holds more than 30 patent deriving from her research. In this role as research area manager, she continuously bridges the business strategies with the research roadmaps.

GeSI

Luis Neves

1 Introduction: GeSI and the Holistic Approach to Sustainability

The Global e-Sustainability Initiative groups around 30 of the world's leading ICT and telecom companies. GeSI was initially created in 2001 at the initiative of UNEP, the United Nations Environment Programme, and its work was initially focused on supply chain and the environment. In 2006 GeSI became an independent association and, as its membership grew, started streamlining its activities and looking at other aspects of sustainability, adopting a holistic approach to CSR. Today its core activities include work on the enabling potential of ICT for climate change and human rights, and on the ICT industry's work for supply chain management.

Today, GeSI is a well-established organization with a large network of members and partners, advocating and raising awareness on the transformative power of ICT to enable the transition to a low-carbon economy.

While it is obvious that, like all economic activities, the ICT sector has an impact on the planet and on society too, it also has a unique potential to enable other industries to conduct their business in a more sustainable manner. And, given its cross-cutting nature, this enabling potential can truly be applied everywhere. About a decade ago, we felt that this positive aspect was being overlooked in discussions on climate change and business responsibilities, and that there was the need to develop a "counter-narrative", or better, a narrative able to provide a more complete picture. That was the driving idea behind the SMART series of reports.

L. Neves (✉)
Global e-Sustainability Initiative, Brussels, Belgium
e-mail: luis.neves@gesi.org

© Springer International Publishing AG, part of Springer Nature 2018 281
R. Altenburger (ed.), *Innovation Management and Corporate Social Responsibility*,
CSR, Sustainability, Ethics & Governance,
https://doi.org/10.1007/978-3-319-93629-1_16

2   The SMART Series Concept: SMART2020

In the early 2000s, discussions on ICT and climate change tended to focus on the negative aspects, predicting that the booming adoption of ICT across the world would inevitably bring a rapid increase in the sector's emissions. The positive aspects of the digital revolution, namely how those same technologies could help the world *reduce* emissions in other sectors, was not taken into account.

GeSI members enthusiastically supported the idea to develop a report addressing precisely these questions: in the fight to climate change, can ICT be a positive force? What exactly is its potential in helping the world reduce its carbon emissions? And what are the most impactful and readily available solutions?

The answers we found to these questions were quite interesting: in 2008, GeSI released *SMART2020—Enabling the low carbon economy in the information age*, the first major study identifying the significant contribution that ICT could make in the transition to a low-carbon economy.

The report found that, far from only being a part of the climate change problem, the ICT sector had the potential to actually be a fundamental part of the solution: while it predicted that the emissions from the ICT sector would represent an estimated 2.6% of the total global emissions by 2020, it also found that ICT solutions would enable other sectors to achieve significant emissions reductions, helping them avoid an estimated 7.8 gigatonnes (Gt) of CO_{2e} emissions by the same year. That represented 15% of the predicted total global emissions in 2020; in other words, the potential CO_{2e} savings the ICT sector could enable were five times larger than its own footprint.

Additionally, the report found that the adoption of ICT solutions in sectors such as energy, buildings, transport, and commerce could save up to €600 billion and create 15 million green jobs around the world, again by 2020.

The biggest opportunities identified at the time included: smart motor systems, where the electricity consumption reduction could save almost 1 $GtCO_{2e}$ in 2020, worth €68 billion; smart logistics, where improvements in the efficiency of transport and storage could save 1.5 $GtCO_{2e}$, with energy savings worth €280 billion; buildings, with a 1.7 $GtCO_{2e}$ and €216 billion to be saved from more energy-efficient living and working spaces; and energy, the largest opportunity identified in the study with its potential savings of 2 $GtCO_{2e}$ through the adoption of smart grids.

At the same time, it was fully acknowledged that the sector was rapidly growing, and that efforts were needed to ensure that the emissions growth linked to this growth was as limited as possible. GeSI members committed to reduce their direct carbon footprint within their respective organizations, and to further develop the so-called "greening by ICT" potential for other sectors through the development of increasingly energy-efficient products and services. A summary of those commitments was included in SMART2020, and a look back at 2008 shows that companies have since met, and in several cases exceeded, those commitments.

As previously mentioned, GeSI strives to communicate a holistic approach to sustainability: that is why, next to their more immediate climate change commitments, a great number of our members also demonstrated their engagement in increased and improved supply chain work, life cycle assessments, and e-waste management.

3 SMARTer2020: How Do We Make This Happen?

The SMART2020 report was extremely well received and inspired a number of follow-up studies focused on the specific potential of ICT in single countries (including Germany, Portugal, Spain, and the United States). GeSI started establishing itself as the "sustainability voice" of the ICT industry globally, and, having begun collaborations with a number of UN bodies and other international organizations, continued working to make the enabling potential message heard in different forums around the world. But that message also evolved quickly.

One of the key characteristics of our sector is, of course, the fact that it has an extremely fast development pace. New products and technologies come to the market and change lives in incredibly short timeframes, making new business models possible and creating entirely new opportunities.

From tablets and increased use of broadband networks to cloud computing and smartphones, ICT has become an even more fundamental part of the global economy. For us, this meant that it was necessary to update the SMART2020 message in order to take into account such developments and their impact, and in order to incorporate more recent data in our analysis.

We decided to maintain the 2020 horizon, while at the same time considering a broader use of ICT in other sectors: *SMARTer2020—The role of ICT in driving a sustainable future* looked at a total of six, from transport to agriculture and land use, from buildings to manufacturing, from energy to consumers and services.

The results we found were at the same time surprising and encouraging: first of all, even though the ICT sector had known a rapid expansion, this was accompanied by a significant adoption of energy-saving solutions within the industry itself, limiting the rise in emissions to the point where the direct emissions figure was lower than in the previous report (from 1.4 to 1.3 $GtCO_{2e}$, or 2.6 to 2.3% of the global figure).

The decrease in the ICT industry's own emissions was and is still driven by several factors, including the switch from PCs to smart mobile devices like phones and tablets; the increased use of virtualization; and a more efficient use of cloud computing. Considering that SMARTer2020 was released in 2012, we of course also know that the economic slowdown of the previous years, especially in developing markets, contributed to a decrease in the estimates on the use of devices in those countries.

It is when comparing the ICT industry performance with its enabling potential that the progress made in just four years becomes evident: in 2008, we had found that

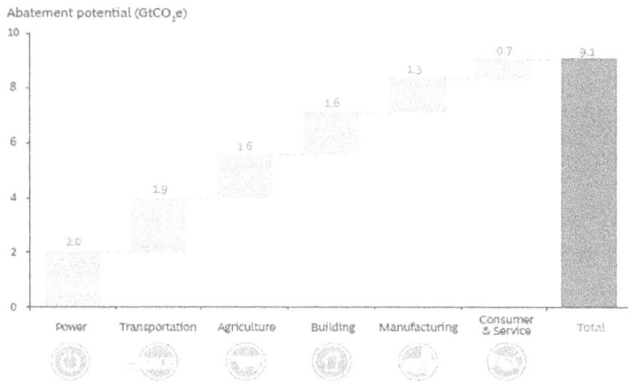

Fig. 1 Summary of estimated abatement potential by sector. Source: Smarter 2020 report

the enabling potential of ICT was five times higher than its own footprint. In 2012, SMARTer2020 quantified the total abatement potential for all the sectors and sub-levers considered in 9.1 $GtCO_{2e}$, from the 7.8 of the previous report. This means that the ICT sector enabling potential was now seven times higher than its own emissions, or 16% higher than originally thought.

As mentioned above, SMARTer2020 looked in detail at the enabling potential of ICT in six sectors, and its findings were very promising for all of them (see Fig. 1).

ICT adoption in the power sector was estimated to be able to yield 2.0 GtCO2e in abatement (or 22% of the total reductions calculated in the report), by playing a critical role in the creation of a more dynamic power market with supply and pricing swiftly responding to changes in demand, and with a greater ability to effectively integrate renewable energy into the power supply.

The emissions reductions in transportation were estimated at 1.9 GtCO2e, or 21% of the total. The increased efficiency in cargo transit through improved logistics networks and fleet management would represent a significant abatement opportunity. In addition, telecommuting and an increased use of video conferencing can reduce transportation needs and emissions.

For agriculture, the emissions reduction estimate was of 1.6 GtCO2e or 18% of the total figure. As the inputs required to grow crops emit large quantities of emissions, ICT solutions allowing farmers to accurately assess how much to irrigate and fertilize their crops can lead to a significant abatement. Systems that reduce the amount of land required to raise livestock and help reduce their methane emissions can also have an important impact. Finally, monitoring equipment can help governments prevent the destruction of rainforests that act as carbon sinks.

In the building sector, ICT applications take the form of systems required to support the generation of renewable energy and its incorporation into the building's power supply, as well as of smart design solutions reducing lighting and heating, ventilation, and air conditioning needs, while at the same time ensuring that those same systems are used efficiently. The combined effect of these applications was found to be potentially able to yield an abatement of 1.6 GtCO2e, that is to say, 18% of the total reductions calculated.

ICT-enabled efficiency in factories and other related applications (for instance, a solution controlling a motor system to better match its power usage to a required output) was estimated to be able to allow for an abatement of 1.2 GtCO2e (13% of total) from the manufacturing sector.

Lastly, the report calculated that the emissions reductions through ICT in the consumer and service sector could reach 0.7 GtCO2e (8% of total). ICT connects consumers to merchants online and allows them to purchase goods without having to physically travel to the store. In addition, ICT-enabled software can develop a more resource-efficient packaging which also generates less waste.

These figures look great on paper. But how do we translate them into reality? During the SMARTer2020 research work, we also discussed the barriers to ICT uptake and what could be done to address them.

While for some ICT solutions the market will drive further uptake, and our main task is therefore to clearly explain the relevant business case, the most significant challenge facing several ICT-based applications was found to be the lack of robust and comprehensive global policies to address climate change. It should also be noted that, while an international framework for climate change was and continues to be needed, it remains clear that effective policies at the national and local level are also needed. National and local governments are best positioned to drive GHG reductions and to enact policies accelerating the use of ICT-enabled solutions as an effective way to facilitate the transition to a low-carbon economy.

For this reason, SMARTer2020 included a deep-dive analysis of the state of play and ICT potential in seven countries, chosen to provide an overview of the various challenges and needs of different world regions: Brazil, Canada, China, Germany, India, the United Kingdom, and the United States; the country analysis also included specific suggestions and policy recommendations to decision-makers on the key actions to undertake in order to address the barriers to ICT uptake.

4   SMARTer2030: The Environment and Beyond

For global climate policy, 2015 was a crucial year: as the world was preparing to meet at the 23rd Conference of the Parties of the UN Framework Convention on Climate Change to discuss and finalize what became known as the Paris Agreement, business leaders were debating the contribution that their respective sectors could bring to the table. Indeed, COP23 was the first such meeting in which the role of business in pushing for an ambitious climate agreement was officially acknowledged and recognized.

Having started championing the enabling role of ICT almost a decade previously, GeSI did not want to miss the opportunity to advocate for our sector in such an important forum. Since 2013, GeSI has been an official partner of the UNFCCC, collaborating to the ICT Pillar of the Momentum for Change Initiative, and participating in COP meetings. In Paris, we wanted to be able to also bring the SMART message—in a renewed form.

SMARTer2030 is the latest and the most ambitious chapter in the SMART series to date. While maintaining the methodology applied in the first two studies, for consistency and in order to be able to better compare the results, we also decided to introduce a few significant novelties.

First of all, the time horizon considered was extended to 2030, to allow to present a longer-term vision; second, the number of sectors considered was brought to eight, with the addition of education and healthcare to those previously considered; third, the analysis considered a broader range of ICT solutions, once again reflecting the fast evolution of this sector; fourth, the report looked at the wider benefits of ICT beyond the most immediate climate change dimension. Thus, carbon emissions reduction was not the only environmental benefit considered, but we also took into account additional metrics such as air quality and resource management; even more importantly, we looked into the social benefits that ICT can deliver, and the growth opportunities it can trigger, thus delivering economic growth and improving people's quality of life. This approach makes all the more sense when we consider how the worldwide growth of the digital economy continues to accelerate, providing the scale necessary to drive greater connectivity and new, disruptive business models. As opposed to the old production-line economy, individuals are firmly at the center of this process, something which we wanted to be reflected in our analysis.

Once again, the findings were deeply significant and encouraging: continuing the trend first shown in the update from SMART2020 to SMARTer2020, the ICT sector's own footprint is projected to continue its decrease as a percentage of global emissions, while its enabling potential is expected to grow.

The report found in fact that ICT can enable a 20% reduction of global CO_{2e} emissions by 2030, holding them at 2015 levels. The sector's own emissions are expected to decrease to 1.97% of the global emissions by 2030, compared to the 2.3% predicted by SMARTer2020. This would result in yet another increase in the overall benefit factor: where SMART2020 found that ICT-enabled reductions in CO_{2e} could be five times higher than its own footprint, and SMARTer2020 raised that figure to seven, the total benefit factor calculated by SMARTer2030 was 9.7: in other words, ICT could enable GHG reductions almost ten times higher than its own emissions by 2030! (See also Fig. 2).

As mentioned above, SMARTer2030 also considered additional environmental benefits beyond emissions reduction: these include a potential increase of agricultural crop yields by 30%, accompanied by yearly savings of over 300 trillion liters of water and 25 billion barrels of oil.

The economic benefits deriving from the adoption of ICT solutions in the eight economic sectors considered (mobility and logistics, manufacturing, agriculture and food, buildings, energy, work and business, health, and learning) could reach $11 trillion per year, or the equivalent of China's GDP in 2015, by 2030. By the same year, ICT can also connect an extra 2.5 billion people to the knowledge economy, giving 1.6 billion of them access to healthcare and access to e-learning tools to half a billion.

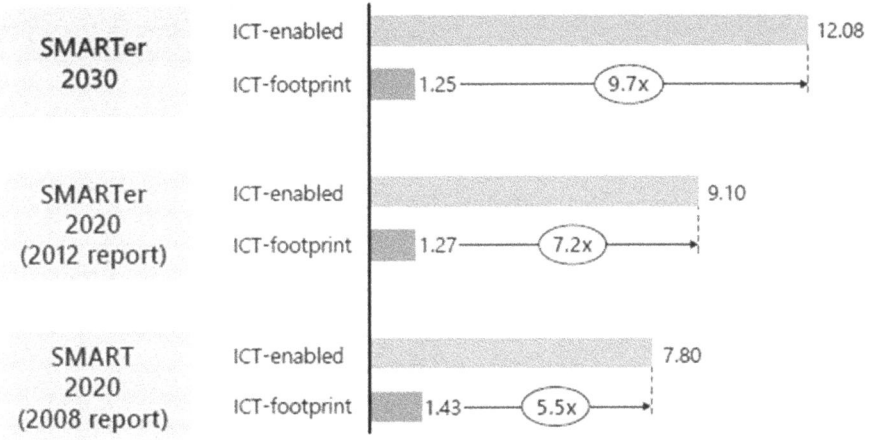

Source: WRI, IPCC, GeSI, SMARTer2020, Accenture analysis & CO₂ models

Fig. 2 Source: SMARTer2030 report

A wider uptake of ICT solutions will of course be needed in order to translate the SMARTer2030 vision into reality. For this to happen, it will be fundamental to address the barriers currently hindering a broader adoption of the technology. SMARTer2030 takes a closer look at such barriers, which can be grouped into three main categories: political, industry-related (i.e. from the supply side), and customer-related (i.e. from a demand perspective).

This means that there are three key stakeholder groups whose involvement is needed to ensure that the barriers to the widespread of ICT solutions are effectively addressed: policy-makers, business leaders, and consumers. Recommendations for each of these groups are included in SMARTer2030, and presented in a more elaborate way in the three playbooks accompanying the main report: addressed to policy-makers, industry, and consumers respectively, the playbooks clearly outline the benefits of ICT for each, and contain calls to action to remove the main barriers to adoption and thus ensure we can reap the full benefits of ICT.

Such actions include, from the policy perspective, correcting the shortcomings of the regulatory environment where inadequate; addressing the lack of investment and financing incentives; and raising awareness of the ICT potential among all relevant stakeholders; from an industry perspective, key actions will include addressing the lack of financing capital; improving cross-sector collaborations and standard integration; and investing in infrastructure. From a consumer point of view, the low affordability of ICT solutions in certain areas must be addressed, as must other country-specific barriers. Finally, we will need to work on the lack of ICT skills and a certain mistrust of the technology, often the result of a combination of scarce knowledge of its functioning and insufficient awareness of its potential benefits.

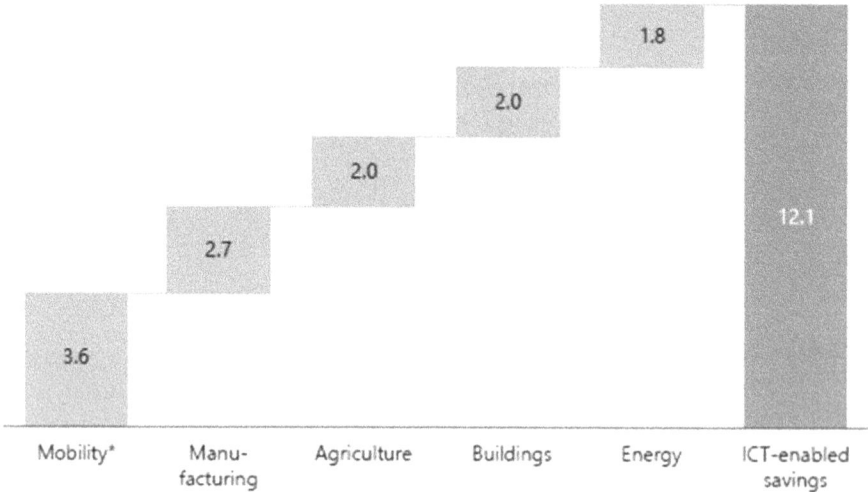

* Mobility solutions consider ICT-enabled improvements to private and commercial mobility and additionally
  consider the reduced need to travel from various sectors, including health, learning, commerce, etc.

Source: WRI, IPCC, World Bank, GeSI, Accenture analysis & CO₂ models

Fig. 3 Source: SMARTer2030 report

And such benefits are not to be overlooked: our updated sector analysis shows that the combined carbon reduction potentially enabled by ICT by 2030 is of 12.1 $GtCO_{2e}$, allocated as shown in Fig. 3.

The key findings by sectors include: for energy, smart grids, analytics solutions and advanced energy management systems can abate 1.8 Gt CO_{2e} and generate \$0.8 trillion in new revenue opportunities. In the agriculture and food sector, smart agriculture can boost yields by 30%, avoid 20% of food waste, and deliver economic benefits worth \$1.9 trillion. At the same time, smart agriculture could reduce water needs by 250 trillion liters and abate 2.0 Gt CO_{2e}.

In the building sector, smart solutions could cut 2.0 Gt CO_{2e} from the housing sector, reducing energy costs by \$0.4 trillion and creating revenue opportunities of \$0.4 trillion. In the mobility and logistics sector, real-time traffic information, smart logistics, intelligent lighting and other ICT-enabled solutions could abate 3.6 Gt CO_{2e}, including abatement from avoided travel.

When it comes to work and business, telecommuting and virtual conferencing can save not only travel but also employees time and money, while at the same time reducing the stress related to commuting. Additional revenues from e-commerce could total \$1.8 trillion and e-work could add \$0.5 trillion while freeing up 100 hours per e-worker annually. In manufacturing, smart solutions including virtual manufacturing, customer-centric production, circular supply chains and smart services could abate 2.7 Gt CO_{2e}.

Finally, ICT could deliver e-health services to 1.6 billion people across the developing and developed world, and create 450 million new e-learning participants in 2030, helping to raise incomes by 11% on average per e-degree.

5   A Case Study from SMARTer2030: Germany

As done in SMARTer2020, SMARTer2030 included a number of deep-dives assessing different ICT solutions in different countries around the world, each with its specific needs and challenges.

With a GDP of close to $4 trillion and a population of 81 million, Germany is the largest national market in the EU, the largest economy in Europe and the fourth largest in the world. The German energy policy aims to reduce GHG emissions by 40% by 2020, and by 80–95% by 2050. With 31%, renewable generation in 2015 had surpassed that of coal (at 26.1%). Wind energy produces 8.9% of total German energy consumption and Germany was by far the biggest market for wind technology in 2014, installing nearly half of new wind farms in the EU. A major policy goal for the country is its commitment to phasing out its nuclear generators by 2022.

Our analysis showed that ICT had an important role to play in helping the country meet its sustainability targets. For example, the buildings sector is responsible for 40% of the primary energy consumption in Germany, and approximately 33% of emissions. The German government has set ambitious energy reduction goals for the buildings sector, including an 80% reduction in primary energy use in buildings by 2050. A high number of private cars means that fuel use is another area where the opportunity to save is high.

In terms of mobile phone penetration, Germany has already reached the saturation point with 121 mobile phones per 100 people. Smartphones meanwhile are picking up fast with a 50% penetration rate in 2014, up from 41% in 2013 and an expected 80% by 2017. In parallel, internet usage through mobile devices has grown to 54% from 40%. Some 843,000 people work in the digital sector in Germany, which is also one of the world's leading exporters of high-tech products.

While the country is progressing quickly towards its goals, there are a number of opportunities across sectors where ICT could play an even bigger role and help bring transformational solutions for sustainable growth.

The sectors in which ICT could have the greatest impact in Germany are smart energy, smart manufacturing, and smart mobility. Our analysis demonstrates that ICT is fundamental to realizing "Energiewende" (the energy transformation program) by enabling the full integration of renewables onto the smart grid. Likewise, smart manufacturing could benefit an economy which still retains a strong manufacturing base, and connecting Germany to the Internet of Things, or Industry 4.0, is an ICT-driven priority with massive growth potential. Last but not least, the German automotive sector can improve its mobility solutions through ICT by scaling up car-sharing solutions and testing fully connected infrastructure with a view to rolling out automated driving.

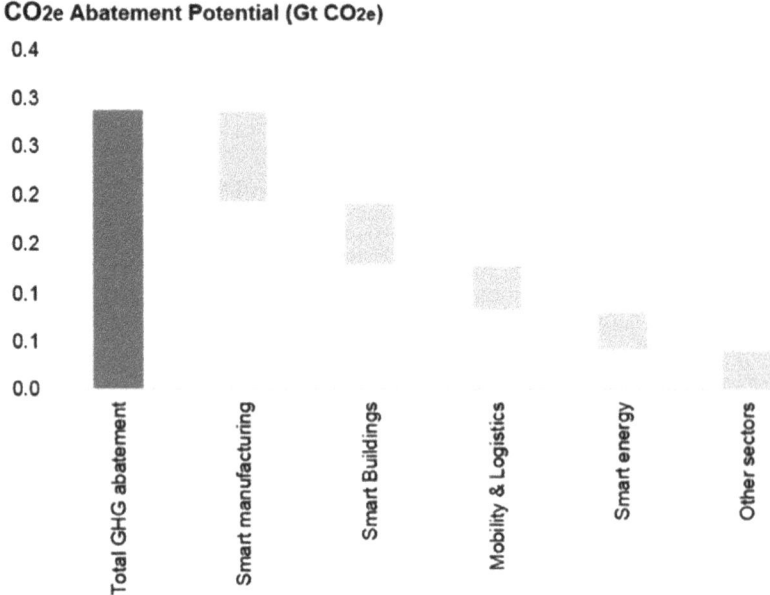

Fig. 4 Germany—Total CO2e abatement potential of ICT across sectors. Source: SMARTer2030 report

Across the eight sectors analyzed in SMARTer2030, our research has identified a total abatement potential of 0.29 GtCO$_{2e}$ for Germany by 2030, with smart manufacturing and smart buildings accounting for more than 60% of these total savings (see Figs. 4 and 5).

6 The Challenge Ahead: Delivering the Sustainable Development Goals

When GeSI was founded back in 2001, its mission referred to the UN Millennium Development Goals as a starting point. It was natural therefore that, as these were replaced by the Sustainable Development Goals (SDGs), GeSI continued to remain involved in the discussion.

In 2014, as the UN General Assembly was preparing to officially adopt the SDGs, GeSI and the International Telecommunications Union (ITU) issued a joint call to policy-makers, inviting them to acknowledge ICT as one of the key implementation means for the achievement of the Goals.

After the release of SMARTer2030, we decided to look at the report findings against the SDGs, to assess where ICT could be most relevant. The #SystemTransformation report, released in 2016, found that digital solutions will

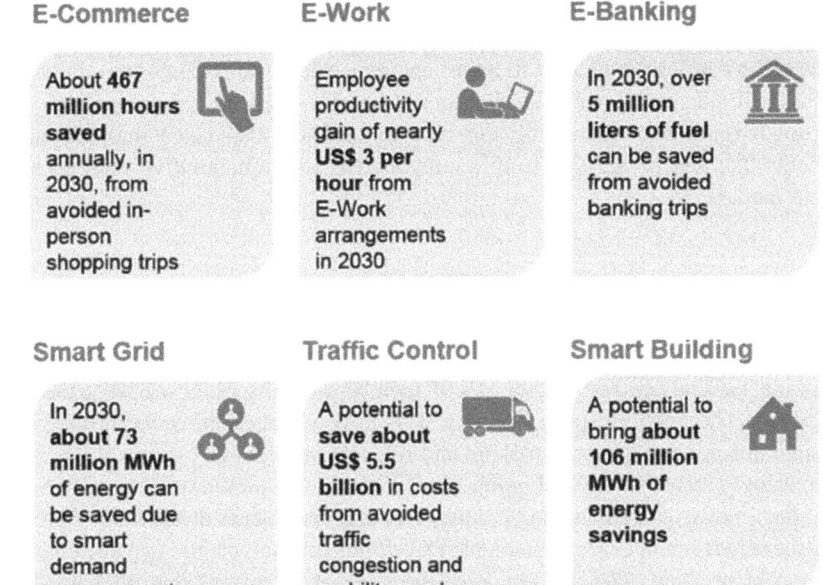

Fig. 5 Germany—ICT can play a significant role in reducing different sector's sustainability impacts. Source: SMARTer2030 report

be indispensable to achieve all 17 SGDs, and more than half of the related 169 targets. The report identified six SDGs where the impact of ICT can be more immediate and significant: these are SDG 3 (good health and well-being), 4 (quality education), 8 (decent work and economic growth), 9 (industry, innovation and infrastructure), 11 (sustainable cities and communities), and 13 (climate action).

With the deadline for the achievement of the SDGs set to 2030, we have an ambitious and difficult task ahead of us. The launch of the #SystemTransformation report was also the occasion for GeSI to officially adopt the SDGs as the overarching framework for our activities up to 2030, in the spirit of that holistic approach to sustainability which has inspired and guided the organization over the last decade. The evolution of that approach is well summarized by the SMART series, initially focused on filling a gap in the narrative about climate change and now developed into a 360°C assessment of the enabling potential of ICT.

In the same spirit, we have also continued our activities on the responsible business side: we are conscious of the fact that, if we want to lead by example and inspire other sectors as well as policy-makers to keep ICT high on their list of transformative solutions, we need to prove our commitment to making the difference.

For this reason, these last few years we have been working with our members to implement and improve a number of tools aimed at supporting them in managing their supply chain and risks thereof, including those related to human rights. Long and complex supply chains like those of ICT companies can represent a continuous challenge and a resource drain, especially when located in developing countries, and need to be carefully managed to avoid negative effects on local communities.

In this field as well, we believe in an opportunity-oriented approach which, while acknowledging and addressing the industry impact, also takes into account the benefits that can be derived from a constructive and collaborative view of supply chain management.

7 Conclusions

Looking back at the last decade, GeSI members can be proud of what they have achieved. The enabling potential of ICT is now a part of the story in the climate change debate, and decision-makers and relevant stakeholders across the world are increasingly aware of the technology potential in meeting the challenges ahead.

Much, however, remains to be done. The implementation of the Paris Agreement and the achievement of the Sustainable Development Goals are the two key challenges the world set itself. The magnitude of the work ahead means that no actor—State, industry, civil society—can deliver it alone. The role of business in supporting climate change work was fully recognized at COP21 and COP22, and it is our intention to continue delivering on our commitments in this field.

From an industry perspective, being sustainable and supporting the transition to a low-carbon economy means that CSR needs to become central and fully integrated into a company's approach; far from being a PR or even a green-washing exercise, the concept of making business sustainably needs to be at the core of how businesses operate. Industry organizations can play a key role in this effort by providing platforms for dialogue, sharing lessons learned, and working together to tackle those challenges that companies alone cannot solve.

8 About GeSI

The Global e-Sustainability Initiative (GeSI) is a strategic partnership of the Information and Communication Technology (ICT) sector and organizations committed to creating and promoting technologies and practices that foster economic, environmental and social sustainability. Formed in 2001, GeSI's vision is a sustainable world through responsible, ICT-enabled transformation. GeSI fosters global and open cooperation, informs the public of its members' voluntary actions to improve their sustainability performance, and promotes technologies that foster sustainable development.

GeSI enjoys a diverse and global membership, representing around 30 of the world's leading ICT companies and partners with over 20 global business and international organizations such as the International Telecommunications Union

(ITU), the United Nations Framework Convention on Climate Change (UNFCCC), the United Nations Environment Program (UNEP), the World Business Council for Sustainable Development (WBCSD), the World Resources Forum Association (WRFA), the Institute of Electrical and Electronics Engineers (IEEE), the Responsible Business Alliance (RBA), the Centre for Sustainable Consumption and Production (CSCP), The Institute for Sustainable Development and International Relations (IDDRI), the Global Climate Forum (GCF), the World Green Building Council (WGBC), the Flemish Institute for Technological Research (VITO), and the International Energy Agency (IEA)—as well as a range of international stakeholders committed to ICT sustainability objectives to share and develop ideas, launch joint initiatives, and collaborate on a broad range of sustainability projects. These partnerships help shape GeSI's global vision regarding the evolution of the ICT sector, and how it can best meet the challenges of sustainable development.

For more information, see www.gesi.org.

Luis Neves was born in Portugal. In 1975 he finished his University degree in History. In May 2004 Luis Neves joined Deutsche Telekom. As from December 1st 2008 he was assigned to the position of Vice President Corporate Responsibility. Effective February 2012 he was appointed to the position of Group Sustainability and Climate Protection Officer, a position that he held until October 2017, when he was appointed Managing Director of the Global e-Sustainability Initiative (GeSI). Luis Neves has been playing a fundamental role in promoting the role of Information and Communication Technology (ICT) in relation to Climate Change and more broadly to Sustainability. In 2006 he was elected Chairman GeSI, which he successfully developed to become a globally recognised organisation in the sustainability area.

In addition, throughout the years Luis held positions in many organizations and initiatives including Steering Committee Member of the United Nations Global Compact Lead Group, Steering Committee Member of the United Nations "Caring for Climate" Initiative, Member of the ERT—European Round Table of Industrialists Climate Change Group, Member of the SAP Sustainability strategy, reporting & assurance Board Advisory Panel, Co-chair of the ICT4EE Forum, Member of the ICC Commission on Environment and Energy and Green Economy Task Force Meetings, Member of the Steering Committee of "Econsense", The German Sustainability Association, Member of the Environment and Sustainability Steering Committee of BITKOM, the German ICT Association, Advisory Board member of the UNFCCC Momentum for Change Initiative, Member of the Leadership Team of the Joint Audit Cooperation (JAC) initiative, Jury member of GreenTec Awards, member of the Advisory Council of B.A.U.M. e.V., and Board Member of the World Resources Forum Association.

Sustainable Entrepreneurship: Family Firms as Sustainability Pioneers

Sylvie Scherrer and Claudia Binz Astrachan

1 Introduction

Future-oriented companies have long recognized the call for sustainability to be an important strategic success factor rather than a short-term trend (Lacy et al. 2010; Nidumolu et al. 2009; Schaltegger et al. 2011). Younger customers in particular are increasingly demanding intelligent and sustainable product and service offerings, and employees prefer working for companies with a proven sustainability track record (Jenkin 2015). First signs of this re-evaluation of values have already become apparent—fair trade fashion has never been so chic, organic food has never so in demand, and sustainable employers have never been more popular than they are today (Zukunftsinstitut 2016).

Current research also shows that a sustainability orientation can benefit a business economically, particularly when combined with entrepreneurial thinking and action, as well as a strong innovation orientation (Schaltegger and Wagner 2011). In contrast to *Corporate Social Responsibility* (CSR) or Corporate Sustainability—two approaches that are already established in research and practice—Sustainable Entrepreneurship is characterized by a strong market-based as well as innovation-oriented approach to sustainability. Young small and medium-sized companies such as the Swiss car-sharing company *Mobility* or bag manufacturer *Freitag* as well as established, large companies like *Philips* or *Unilever* demonstrated just how successful a sustainable entrepreneurship strategy can be. While these businesses differ in many aspects, they are all considered pioneers in their respective industry—a status they gained by continuously and consciously recognizing sustainability challenges as a source of inspiration for the development of new products, services and business

S. Scherrer (✉) · C. B. Astrachan
Lucerne Business School, Luzern, Switzerland
e-mail: sylvie.scherrer@hslu.ch; claudia.astrachan@hslu.ch

© Springer International Publishing AG, part of Springer Nature 2018
R. Altenburger (ed.), *Innovation Management and Corporate Social Responsibility*,
CSR, Sustainability, Ethics & Governance,
https://doi.org/10.1007/978-3-319-93629-1_17

models. These diverse company types thereby reflect one of the key challenges the novel research field of Sustainable Entrepreneurship struggles with, namely the *question of what both innovative start-ups that were born with a strong sustainability orientation, as well as established, large multinationals that have to radically change in order to integrate a sustainable mindset in their way of thinking, are able to contribute to the sustainability transition* (Hockerts and Wüstenhagen 2010).

Even though Sustainable Entrepreneurship may seem to be an intuitively appealing and easily comprehensible concept, the implementation of such a strategy is nevertheless challenging. The implementation of a Sustainability Entrepreneurship strategy usually leads to fundamental structural and cultural changes within the organization. The process of discovering, developing and utilizing new opportunities that generate an economic, social and environmental value is also complex, uncertain and risky (Cohen and Winn 2006; Dean and McMullen 2007; Shepherd and Patzelt 2011).

Against this background, it is perhaps not surprising that such a young field of research has hardly dealt with the idiosyncrasies of family businesses. For readiness to change, risk-taking or innovation capacity are concepts that are not frequently associated with family businesses. Instead, family-owned companies are often said to be rather change resistant (Miller and Le Breton-Miller 2005). Yet, many family firms have longstanding traditions in social responsibility (Dyer and Whetten 2006), and upon closer consideration, many family firms are indeed pioneers of sustainable entrepreneurship in their regions and industries.

Amongst other things, this may be caused by the inextricable connection between the family and the business, which fosters the enterpreneurial family's interest in social engagement and their desire to consciously and responsibly assume their role as well as the role of their company in society (Niehm et al. 2008). Prior research finds, for example, that family-run companies are significantly more involved in social initiatives than non-family businesses (Bingham et al. 2011) and behave more sustainably than non-family companies (Campopiano and De Massis 2015; Dyer and Whetten 2006). The family's social engagement can serve to protect the reputation of the family and the business; and it may satisfy altruistic motives, for example, to serve a higher purpose than money-making (*"spiritual wellness"*, Dyer and Whetten 2006, p. 789). Families that strongly identify with the company are particularly interested in reiterating of a positive self-image: hence, they want their company to be perceived as particularly responsible and sustainable. Furthermore, qualities such as financial independence, a long-term horizon, the absence of a quarterly thinking driven by shareholder interests, and the ability to quickly take strategic decisions make family companies—at least in theory—an ideal breeding ground for Sustainable Entrepreneurship.

This article therefore considers the concept of Sustainable Entrepreneurship in the context of family businesses. We focus on those family businesses that have a long and successful history, and at the same time have the objective to hand over a successful company to the next generation. Based on qualitative interviews with the owning families of three long-standing Swiss family businesses, we illustrate how different forms of sustainability-oriented entrepreneurship can be promoted success-fully across generations.

2 Sustainable Entrepreneurship: A Multidimensional Concept

Sustainable entrepreneurship suggests that profitable opportunities for for sustainabil-ity often arise when markets fail (Dean and McMullen 2007). Therefore, inefficient companies, externalities, faulty price mechanisms and information asymmetries—if recognized—provide valuable opportunities (Cohen and Winn 2007) that resourceful entrepreneurs and organizations can exploit by using their innovation capacity, proactivity and willingness to carry risk to successfully develop new solutions to existing sustainability issues (York and Venkataraman 2010).

Today, researchers from different research fields agree that sustainable entrepre-neurship may be one of the most promising drivers for a more sustainable future (Pacheco et al. 2010). Yet, it remains unclear how to define the versatile and multi-dimensional concept of Sustainable Entrepreneurship—something that is indeed characteristic for an emerging and multi-disciplinary field (Pacheco et al. 2010; Shepherd and Patzelt 2011).

Existing definitions of sustainable entrepreneurship, on the one hand, vary in their understanding of sustainability. Although the majority of work represents a holistic, integrated understanding of sustainability, definitions usually focus on partial aspects of sustainability, such as environmental sustainability (Dean and McMullen 2007; Hockerts and Wüstenhagen 2010) or social justice (Wagner 2010). On the other hand, current literature attributes different roles to to the concept, at different levels: e.g., Sustainable Entrepreneuship is described as a driver of social change towards sus-tainability (Hörisch 2015), as the changing force of whole industries and markets (Hockerts and Wüstenhagen 2010; Shepherd et al. 2013), as an innovation driver in existing companies (Adams et al. 2016; Schaltegger and Hansen 2013; Schaltegger and Wagner 2011), or as a general value-orientation in organizations. In an effort to integrate these different perspectives, Shepherd and Patzelt (2011) suggest that Sustainable Entrepreneurship "*is focused on the preservation of nature, life support, and community in the pursuit of perceived opportunities to bring into existence future products, processes, and services for gain, where gain is broadly construed to include economic and non-economic gains to individuals, the economy, and society*" (p. 142).

However, the process of sustainability-oriented entrepreneurship differs greatly between established firms and new ventures. While new companies are often founded with a strong sustainability vision at their heart, the introduction of a *Sustainable Entrepreneurship* mindset in established, profit-oriented companies may be more complex, since a conscious effort must be made to make sustainability an integral part of the innovation process, which often demands a fundamental change in corporate philosophy as well as in products, processes, and practices (Adams et al. 2016). Sustainable Entrepreneurship in established companies therefore not only means incremental improvements in processes and products, but also a fundamental rethinking and adaptation of the business model, and more broadly of the role of the company in society (Schaltegger and Hansen 2013).

This might explain why current research on Sustainable Entrepreneurship focuses either on new ventures that were strongly oriented towards sustainability from their beginnings (e.g., examples from the fashion industry include *Mud Jeans, Carpasus,* or *Fabric Denim*; examples from the food industry include *Winnow* or *Original Unpacked*), or on established large companies that aim to transform their business model towards sustainability (e.g., *GE, Philips, Walmart, IKEA*). It is striking, however, that Sustainable Entrepreneurship has hardly been investigated in the context of family-owned enterprises thus far, even though many sustainability pioneers—both younger and smaller as well as older and larger companies—are family-owned. Given the ubiquity and idiosyncracies of family firms, we suggest that the exploration of Sustainable Entrepreneurship in family firms may provide interesting insights.

3 Family Businesses: Pioneering Sustainable Entrepreneurship

Family enterprises are usually little-known for their innovative strength, risk-taking and proactivity. They are often accused reluctance to change and low risk-taking, which can significantly hinder entrepreneurial activity (Zahra 2005). On the other hand, family businesses can rely on specific characteristics and resources, such as the rapid decision-making, financial independence, reputation, sound market knowledge and a long-term time horizon (Miller et al. 2016; Chrisman et al. 2015; De Massis et al. 2015), which may foster entrepreneurship (Aldrich and Cliff 2003; McCann et al. 2001; Miller et al. 2016) and innovativeness (Chrisman et al. 2015; De Massis et al. 2015; Duran and Kammerlander 2015; König et al. 2013).

Accordingly, the innovation process in family enterprises tends to differ from non-family firms in that they prefer relatively safe and foreseeable innovation processes, which they can finance with their own resources (Alberti and Pizzurno 2013; Chrisman and Patel 2012; De Massis et al. 2015). This preference can partly be explained by their desire to retain long-term family control and the remain independent of external financiers (Gomez-Mejia et al. 2011). However, if the family business' longevity is at risk, family businesses often choose selected, radical innovation projects which they pursue quickly and consistently (Allison et al. 2014; König et al. 2013), and that yield a high innovation output with a comparably low innovation input (Duran and Kammerlander 2015). Paradoxically, family businesses often have better skills, but a lower willingness to innovate than non-family businesses (Chrisman et al. 2015). Similarly, recent research shows that family enterprises tend to be very successful in translating their sustainability commitment into successful innovations (Craig and Dibrell 2006; Wagner 2010). This further suggests that family businesses may be able to build on unique success factors to foster Sustainable Entrepreneurship and create an environment for innovation.

Subsequently, we present a series of case studies with long-lived Swiss family businesses that have all been recognized as particularly innovative in their respective industries, and which acted in different ways as pioneering sustainability entrepreneurs. The case studies serve to showcase how family businesses can create an ideal environment for Sustainable Entrepreneurship.

4 Enabling Sustainable Entrepreneurship in Family Businesses

For this contribution, we interviewed several traditional Swiss family businesses that are generally regarded as particularly innovative in their respective markets. The interviews with members of the owning family as well as management-level employees lasted between 50 and 80 minutes, and were analyzed by means of qualitative content analysis (Yin 2003).

4.1 Family Values as the Foundation of Sustainable Entrepreneurship

Few family businesses have a well-defined Corporate Social Responsibility (CSR) vision or a formal strategy. It seems as though for most family firms, being a good corporate ciziten and 'doing right' by their values as well as their stakeholders is something that comes to these companies naturally—as they strive to protect and maintain a positive family and business reputation (Dyer and Whetten 2006). Values such as decency, transparency, honesty, trustworthiness and reliability then serve as a moral compass for socially responsible and sustainability-oriented behavior.

Family businesses have long-term goals that span across several generations; on the one hand, these goals ensure that the ancestors' heritage is preserved, on the other hand, they ensure the firm's ability to succeed for future generations. This perspective creates an environment that fosters the desire to creative innovative, sustainable, socially responsible and resource-friendly solutions. The subsequent case of *Victorinox AG* shows how a value-oriented family business can remain true to their values as a socially minded employer, despite a demanding market environment.

Case 1: Victorinox AG

In 1884, Karl Elsener founded a knife shop in a remote region of Central Switzerland to counteract the increasing emigration to the city. Until today, the world-famous red Swiss Army Knife with the white cross is produced in Ibach, a village in the small mountain canton of Schwyz. Over the last four generations the small knife shop has been transformed into a globalized and diversified company with approximately 1700 employees worldwide, including 900 in Ibach. While the company is performing well today, the family has faced difficult periods over the course of the last 130 years: in the aftermath of September 11, 2001, and the resulting ban on pocket knives in air travel, the company's sales fell by more than 30%. The survival of the Swiss production site was suddenly in danger. Although the Elsener family knew that the adaption to this new market situation would take time, they decided that the dismissal of their employees for economic reasons was not an option. As a consequence, the company 'lent' around 60 *Victorinox* employees to other companies in the area, and even organized a bus transfer to their temporary workplace. Carl Elsener Jr., CEO of the fourth generation, is convinced that this experience has further reinforced the cohesion of the *Victorinox* family. It further exemplifies how the company regularly renounces on short-term profits to stay true to its social mission. Strongly anchored in Christian values, the management of *Victorinox* often relies on the Bible in difficult situations. These values, combined with a deep connection to nature, also define *Victorinox's* sustainable approach to natural resources. Thanks to large investment in process innovations, the recycling pioneer is now able to filter around 600 tons of steel per year—about a quarter—from the cooling water. The heated cooling water is in turn used to heat the business buildings and the apartments of the employees. In addition to process innovations, the *Victorinox* concept of sustainability also leads to new products, such as the pocket knives launched in June 2016 that are made from recycled Nespresso capsules.[1] Lastly, since 2000, 90% of the company belongs to the company's foundation, which intends to prevent the company from being financial risks related to inheritance and succession issues. 10% of the company shares are held in a non-profit foundation that serves philanthropic purposes.

The strong value orientation and long-term time horizon provide the Elsener family with much needed support, courage, and stamina to accept profit setbacks, while actively seeking solutions that are in-line with their values.

Today, *Victorinox* is flourishing—according to Carl Elsener, not least because of the employee's commitment. The family aims to make the employees feel like being part of the *Victorinox* family, partly through their own modesty: There is no special

---

[1]https://www.victorinox.com/ch/de/%C3%9Cber-uns/Unternehmen/Nachhaltigkeit/Alles-%C3%BCber-Nachhaltigkeit/cms/infotainment

treatment for the management level, no bonuses, and the CEO earns a maximum of five times more than the employee with the lowest salary. Their efforts seem to be appreciated by employees that care about the company and indeed, many of *Victorinox'* innovations stem from employees' ideas and suggestions.

The value-orientation of the owning family—shared by the workforce—can be a driver and a central resource for entrepreneurial sustainability: family firms that uphold values (such as modesty or fairness) that may be at odds with a general for-profit orientation may be more likely to use their abilities and resources in an innovative way towards promoting socially responsible and environmentally sustainable solutions.

4.2 Long-Term Orientation as a Driving Force Behind Sustainable Entrepreneurship

Family businesses are known to think in generations rather than quarters—a distinct long-term orientation is hence a central characteristic associated with family ownership. The desire to hand over a successful and future-oriented company to the next generation strongly affects the entrepreneurial behavior of both the business and the owning family. For example, investments in new products, processes or business lines are decided against the background of their contribution to the company's long-term viability. Long-term orientation also drives entrepreneurial families to maintain stable, long-term relationships with their various stakeholders such as customers and suppliers: these strong partnerships again help in the development of future-oriented products and services.

However, the *conditio sine qua non* for the long-term existence of a family business is the willingness of the next generation to continue the company. This requires the development of a cross-generational vision for family and enterprise that meets the needs of all generations involved (Cohen and Sharma 2016). An integral part of such a process is the definition of compelling goals for the family that go beyond the mere ownership of the organization. In many cases, these goals revolve around long-term family cohesion or the desire to use economic success for a good cause that is dear to the family (e.g., cancer research, development assistance). The joint pursuit these goals that go beyond profit aspirations promotes the bond that glues the family together, and it also strengthens the family's identification with the company. The company may therefore serve as a vehicle to achieve these non-financial goals.

The case study of *UMB AG Bottighofen* shows how sustainability-oriented entrepreneurship can be used not only for economic success, but also to contribute to the family cohesion.

Case Study 2: UMB AG Bottighofen
In 1868, the Munz family started their mill operation at Lake Constance. More than 120 years later the entrepreneurial family was forced to close the factory for economic reasons. An industrial reorientation was difficult, because in the meantime, the surroundings of the mill had become a protected residential and recreation area. For the family, however, it was clear that they their entrepreneurial endeavors to continue. The Munz family therefore founded a real estate company and converted the former production site into a high-quality residential complex, thus re-cultivating the land near the lake. In addition to the carefully restored buildings and a museum, the façades of the core buildings are reminiscent of the mill's 1000-year history. With the advent of the seventh generation and its desire to contribute actively to regional development, the young generation decided to revive the former space in a more environmentally friendly and social orientation, hence opening a restaurant on site and creating a co-working space for local entrepreneurs and creatives. These two initiatives had a positive impact on the public perception of the family and the company. However, they had the greatest impact on the owning family: as members of different generations worked on the realization of these projects, they once again felt involved in the family business. The perception of the family business within the family also changed—suddenly the crisis of the nineties was no longer the focus, but instead the possibility to extend the almost 150-year-old heritage of the ancestors.

It often seems to be the case that the younger generation acts as the driving force behind the development of new, sustainability-oriented offers (Delmas and Gergaud 2014). The case of *UMB AG Bottighofen* exemplifies how working together on a project that all family members feel passionate about fosters cohesion in the family. This is particularly the case if the project work not only serves financial objectives, but also creates an active, value-generating contribution for employees, customers, and the environment. Thus, sustainability-oriented entrepreneurship can serve as an effective strategy for securing non-financial, cross-generational goals.

4.3   Being Innovative as a Way of Life

Successful family businesses build on existing strengths, and often recollect the pioneering and entrepreneurial spirit of their ancestors. This leads to an innovation-affirming attitude within the family, and a strong innovation culture within the company. It is the combination of the long-term perspective and the strong value orientation that leads these entrepreneur families to focus on the development of new business opportunities rooted in sustainability. Furthermore, their strategic flexibility and financial independence allows most family businesses to pursue long-term innovation projects that require high investments, yet may not produce results in the short term. This may explain why certain family businesses—like *Trisa* in the following case—can still be successful in highly competitive markets. The only chance to face the pressure of the global competitors is the use of unique technologies, and the development and maintenance of a strong innovation culture.

Case Study 3: Trisa AG

The founding of Trisa AG in 1887 was characterized by a great deal of pioneering spirit and trust in a promising future. Today, Trisa AG is a fourth-generation manufacturer, and one of the leading suppliers of brush products in the world of oral, hair and personal care. Around 1100 employees produce more than one million toothbrushes per day, in Switzerland. Around 95% of all products are exported. The entrepreneurial Pfenninger family recognized early on that any company with a Swiss production site would only remain globally successful by virtue of its innovative power and technological market leadership. In the middle of the 1960s, the owning family therefore decided to take a drastic step and began to heavily invest in high-tech production facilities. However, the family felt that it was even more important to secure the innovative power and commitment of its employees. As a result, 30% of the family's shares were sold to employees, who from then on also constituted half of the Supervisory Board. The transfer of the family's innovative power to the workforce laid the foundation for a participatory, innovative and performance-oriented corporate culture. The long-term orientation of the broad owners' circle allows for a comprehensive, innovation-oriented understanding of sustainability, which is now part of the *Trisa* brand. This basic understanding promotes an environment in which sustainability-oriented, innovative product can flourish.

The desire to serve the customer and at the same time protect the family's reputation leads to a strong quality and customer orientation in many family firms, which allows these companies to fulfill their promise to customers, which in turn strengthens customer trust and facilitates the development of stable and long-term customer relationships. In combination with sound knowledge of market and consumers, this makes it easier for the company to anticipate customer needs. Today, customers are increasingly asking for resource-conserving, long-lasting products and service solutions, that possible also serve a social purpose (Adams et al. 2015)—which is is in-line with the values many family businesses uphold. The *Trisa* example illustrates how sustainability can be a key competitive advantage for a company as a dimension of an innovative and high-quality service.

5 Concluding Thoughts

Traditional value-oriented family businesses are experiencing a renaissance in the twenty-first century. After having largely been ignored in scientific research and portrayed as an outdated business model in popular magazines over the past decade, they are now seen as future-oriented model companies, and as an inspiring alternative to the profit- and short-term oriented public companies. The question therefore arises as to what new perspectives are opened by researching family businesses that have successfully endured for generations, and for whom sustainability and innovation have always been an integral part of their corporate culture.

Firstly, family businesses use their strong values consciously (and courageously) as a reliable moral compass for sustainability-oriented decisions. In turbulent times, values provide continuity and stability. Furthermore, decision-making—which often requires companies and families to make sacrifices—also helps to maintain these values by the willingness and ability to develop creative and new solutions with a social added value. However, a strong value orientation is not a family business idiosyncracy. *Each company has their own values that serve as guidelines for action for employees and which can foster a climate that promotes innovation and sustainability. It is however essential that the management does not only communicate these values, but lives the values authentically and consistently in everyday life—which is something that many family-owned businesses excel at.*

Second, transgenerational thinking is part of family business DNA. A transgenerational outlook—a central aspect of a holistic sustainability approach—is deeply embedded in family businesses. The company vision is developed and grown with the desire to build and maintain a successful, sustainable company that future generations can also be proud of. *A departure from mere short-term profit-making, with the aim of realizing a vision that transcends pure shareholder value, is an important step towards a sustainable company.*

Thirdly, family enterprises rely on their ability to innovate to achieve this vision of sustainable success. In the development of innovative and sustainability-oriented solutions, family enterprises can count on stable and trust-based relationships with their stakeholders. *Successful family businesses often promote and cultivate a participatory innovation culture, which benefits from the loyalty and motivation of the workforce as well as strong relationships with important stakeholders such as customers and suppliers.*

Ultimately, because of the inextricable connection between the family and enterprise systems—both of which serve a different purpose—family businesses are experienced in dealing with conflicting situations: they regularly trade off short-term profits for long-term success, or make decisions in favor of family harmony instead of business profits. *Successful family businesses excel in exploiting these inherent paradoxes as sources for new entrepreneurial solutions.*

References

Adams, R., Jeanrenaud, S., Bessant, J., Denyer, D., & Overy, P. (2016). Sustainability-oriented innovation: A systematic review. *International Journal of Management Reviews, 18*(2), 180–205. https://doi.org/10.1111/ijmr.12068

Alberti, F. G., & Pizzurno, E. (2013). Technology, innovation and performance in family firms. *International Journal of Entrepreneurship and Innovation Management, 17*(1/2/3), 142. https://doi.org/10.1504/IJEIM.2013.055253

Aldrich, H. E., & Cliff, J. E. (2003). The pervasive effects of family on entrepreneurship: Toward a family embeddedness perspective. *Journal of Business Venturing, 18*(5), 573–596. https://doi.org/10.1016/S0883-9026(03)00011-9

Allison, T. H., McKenny, A. F., & Short, J. C. (2014). Integrating time into family business research: Using random coefficient modeling to examine temporal influences on family firm ambidexterity. *Family Business Review, 27*(1), 20–34. https://doi.org/10.1177/0894486513494782

Bingham, J. B., Dyer, W. G., Smith, I., & Adams, G. L. (2011). A stakeholder identity orientation approach to corporate social performance in family firms. *Journal of Business Ethics, 99*(4), 565–585. https://doi.org/10.1007/s10551-010-0669-9

Campopiano, G., & De Massis, A. (2015). Corporate social responsibility reporting: A content analysis in family and non-family firms. *Journal of Business Ethics, 129*(3), 511–534. https://doi.org/10.1007/s10551-014-2174-z

Chrisman, J. J., & Patel, P. C. (2012). Variations in R&D investments of family and nonfamily firms: Behavioral agency and myopic loss aversion perspectives. *Academy of Management Journal, 55*(4), 976–997. https://doi.org/10.5465/amj.2011.0211

Chrisman, J. J., Chua, J., Massis, A., de Frattini, F., & Wright, M. (2015). The ability and willingness paradox in family firm innovation. *Journal of Product Innovation Management, 32*(3), 310–318.

Cohen, B., & Winn, M. (2007). Market imperfections, opportunity and sustainable entrepreneurship. *Journal of Business Venturing, 22*(1), 29–49. Retrieved from http://www.researchgate.net/profile/Boyd_Cohen/publication/4968082_Market_imperfections_opportunity_and_sustainable_entrepreneurship/links/544159cb0cf2a76a3cc7e006.pdf

Cohen, A., & Sharma, P. (2016). *Entrepreneurs in every generation: How successful family businesses develop their next leaders.* Oakland, CA: Berrett-Koehler.

Craig, J. B., & Dibrell, C. (2006). The natural environment, innovation, and firm performance: A comparative study. *Family Business Review, 19*(4), 275–288. https://doi.org/10.1111/j.1741-6248.2006.00075.x

De Massis, A., Frattini, F., Pizzurno, E., & Cassia, L. (2015). Product innovation in family versus nonfamily firms: An exploratory analysis. *Journal of Small Business Management, 53*(1), 1–36. https://doi.org/10.1111/jsbm.12068

Dean, T. J., & McMullen, J. S. (2007). Toward a theory of sustainable entrepreneurship: Reducing environmental degradation through entrepreneurial action. *Journal of Business Venturing, 22*(1), 50–76. https://doi.org/10.1016/j.jbusvent.2005.09.003

Delmas, M. A., & Gergaud, O. (2014). Sustainable certification for future generations: The case of family business. *Family Business Review, 27*(3), 228–243. https://doi.org/10.1177/0894486514538651

Duran, P., & Kammerlander, N. (2015). Doing more with less: Innovation input and output in family firms. *Academy of Management Journal, 59*(4), 1–75.

Dyer, W. G., & Whetten, D. A. (2006). Family firms and social responsibility: Preliminary evidence from the S&P 500. *Entrepreneurship Theory and Practice, 30*(6), 785–802. https://doi.org/10.1111/j.1540-6520.2006.00151.x

Gomez-Mejia, L. R., Cruz, C., Berrone, P., & De Castro, J. (2011). The bind that ties: Socioemotional wealth preservation in family firms. *The Academy of Management Annals, 5*(1), 653–707. https://doi.org/10.1080/19416520.2011.593320

Hockerts, K., & Wüstenhagen, R. (2010). Greening Goliaths versus emerging Davids – Theorizing about the role of incumbents and new entrants in sustainable entrepreneurship. *Journal of Business Venturing, 25*(5), 481–492. https://doi.org/10.1016/j.jbusvent.2009.07.005

Hörisch, J. (2015). The role of sustainable entrepreneurship in sustainability transitions: A conceptual synthesis against the background of the multi-level perspective. *Administrative Sciences, 5*(4), 286–300. https://doi.org/10.3390/admsci5040286

Jenkin, M. (2015). Millennials want to work for employers committed to values and ethics. *The Guardian*. Retrieved from https://www.theguardian.com/sustainable-business/2015/may/05/millennials-employment-employers-values-ethics-jobs

König, A., Kammerlander, N., & Enders, A. (2013). The family innovator's dilemma: How family influence affects the adoption of discontinuous technologies by incumbent firms. *Academy of Management Review, 38*(3), 418–441. https://doi.org/10.5465/amr.2011.0162

Lacy, P., Cooper, T., Hayward, R., & Neuberger, L. (2010). A new era of sustainability. *UN Global Compact – Accenture CEO Study*. Retrieved from http://www.unglobalcompact.org/docs/news_events/8.1/UNGC_Accenture_CEO_Study_2010.pdf

McCann, J. E., Leon-Guerrero, A. Y., & Haley, J. D. Jr. (2001). Strategic goals and practices of innovative family businesses. *Journal of Small Business Management, 39*(1), 50–59. Retrieved from http://www.scopus.com/inward/record.url?eid=2-s2.0-0035531260&partnerID=40&md5=c3c297f8533764228bc55232f8bfe80a

Miller, D., & Le Breton-Miller, I. (2005). Management insights from great and struggling family businesses. *Long Range Planning, 38*(6), 517–530. https://doi.org/10.1016/j.lrp.2005.09.001

Miller, D., Steier, L., & Le Breton-Miller, I. (2016). What can scholars of entrepreneurship learn from sound family businesses? *Entrepreneurship: Theory and Practice, 40*(3), 445–455. https://doi.org/10.1111/etap.12231

Nidumolu, R., Prahalad, C., & Rangaswami, M. (2009). Why sustainability is now the key driver of innovation. *Harvard Business Review, 87*(9), 55–64.

Niehm, L. S., Swinney, J., & Miller, N. J. (2008). Community social responsibility and its consequences for family business performance. *Journal of Small Business Management, 46*(3), 331–350. https://doi.org/10.1111/j.1540-627X.2008.00247.x

Pacheco, D. F., Dean, T. J., & Payne, D. S. (2010). Escaping the green prison: Entrepreneurship and the creation of opportunities for sustainable development. *Journal of Business Venturing, 25*(5), 464–480. https://doi.org/10.1016/j.jbusvent.2009.07.006

Schaltegger, S., & Hansen, E. G. (2013). Unternehmerische Nachhaltigkeitsinnovationen durch nachhaltiges Unternehmertum. In R. Altenburger (Ed.), *CSR und Innovationsmanagement* (pp. 19–30). Berlin: Springer.

Schaltegger, S., & Wagner, M. (2011). Sustainable entrepreneurship and sustainability innovation: Categories and interactions. *Business Strategy and the Environment, 20*(4), 222–237. https://doi.org/10.1002/bse.682

Schaltegger, S., Lüdeke-Freund, F., & Hansen, E. G. (2011). Business cases for sustainability and the role of business model innovation: Developing a conceptual framework. *SSRN Electronic Journal* (January 2016). https://doi.org/10.2139/ssrn.2010506

Shepherd, D. A., & Patzelt, H. (2011). The new field of sustainable entrepreneurship: Studying entrepreneurial action linking "What Is to Be Sustained" with "What Is to Be Developed". *Entrepreneurship Theory and Practice, 35*(1), 137–163. https://doi.org/10.1111/j.1540-6520.2010.00426.x

Shepherd, D. A., Patzelt, H., & Baron, R. A. (2013). "I Care about Nature, but . . .": Disengaging values in assessing opportunities that cause harm. *Academy of Management Journal, 56*(5), 1251–1273. https://doi.org/10.5465/amj.2011.0776

Wagner, M. (2010). Corporate social performance and innovation with high social benefits: A quantitative analysis. *Journal of Business Ethics, 94*(4), 581–594. https://doi.org/10.1007/s10551-009-0339-y

Yin, R. K. (2003). *Case study research: Design and methods* (3rd ed.). Thousand Oaks: Sage.

York, J. G., & Venkataraman, S. (2010). The entrepreneur-environment nexus: Uncertainty, innovation, and allocation. *Journal of Business Venturing, 25*(5), 449–463. https://doi.org/10.1016/j.jbusvent.2009.07.007

Zahra, S. A. (2005). Entrepreneurial risk taking in family firms. *Family Business Review, 18*(1), 23–40. https://doi.org/10.1111/j.1741-6248.2005.00028.x

Zukunftsinstitut. (2016). *Authentisch und ästhetisch: Nachhaltigkeit 2.0*. Retrieved from https://www.zukunftsinstitut.de/artikel/tup-digital/04-next-economy/01-longreads/nachhaltigkeit-20/

Sylvie Scherrer is a lecturer and research assistant at Lucerne University of Applied Sciences and Arts in Switzerland, where she focuses family businesses, entrepreneurship and sustainability. She also is a Ph.D. Candidate in Business Administration at the Centre for Family Enterprise and Ownership at Jönköping University in Sweden.

Claudia Binz Astrachan is a lecturer and research assistant at Lucerne University of Applied Sciences and Arts in Switzerland, where she teaches seminars on family business and entrepreneurship and runs an annual programme for family business owners. She also serves as a chair of the Special Interest Group Family Business Research of the European Academy of Management (EURAM), and as the Consortium Chair of the International Family Enterprise Research Academy (IFERA). She earned her Ph.D. at the Witten Institute for Family Business at Witten/Herdecke University (Germany).

Socially Driven Stakeholder Networks of German Family-Owned Companies as Enablers of Economic Success: A Theoretical and Empirical Study

Henry Schaefer and Friedrich Voelker

1 Introduction

In Germany in many industries like machine building, chemical or automotive family-owned companies stand for above average R&D activities, high quality products, close ties with their customers and outstanding agility. Many of those companies are also well-known for long-lasting and valuable activities in corporate citizenship, mainly focused on their direct neighborhoods and particular stakeholders. The financial success of those companies is often grounded on their networks, as it is typical for the German automotive industry with their close ties to OEMs (see Schäfer and Baumann 2013, pp. 1–2). Since the economic role of such networks for the success of companies in general is well understood, the study of the relevance of stakeholder networks embedded in firm-specific understanding and practice of Corporate Social Responsibility (CSR), the role of in such a way created social capital and especially the links to family-owned companies are just at the beginning and only few empirical studies on that issue exist.

Today the majority of CSR research is focused on public listed companies, owned by shareholders and operated by an external management. An increasing number of investors operate on capital markets according to principles for responsible investment (e.g. the UN PRI, see Schäfer 2012). With the help of rating agencies they monitor the CSR performance of companies and try to enforce their notion of a 'good' environmental, social and governance related policy which a company's management has to execute (see Schäfer 2009). Only of minor importance for CSR related research and of low awareness in the public seem to be 'family-owned companies'

H. Schaefer (✉) · F. Voelker
Institute of Business Administration, Dept. III (Corporate Finance), University of Stuttgart, Stuttgart, Germany
e-mail: h.schaefer@bwi.uni-stuttgart.de

© Springer International Publishing AG, part of Springer Nature 2018
R. Altenburger (ed.), *Innovation Management and Corporate Social Responsibility*,
CSR, Sustainability, Ethics & Governance,
https://doi.org/10.1007/978-3-319-93629-1_18

(abbreviated as 'FoC' in the following). Generally FoCs are understood as representatives of more long-term, sustainable and conservative strategies—attributes that fit very well with the postulate of inter- and intra-generational thinking of the concept of sustainable development (see WCED 1987, p. 43). Under those circumstances, a FoC often develops a specific understanding of the social and ecological responsibility of the company even towards its stakeholders (see Schäfer 2010, p. 18). Such roots of CSR self-understanding lead to specific ways and styles of how FoC owners and managers practice CSR. As (Schäfer and Goldschmidt 2010) discussed, CSR of FoCs often underlies specific motivations of their owners and owner-managers.[1]

1.1 The Uniqueness of FoCs

The ownership structure of FoCs is very often represented by a family. It represents an institution that dominates the self-understanding of the company, its cultural values, the strategic course and the rules of conduct of the company (see Klein 2010, p. 68 ff.). From a sociological point of view a family includes a dynastic aspect which means that different members of one or multiple 'core families' (mother, father and children) have a strong common feeling of belonging together (see Klein 2010, p. 68). This motivates dynastic families to accumulate a joint family-wealth and pass it on from generation to generation (see Le Breton-Miller and Miller 2006, p. 732 f.; Naldi et al. 2007, p. 36). When trying to find an established definition of the term 'family-owned company' one is faced with many different approaches (see Chrisman et al. 2005, p. 556 f.; Chua et al. 1999, p. 21; Flören 2002, p. 15 ff.). Litz (1995, p. 71) even describes this as a considerable definitional confusion. Nevertheless almost all of the existing definitions can be attributed to either the 'components-of-involvement approach', the 'essence approach' (Chrisman et al. 2005, p. 556 f.) or a mixture of the two such as the 'F-PEC Scale' (see Astrachan et al. 2002).

The common theoretical implication of these approaches is that the owning family can always directly or indirectly exert control over the company and therefore influence corporate decision-making processes, driven by their individual goals,

---

[1]Definitions and understandings of CSR vary enormous (see Dahlsrud 2008). A common denominator is the understanding of CSR as an attempt to cause a fair internalization of negative external effects of an enterprise's activities. This is highly relevant for stakeholders on the one hand side (see Carroll 1999). On the other hand side stakeholders can also be receiver of positive external effects of an enterprise's operations (see Eells and Walton 1974). Practically seen, CSR is understood beyond legal obligations ('the law to operate') as it is determined by voluntarily undertaken business activities. They enable or ensure that stakeholders facilitate a firm's operations or refrain to threaten them. This describes CSR according to the license to operate paradigm (see Gunningham et al. 2004). With the concept of the Sustainable Development as first offered by the Brundtland-Commission in 1987 and worked out on the UN-level, the anthropogenic form of the sustainability concept has been crystallized at the firm level as 'Corporate Sustainability'. With a Corporate Sustainability approach, the complete entrepreneurial behaviour becomes explicitly collimated in the triad of economic, social and ecological dimensions (see WCED 1987).

values and personalities (see Schäfer 2010, p. 4). This makes FoCs unique from the viewpoints of the principal-agent theory as well as of the resource-based view—the most important theoretical streams within FoC research. Indeed FoCs often have access to specific resources that are not available for non-family owned companies. This characteristic bundle of resources is called 'familiness' (see Habbershon and Williams 1999, p. 11). According to the inter-generational impetus of the economic activities of families, they are often described as long-term investors that are more interested in sustainable growth of market shares and the equity of the firm than in short-term profit maximization (see Le Breton-Miller and Miller 2006, p. 732 f.; Naldi et al. 2007, p. 36; Habbershon and Williams 1999, p. 4).

1.2 FoCs in the Light of CSR

From the viewpoint of networks, the CSR of FoCs can be described as the intersection of two circles: The first circle represents the interior network of the owners of the FoC, i.e. in most cases the family. It consists of direct networks of owners who drive the company as managers or as members in supervisory boards. Corresponding to this is the indirect network which represents the sole ownership structure with the bulk of owners. They are either not interested in a role as owner-managers or excluded from management by inter-family contracts. Although they are non-operating branches of the entire family their tradition and heritage can strongly influence the FoC's understanding of CSR.

The second circle to be distinguished is the network of collaborations between the FoC's management (i.e. partly with the family members who own the FoC) and stakeholders that are critical in the sense of Freeman (1984) for the FoC. In the following the focus lies on that second circle, i.e. the collaboration of a FoC as an entity with stakeholders. In general, little research has been done on the intricacies of CSR in such networks (see Dyer and Whetten 2006, p. 785; Uhlaner et al. 2004, p. 186). The few existing empirical studies suggest a link between family firm size and the characteristics of such stakeholder-networks: Schäfer (2010) and Schäfer and Goldschmidt (2010) show that small and medium-sized FoCs (up to an annual turnover of 50 Mio Euros) in Germany exhibit a dominant focus on customers in their stakeholder related CSR activities, whereas FoCs with an annual turnover above 50 Mio Euros address CSR mainly to their employees. Another empirical study by Keese et al. (2010, p. 215) supports these findings as it concludes that for German FoCs a positive trade-off exists between company size and the amount invested into education and trainings of their employees. For Dutch FoCs Uhlaner et al. (2004, p. 189) could identify employees and customers as the most prominent stakeholders. Looking at European small and medium-sized companies, the Commission of the European Union (2002, p. 28) stated that the CSR activities of smaller companies is rather motivated by a potential increase of customer loyalty than employee satisfaction. These findings suggest that there seems to be a special relationship between the size of a FoC and its preferred stakeholders to which CSR activities of FoCs are directed to. What is so far unexplored, are the reasons for such size-dependent CSR activities addressed to stakeholders. The following study sticks to that question and focuses on employees and customers, i.e. primary stakeholders with contractual claims against a FoC (see Wheeler and Sillanpää 1997, p. 5). This study contributes to that presumption and the related academic work in two important ways. First and based on the cited current empirical research on FoCs, the paper derives several theoretical explanations by introducing the notion of stakeholder related social activities of a FoC as a type of idiosyncratic, strategic investment into resources and competences. Special emphasis is put on how CSR can leverage FoC-stakeholder networks and create social capital that contributes to the economic success of a FoC. In our paper CSR activities of a FoC are understood as activities towards stakeholders to foster their collaboration with a FoC. We will distinguish between vertical collaboration, i.e. those of the management and employees (relevant for large FoCs) and horizontal collaborations, which describe the interactions of customers and the management (of small and medium-sized FoCs).

Based on the first part, the study is followed by the formulation of a set of hypotheses that are deduced from the previous theoretical research (stakeholder theory, models of the resource-based and dynamic capability approaches, moral intensity model). In the second part of our study those hypotheses are empirically tested to proof whether and to what extent they are relevant to explain CSR activities of FoCs that are addressed to one of the two stakeholder groups. Although the empirical study is done in the country-specific context of Germany, the outcome of the study allows to a certain degree the derivation of fundamental insights that can be generalized and transferred to other countries as well.

The paper is organized as follows: It starts with the explanation of stakeholder related CSR activities as a type of idiosyncratic, strategic investment into strategic resources of a FoC. For the research goals of this paper, the FoC definition of the Commission of the European Union (2009, pp. 8–10) seems most suited since it can easily be translated into statistically measurable variables while also being routed in theory and well established in literature. According to this definition, a company qualifies as a FoC if the majority of decision-making rights is held by the owning family (or related families) and at least one member of the family is directly involved in managing or supervising the company. By analyzing the business environments of small and large FoCs as defined by their annual revenues (below 50 Mio Euros), two size-dependent variables are identified: the degree of internationalization and the competitiveness of a FoC. Subsequently, the effects of these variables and the firm-size itself on addressing CSR related investments to either employees or customers are discussed and according hypotheses are derived. These hypotheses are than tested by statistically analyzing self-completion questionnaires sent out to owners and managers of German FoCs.

2 Research Model and Hypotheses

There seems to be an obvious inherent fit between the long-term horizon of FoCs and CSR since many of their CSR activities can be understood as investments into intangible resources such as human, social and reputational capital with a long level of maturity. If a FoC has created them strategically, they have the potential to lead to economic competitive advantages over time (see Branco and Rodrigues 2006, p. 112; McWilliams et al. 2006, p. 3; Perrini 2006, p. 306; Van de Ven and Graafland 2006, p. 114; Weber 2008, p. 251). Such a competitive advantage will be reasoned by CSR activities that create context dependant knowledge which can serve as an 'isolating mechanism' (see Mahoney and Pandian 1992, p. 371): It is heterogeneous compared to competitors, unique with reference to a specific firm and immobile as it cannot be exchanged (see Kyläheiko and Sandström 2007, p. 969). Under those circumstances it represents a kind of knowledge based assets, that cannot be simply imitated by competitors as it shares attributes like patents, brands or copy rights with the exception that they are protected by law. Although CSR related asset formations have no exclusive property rights the above mentioned attributes work as substitutes and allow the exploitation of economic rents. In order to qualify a CSR activity as such an investment, social, environmental and governance related activities of a FoC must have a positive economic impact on so-called 'VRIN-resources' (see in general Barney 1991, p. 101). CSR activities can be clustered as VRIN if they contribute to the financial success of a company and if they qualify as resources with the attributes $v$aluable, $r$are, in-imitable, $n$on-substitutable. From a dynamic perspective VRIN resources cannot by their economic nature completely be specified ex ante due to their entire and often uncertain future outcomes.

2.1 Research Model

As the modern knowledge based theory of the firm states, nowadays the exploration and exploitation of knowledge drives economic success of companies in most industries. Knowledge exchange is not restricted to the boundaries of companies but often requires openness to the surrounding environment of the company, especially to its stakeholders (Kogut and Kulatilaka 2001, p. 747). Knowledge in the sense of a VRIN resource is highly intangible, often emerges outside markets and (companies') hierarchies and is grounded on long-lasting co-operations (Conner and Prahalad 1996, p. 477). The attributes of such knowledge based resources can be expressed in terms of networks, highly flexible and plastic resources, intangibles (like brands, customer value or employee qualification), new risk dimensions (structural and procedural uncertainties), significant role of implicit and incomplete contracts and the generation of tacit knowledge in conjunction with stakeholders (see Halal 2001, S. 31). According to the resource dependence theory (see Pfeffer and Salancik 1978) companies cannot generate all of their VRIN-resources by themselves and are

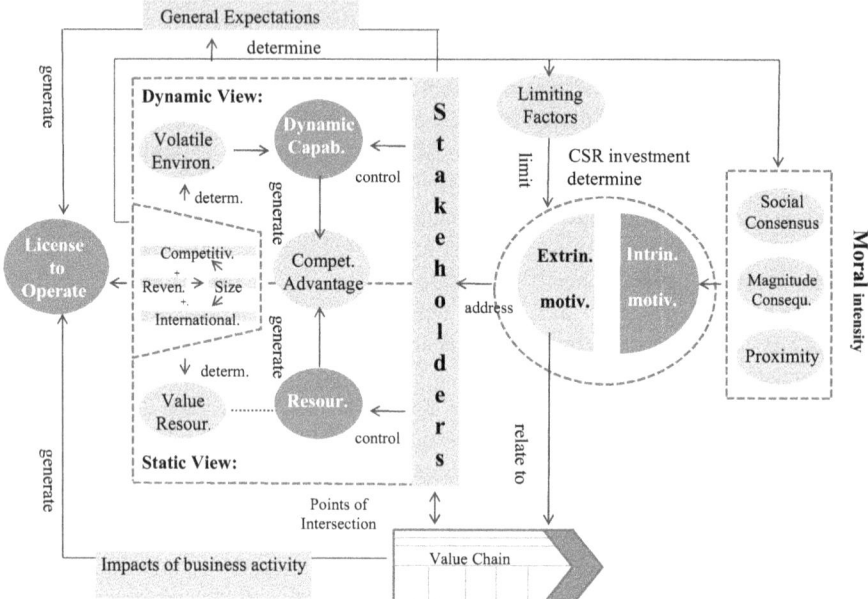

Fig. 1 Research model: stakeholder theory in conjunction with the resource based view and the dynamic capabilities approach

thus depending on exchange relationships with the owners of such resources or have to develop such resources jointly with collaborators outside or inside the company. A well installed and functioning collaboration of FoCs with economically critical stakeholders bears the potential to create a very specific kind of intangible resource—namely social capital. It is of superior economic value for a FoC, if its operating environment is characterized by growing dynamics, ambiguities and turmoil, requires steady learning and discovery processes and is depending on coordination mechanisms for multiple stakeholder relations. Those relationships can either have a relational (see Galbreath 2005, p. 982) or a reputational (see Zoeteman 2012, p. 163) resource character. In the following we focus on the ability of social capital to serve as a VRIN resource that renders services for the FoC which increases its financial profits (see Cornell and Shapiro 1987 and Fig. 1).

As the creation of knowledge is justified by its ability to generate innovation rents, the collaboration with stakeholders under those circumstances can be interpreted as an innovation network. By exploiting innovation capacities the network can lead to a competitive advantage if it is characterized as a resource with causal ambiguity, social complexity (see Barney 1991, pp. 108–110) and path-dependency according to trust based social co-operations between the FoC and relevant stakeholders (see Sharma and Vredenburg 1998, p. 740).

Stakeholders can play a role as bearers of specific knowledge, competences, cultural identities etc. that could be used by a company's management as an ingredient in its value creation process. The exchange of knowledge presumes either articulation or codification. In both cases social interaction is a necessary precondition and networks serve as adequate institutions. The more relevant an in such a way created resource for the economic value creation and financial success is, the more the management has an incentive to collaborate with stakeholders. If such transactions occur repeatedly, over a longer time horizon and become integral part of the value creation function of a company, management has further incentives to hold strong ties to appropriate stakeholders. Formal routines, regulated collaboration etc. would than follow and build up a network over time (see Arora and Petrova 2010, p. 5). Such a FoC-stakeholder network develops special structures of knowledge exchanges and knowledge generations, if a FoC has installed procedures and architectures that deliver to stakeholders exclusive benefits and make it worth for them to join the network and to stay in it (see Russo and Fouts 1997, p. 539). The installment of such prerequisites share many attributes of investments into intangible capital. In the case of employees social benefits are the most important incentives for joining the network. Customers often stay in networks if they can expect tailor-made products and services with an extraordinary quality, reliability and accuracy. The FoC has an incentive to protect the stakeholder network if the generation of knowledge and the screening and diffusion of information leads to superior innovation, reduces transactions costs (e.g. costs to coordinate) and increases the efficiency of the whole value chain and its output. In that sense a FoC-stakeholder network is able to create intangible resources that share many attributes of social capital (see Nahapiet and Ghoshal 1998, S. 245–252).

According to Penrose we can interpret the social capital character of such a FoC-stakeholder network as an organization's resource which offers specific economic services that can be rendered by a FoC (see Penrose 1959, p. 137). From a dynamic point of view FoC-stakeholder networks should over time allow the combination and exchange of new knowledge. In such a way the network generates a collective mind (see Weick and Roberts 1993, p. 357) with which in- and outside a FoC complementary operations between management and employees or customers, can lead to organizational routines (see Levinthal 2003, p. 366) and build up intellectual capital that increases the adaptive efficiency of a FoC. If it is network-specific, it cannot be imitated by competing companies or other networks. And if it works successful it allows for competitive product innovation that can exploit 'Schumpeter rents' (see Tsai and Ghoshal 1998, p. 467). The creation of such social capital represents an intangible resource that is best described as a dynamic capability (see Teece et al. 1997; Teece 2009; Kogut and Zander 1993). It represents a 'bundle of real options' for future strategic choices (see Bowman and Hurry 1993, p. 762) developed by the FoC's former stakeholder network-based investments into knowledge and dynamic capacities. In most cases, such investments have only to a limited extent direct positive cash flow effects or cannot be unambiguously identified as the cause of positive cash flow effects (Kyläheiko et al. 2002). Nonetheless, they provide flexibility for potential subsequent investments in the form of options (see Amram and Kulatilaka 1999, p. 3; Trigeorgis 1996, p. 4), i.e. they represent flexibility-generating investments which are valuable in environments of high uncertainty and the threat of management decisions with irreversible consequences (see Dixit and Pindyck 1994).

From a financial point of view FoCs then should address their CSR towards those stakeholders who control VRIN-resources that are of highest importance for a company. They would allow for the leveraging of dynamic capabilities as it operates with already installed assets in place and aims to yield returns by the exploitation of such VRIN-resources. From a strategic point of view the implementation of stakeholder driven dynamic capabilities has to be understood as platforms or 'institutions-as-reserves' (see Foss 1998, p. 12) that allow special growth and flexibility operations due to unforeseeable future changes in the operating environment of a FoC.

However, FoCs do not only engage in CSR since they want to invest in their intangible resources and thus gain a competitive advantage. Often, they are simply interested in 'doing good' without thinking about any financial return (see Van de Ven and Graafland 2006, p. 119; Perrini and Minoja 2008, p. 59; Schäfer 2010, p. 17; Uhlaner et al. 2004, p. 186). In fact, many authors argue that this kind of selfless CSR can have negative effects on the financial success of the company (see Elahuge 2005, p. 14; see Waddock and Graves 1997, p. 305).

2.2 Derivation of Hypotheses

In the following we take a closer look at FoCs in Germany and their CSR activities. Based on Schäfer (2010) and Schäfer and Goldschmidt (2010) we distinguish between small and medium-sized companies with an annual turnover up to 50 Mio Euros and large family-owned companies with an annual turnover above 50 Mio Euros. As the mentioned empirical studies have exhibited, the dominant focus of the CSR activities of small and medium-sized enterprises lies on customers. CSR activities of large family-owned companies, however, are mainly addressed to their employees. During the following part of the paper, reasons for such observed size-dependent differences in addressing CSR towards employees and customers are discussed and according hypotheses are derived. For that purpose we will distinguish the nature of CSR activities into two motivation clusters: extrinsic and intrinsic. We begin by taking a closer look at extrinsic motivated CSR, i.e. activities that are motivated by achieving some form of financial benefit (or by avoiding financial loss) for the FoC.

Smaller companies are often described as generating the vast majority of their revenue with only a few or even just one customer (see Hankinson et al. 1997, p. 173). This significantly weakens their negotiation position towards these customers and makes them highly dependent (see Jenkins 2004, p. 44; Lepoutre and Heene 2006, p. 264; Spence 2007, p. 534). Hence large customers (measured in their profit share or revenue share of a FoC) can easily force their own interests onto small companies including strict CSR-guidelines that are declared as a pre-condition for conducting any (further) company (e.g. so-called pre-qualification guidelines). Smaller FoCs thus have little choice but to abide by these guidelines. These considerations lead to the first hypothesis:

H1 *'CSR activities of small FoCs are influenced more strongly by customer-pressure than CSR activities of large FoCs. Therefore, small FoCs engage more often in customer-focused CSR than large FoCs.'*

The efficiency of the transfer of knowledge within FoCs strongly depends on existing network structures between employees. We argue that due to their sheer organizational size, larger FoCs experience a more inefficient transfer of knowledge. This is mainly because the number of geographically distant and functionally distinct departments increases with firm size (see Jenkins 2004, p. 41). Communicating tacit knowledge between these departments is associated with transaction costs that can be prohibitively high (see Kogut and Zander 1993, p. 629 f.; Teece et al. 1997, p. 525). Hierarchy also tends to increase with firm size (see Delmastro 2002, p. 120) and hinders the vertical flow of information (see Thiessen et al. 2007, p. 229). These factors decrease the absorptive capacity of large FoCs and thus their ability to generate dynamic capabilities. If decision-makers of large FoCs are aware of this, it is to be expected that they take actions to compensate for this disadvantage. In fact, employee-focused CSR can be used to foster the transfer of knowledge by increasing employee loyalty (see McWilliams and Siegel 2001, p. 112) and thus maintaining or even strengthening the (vertical) network structures within a FoC. Hence we formulate hypothesis

H2 *'In order to generate dynamic capabilities, large FoCs have to be more pro-active about maintaining the network structures between their employees than small FoCs. Therefore, large FoCs engage more often in employee-focused CSR activities than small FoCs.'*

Research shows a strong correlation between (family) company size and degree of internationalization (see Jenkins 2004, p. 41; KfW Bankengruppe 2006, p. 10 f.; Mittelstaedt et al. 2003; Schäfer 2010, p. 7). This means that large FoCs are confronted with a more complex business environment and challenges of stakeholders than small FoCs since they have to operate within different cultures, judicial systems, and languages (see Klein 2008, p. 8; Luo 2000, p. 361). This complexity may pose a threat but can also be an opportunity for companies to differentiate themselves from their competitors. Therefore, intercultural competence is a valuable resource as defined by Barney. Intercultural competence is a form of social and human capital as it is determined by the knowledge and experience of employees. Employee-focused CSR can be used to build and maintain human capital by: (1) improving employee retention, (2) building a reputation for being an attractive employer and thus attracting skilled/experienced staff and (3) directly increase the knowledge of employees through education and qualification programs (see Arora and Petrova 2010, p. 5; Branco and Rodrigues 2006, p. 121; Eweje and Bentley 2006, p. 10; Sharma et al. 2009, p. 210 f.). We summarize these findings in hypothesis

H3 *'Large FoCs are more dependent on their intercultural competence than small FoCs. Therefore, large FoCs engage more often in employee-focused CSR than small FoCs.'*

It could also be argued that increasing levels of internationalization lead to a higher degree of overall environmental volatility since international companies are not only faced with political, competitive, technological, economical, etc. changes in one but in many countries. If it becomes relevant, large international FoCs should be more depending on dynamic capabilities than small local FoCs since the economic value of dynamic capabilities increases the ability to response to environmental uncertainty (see Augier and Teece 2009, p. 126 f.). There is only rare literature that describes the link between dynamic capabilities and CSR (see Lattemann et al. 2007, p. 8), but for employee-focused CSR some authors (see Carter 2005, p. 182; Cramer 2005) have proven that a management's employee-focused CSR can foster the creation of dynamic capabilities. Hence for the fourth hypothesis we formulate:

H4 *'Large FoCs are faced with a more volatile environment than small FoCs and react by taking active measures to build dynamic capabilities. Therefore, large FoCs engage more often in employee-focused CSR than small FoCs.'*

Due to their limited organizational size and distinctive local or regional focus, small FoCs often have very personal, trusting and non-bureaucratic relationships with their customers (see Graafland et al. 2003, p. 46; Spence 2007, p. 537; Vives 2010, p. 110). This can allow the establishment of long-term networks between small FoCs and their customers that are characterized by cooperation rather than market transactions. A typical example for such networks can be found in the German automotive supplier industry (see Schäfer and Baumann 2013). Such close customer-relationships can be interpreted as a valuable resource (see Wölfer 2010, p. 399). They are often the only way for small FoCs to compete with and differentiate themselves from larger companies which operate with economies of scale, large financial reserves and extensive R&D capacities (see Acs and Audretsch 1987, p. 567; Spence 2007, p. 537). Based on the assumption that customer-focused CSR can improve relationships with customers (see Knox and Maklan 2004, p. 510; Weber 2008, p. 249; Vlachos et al. 2008, p. 170 ff.) we deduce the next hypothesis:

H5 *'Close relationships with customers are more important for the capability to survive and the competitiveness of small FoCs than for large FoCs. Therefore, small FoCs engage more often in customer-focused CSR than large FoCs.'*

It is a common understanding in management theory and practice that innovations drive a firm's financial success by stimulating employees to create new ideas (see Van de Ven 1986). In a recent survey for Germany, the outstanding relevance of R&D for the product and process innovations of large FoCs, together with a above average R&D ratio was confirmed. Another outcome of the survey highlights, that the most important bottleneck for sustaining and extending R&D is the shortage of qualified staff (BDI/Deutsche Bank 2016, pp. 8–12). Brown and Duguid (1991), Howells (2002), Nonaka (1991) provide clear evidence that in a firm the capacity to innovate can be increased by social interactions and knowledge sharing among idea providing employees. Intra-firm social networks, cooperative behavior between employees and the perceived fairness among network members influence the effectiveness of innovation capabilities positively as it creates an environment for knowledge sharing and creativity (see Frank et al. 1996; Gächter et al. 2010). Participation behavior and knowledge sharing are two decisive conditions for the innovative capacity of a firm, the attractiveness of its products, its competitive strength and at last of its financial success. Brown and Duguid (1991) and Howells (2002) could proof that relationships between innovation idea value and social interaction as well as between innovation idea value and knowledge sharing among idea providers are most important factors for innovation outcomes (see also Thieme et al. 2003), Such social interactions are not self-evident in firms. If a firm's management does network-building not leave up to a spontaneous process but intends to strive for a managerial and goal-oriented process, it has to implement formal institutions (see Björk and Magnusson 2009). Such measures are investments and contribute to a firm's social capital. Moreover it is often accompanied by fostering the human capital inside the firm. Social and human capital are elementary parts of the modern understanding of corporate social responsibility and should contribute to the financial success of a firm.

In his empirical analysis of the CSR of FoCs Schäfer (2010, p. 7) came to the conclusion that large FoCs have powerful financial and human resources at their disposal which increases their ability to innovate (see Mueller and Tilton 1969, p. 571). For FoCs such innovations are addressed to products and services as well as to operative processes within the FoC (see Utterback and Abernathy 1975). Depending on their strategic focus, large FoCs can thus use innovation not only to be a technological frontrunner but also a quality (see Prajogo and Sohal 2003, p. 912) or price leader in their respective markets (see Jácome et al. 2002, p. 190 f.). It enables large FoCs to build significant barriers to entry (see Siegfried and Evans 1994, p. 132, 140). It should protect large FoCs from new entrants and strengthen their oligopolistic positions they often operate in. Since human capital and social capital significantly determine the innovative capacity of a company (see Subramaniam and Youndt 2005, p. 450 ff.), they are extremely valuable resources for large FoCs given their competitive environment. As we argued before, employee-focused CSR can be used to build human and social capital. We thus formulate:

H6 *'Investing in human and social capital to increase their innovativeness is more important for the survival capability and competitiveness of large FoCs than for small FoCs. Therefore, large FoCs engage more often in employee-focused CSR than small FoCs.'*

The highly innovative environment of large FoCs is characterized by a high rate of ongoing (technological) change with structural and procedural uncertainty. Technological changes occur in many sectors as rapid developments with structural effects and often unforeseeable trends. Such an uncertain and volatile environment requires company agility and flexibility, i.e. the development and installment of dynamic capabilities (see Augier and Teece 2009, p. 126 f.). To cope successfully with those uncertainties is prominent for the competitive strength and economic sustainability of large FoCs (see Özsomer et al. 1997, p. 403). Since employee-focused CSR can be used to build dynamic capabilities, we formulate hypothesis

H4A *'Large FoCs are faced with a more volatile environment than small FoCs and react by taking active measures to build dynamic capabilities. Therefore, large FoCs engage more often in employee-focused CSR than small FoCs.'*

As we stated earlier, CSR can be motivated by purely altruistic reasons. We call this in the following 'intrinsic CSR'. It is not motivated by achieving financial gains or preventing of financial losses but, as any ethically motivated action, by the moral intensity (see Jones 1991). Moral intensity itself is a function of three variables: (1) probable magnitude of consequences, (2) proximity, and (3) social consensus (see McMahon and Harvey 2006). The premise is that FoCs will focus their CSR activity on stakeholder groups that find themselves in the situation with the highest moral intensity.

The vast majority of internationally operating large FoCs is not engaging in any foreign direct investment but is relying on exports (see IfM Bonn 2007, p. 105). This means that in most cases no or only few FoC employees are stationed abroad. They live and work geographically close to FoC management and thus within the same cultural area. Customers of international FoCs, however, by definition are spread over different countries and cultures. The more regional FoCs become, the closer customers live to the residential of the company, their owners and the management. Thus the proximity of customers (and therefore their moral intensity) should increase as FoCs decrease in their degree of internationalization. This brings us to the next hypothesis.

H7 *'The proximity between small FoCs and their customers is higher than the proximity between large FoCs and their customers. Therefore, small FoCs engage more often in customer-focused CSR than large FoCs.'*

3 Empirical Analysis

In order to test our hypotheses, a self-completion questionnaire was designed by breaking down the hypotheses into a total of 38 necessary and sufficient conditions and translating them into statistical variables. It was pretested by interviewing managers and owners of 14 FoCs in Germany in May 2012. The sample for the final questionnaire was selected by using the network of the German Chambers of Commerce (Industrie- und Handelskammern) to reduce the problem of below-average response rates that is commonly characteristic to FoC research. The collaboration with the Chambers of Commerce allowed us to use their address data bases, their official postal codes and should therefore facilitate to reach the survey targets and to increase their willingness of FoCs to respond. Out of the 80 German Chambers, 15 of them forwarded the questionnaire to a total of 802 FoCs. The questionnaire was also included in general monthly electronic newsletters that were sent out to approximately 3,700 FoCs.

3.1 Results

From September to November 2012, a total of 105 questionnaires were returned to us that were both complete and met the FoC definition criteria of this paper. 73% of respondents are small or medium sized FoCs (revenues below or equal to 50 Mio Euros), 27% are large FoCs (revenues above 50 Mio Euros). The responding FoCs are based in 10 of the 16 German states with a strong focus on Baden-Wuerttemberg (61 FoCs) as the center where FoCs in Germany are situated. The sample includes a wide variety of different industries with services, trade and mechanical engineering being the two most common. In order to test our hypotheses, we compared the answers given by large FoCs to those of small FoCs. Whenever possible, we used Student's t-test to check for differences in mean values for statistical significance. Only when levels of measurement did not allow for this (i.e. when answers are nominal or ordinal), Pearson's χ^2-test was used.

3.1.1 Regression Results

Testing of H1 In a first step we could confirm the expected significant negative relationship between firm-size and the percentage of overall revenues, generated with the largest 5 customers. According to Table 1 we interpret the result, that small FoCs are more depending on a smaller number of customers compared to large FoCs.

However, this dependency does not seem to translate into small FoCs getting forced into complying with the CSR-standards of their customers more often than large FoCs: the quota of customer requests for CSR that are actually implemented does not significantly vary with FoC size (see Table 2). We therefore have to dismiss H1.

Table 1 t-Test of customers' power

	FoC Size	Mean	p (Sig. 2-sided)	p/2 (Sig. 1-sided)
Number of customers	$> = €$ 50 Mio	32.999	0.575	0.2875
	$< €$ 50 Mio	62.897		
Revenue with the largest 5 customers	$> = €$ 50 Mio	2.86[a]	0.122	0.061
	$< €$ 50 Mio	3.64[a]		

[a]The values represent intervals on a %-scale (1 ~ "<10%", 2 ~ "10–19%", 3 ~ "20–29%", etc.)

Table 2 t-Test of H1, a FoC's CSR requested by customers

	FoC Size	Mean	p (Sig. 2-sided)	p/2 (Sig. 1-sided)
CSR requested by customers	> = € 50 Mio	11.68	0.824	0.412
	< € 50 Mio	10.55		
There of implemented in a FoC	> = € 50 Mio	6.89	0.996	0.498
	< € 50 Mio	6.91		
Quota of such implemented CSR	> = € 50 Mio	74.55%	0.676	0.338
	< € 50 Mio	70.45%		

Testing of H2 Both large and small FoCs regard the exchange of knowledge among their employees as a very critical factor for the company's economic success. They also strongly agree that personal relationships between their employees determine the effectiveness of the exchange of knowledge.

However, large FoCs rely significantly more often on formal tools of communication (such as meetings, conferences etc.) than small FoC's. This can be interpreted as FoCs increasingly taking active measures against inefficiencies in the exchange of information as they grow larger. Large FoCs also identify the resignation of employees as more likely to negatively impact the exchange of knowledge than small FoCs. It makes sense since large FoCs are more dependent on single individuals to maintain the flow of information due to informal contacts between different departments of the company—so called 'gatekeepers'. Surprisingly though, large FoCs do not try to increase employee retention in order to maintain network structures between employees und thus increase the exchange of knowledge. Therefore we have to reject H2. Table 3 summarizes the statistical results.

Testing of H3 t-Test results show a highly significant relationship between FoC size and the percentage of revenue generated outside of Germany. This difference in the geographical reach of small and large FoCs goes hand in hand with a significantly higher appreciation for the importance of intercultural competences of employees in large FoCs (see Table 4).

The international orientation of large FoCs seems to affect their CSR activities. On the one hand, they use significantly more tools to help their employees to develop intercultural competences (e.g. trainings, seminars, working abroad). On the other hand, they try to retain and recruit inter-culturally competent employees significantly more often by using employee-focused CSR (see Table 5). We therefore accept H3.

Testing of H4 H4 suggests a positive link between the degree of internationalization of FoCs and the (perceived) volatility of their environment. The empirical test does not confirm a significant difference in how small and large FoCs perceive the rate of structural and procedural changes in different parts of their operating environment (e.g. competitive situation, economy, technological development, values and norms). Therefore we have to dismiss H4. Interestingly though, large FoCs feel significantly more dependent on the abilities of their employees to be prepared for such changes. They also take significantly more measures (e.g. trainings, qualification programs) to

Table 3 t-Test of H2 concerning the management's knowledge exchange with employees

	FoC Size	Mean	p (Sig. 2-sided)	p/2 (Sig. 1-sided)
Importance of knowledge	> = € 50 Mio	4.54[a]		
Exchange for a FoC's financial profit	< € 50 Mio	4.31[a]		
Importance of personal contact for	> = € 50 Mio	4.25[a]		
knowledge exchange	< € 50 Mio	4.30[a]		
Degree of formalization of knowledge	> = € 50 Mio	3.18[b]	0.05	0.025
exchange	< € 50 Mio	2.71[b]		
Disruption of personal contacts through	> = € 50 Mio	3.61[c]	0.087	0.0435
fluctuation	< € 50 Mio	3.19[c]		
Retention employees to foster knowl-	> = € 50 Mio	3.82[c]	0.899	0.4495
edge exchange	< € 50 Mio	3.79[c]		

[a]The values represent choices on a Likert-scale: "1 = Not important at all" and "5 = Very important"
[b]The values represent choices on a Likert-scale: "1 = Very informal" and "5 = Very formal"
[c]The values represent choices on a Likert-scale: "1 = Not true at all" and "5 = Very true"

Table 4 t-Test of inter-cultural competences of employees

	FoC Size	Mean	p (Sig. 2-sided)	p/2 (Sig. 1-sided)
Relation of revenue abroad to revenue	> = € 50 Mio	5.61[a]	0.000	0.000
earned in Germany	< € 50 Mio	2.16[a]		
Importance intercultural competences	> = € 50 Mio	3.96[b]	0.000	0.000
employees	< € 50 Mio	2.77[b]		
Improvement of intercultural compe-	> = € 50 Mio	3.79[c]	0.000	0.000
tences of employees	< € 50 Mio	2.53[c]		
Retention employees with intercultural	> = € 50 Mio	4.00[c]	0.000	0.000
competences	< € 50 Mio	2.82[c]		
Intercultr. competences counts when	> = € 50 Mio	3.36[c]	0.004	0.002
hiring	< € 50 Mio	2.6[c]		

[a]The values represent intervals on a %-scale (1 ~ "<10%", 2 ~ "10–19%", 3 ~ "20–29%", etc.)
[b]The values represent choices on a Likert-scale: "1 = Not important at all" and "5 = Very important"
[c]The values represent choices on a Likert-scale: "1 = Not true at all" and "5 = Very true"

equip their employees with knowledge and abilities that help them to effectively react to environmental uncertainty (see Table 6).

Testing of H5 The results of a Pearson's χ^2-test show that small FoCs perceive tight relationships with their customers as significantly more important to their competiveness than their ability to innovate. However, this does not cause them to use more customer-focused CSR (see Table 7). Therefore we have to reject H5.

Table 5 t-Test of H3, CSR as a tool for retaining employees and labor market positioning

	FoC Size	Mean	p (Sig. 2-sided)	p/2 (Sig. 1-sided)
CSR to retain employees	> = € 50 Mio	3.96[a]	0.006	0.003
	< € 50 Mio	3.25[a]		
CSR to attract employees	> = € 50 Mio	3.82[a]	0.002	0.001
	< € 50 Mio	2.95[a]		

[a]The values represent choices on a Likert-scale: "1 = Not true at all" and "5 = Very true"
[b]The values represent intervals on a %-scale (1 ~ "<10%", 2 ~ "10–19%", 3 ~ "20–29%", etc.)
[c]The values represent choices on a Likert-scale: "1 = Not important at all" and "5 = Very important"

Table 6 t-Test of FoC's employees related measures to be prepared for future changes in the business environment

	FoC Size	Mean[a]	p (Sig. 2-sided)	p/2 (Sig. 1-sided)
Employees as important enabler to react to unforeseeable changes in the business environment	> = € 50 Mio	4.43	0.074	0.037
	< € 50 Mio	4.01		
Measures of a FoC to be prepared to cope with future changes in the business environment	> = € 50 Mio	3.96	0.307	0.154
	< € 50 Mio	3.75		
Employee qualification as a measure for a FoC's fitness to to cope with future changes	> = € 50 Mio	4.07	0.026	0.013
	< € 50 Mio	3.53		
Employee retention as a measure for a FoC's fitness to cope with future changes	> = € 50 Mio	4.29	0.432	0.216
	< € 50 Mio	4.13		
Employee hiring as enabler to be prepared for the future change	> = € 50 Mio	3.93	0.483	0.242
	< € 50 Mio	3.78		

[a]The values represent choices on a Likert-scale: "1 = Not true at all" and "5 = Very true"

Testing of H6 According to the χ^2-test mentioned before, large FoCs evaluate their ability to innovate significantly higher than tight relationships to their customers (see Table 8).

This observation is supported by a significantly positive relationship between the size of FoCs and their innovation quota (R&D expenditure/revenue): its amounts to 4.40% for small FoCs and to 6.39% for large FoCs (Table 9).

The latter strongly agree that their ability to innovate depends on the knowledge and experience of their employees and that it affects their CSR activity. They use employee qualification measures (such as trainings and education) aiming at increasing the ability of the workforce to innovate significantly more often than small FoCs. They also try to retain and recruit employees with R&D skills more often by using employee-focused CSR (see Table 9). We therefore accept H6.

Testing of H4A One condition of H4 is, that a steadily high innovation rate of oligopolies to which many large FoCs belong to, leads to environmental volatility. While large FoC's do in fact show higher innovation rates than small FoCs, they do

Table 7 t-Test active measures and CSR to build customer relationships

	FoC Size	Mean	p (Sig. 2-sided)	p/2 (Sig. 1-sided)
Take active measures to build tight cus-tomer relationships	$\geq$ € 50 Mio	4.21[a]	0.509	0.255
	< € 50 Mio	4.06[a]		
CSR to build tight customer relationships	$\geq$ € 50 Mio	3.46[a]	0.953	0.477
	< € 50 Mio	3.48[a]		

[a]The values represent choices on a Likert-scale: "1 = Not true at all" and "5 = Very true"

Table 8 χ^2-test concerning innovation and customer relationships

		Competitiveness depends on which critical factor?		
		Relationships with customers	Capability to innovate	Total
FoC Size	< 1 Mio. €	12	1	13
		92.3%	7.7%	100.0%
	1–50 Mio. €	53	11	64
		82.8%	17.2%	100.0%
	> 50 Mio. €	13	15	28
		46.4%	53.6%	100.0%
Total		78	27	105
		74.3%	25.7%	100.0%
	Value	df	p (Sig. 2-sided)	p/2 (Sig. 1-sided)
Chi-squared (Pearson)	16.021[a]	2	0.000	0.000

[a]The values represent intervals on a %-scale (1 ~ "<10%", 2 ~ "10–19%", 3 ~ "20–29%", etc.)
[b]The values represent choices on a Likert-scale: "1 = Not true at all" and "5 = Very true"

not perceive their environment as being significantly more bound to change. We therefore dismiss H4A.

Testing of H7 H7 states that the more narrow geographical focus of small FoCs leads to closer proximity to their customers. In fact our statistical tests show that the percentage of customers living abroad decreases significantly with firm size. However, this is also the case for employees. We therefore cannot argue that the high proximity of customers explains why small FoCs focus their CSR activity on them as opposed to employees and reject H7 (see Table 10).

3.1.2 Discussion of Results

Out of the 38 tested conditions, 23 were confirmed while 15 had to be dismissed. Accordingly, out of 9 hypotheses (7 primary and 2 alternative) 2 were confirmed and 7 rejected. It was confirmed by the survey that large FoCs are more international,

Table 9 t-Test of H6 concerning the importance of employees for a FoC's innovation

	FoC Size	Mean	p (Sig. 2-sided)	p/2 (Sig. 1-sided)
R&D/Revenue	> = € 50 Mio	6.39%	0.078	0.039
	< € 50 Mio	4.40%		
Employees determine innovation	> = € 50 Mio	4.5[a]	0.001	0.001
	< € 50 Mio	3.79[a]		
Measures in order to improve innovation	> = € 50 Mio	4.25[a]	0.002	0.001
	< € 50 Mio	3.57[a]		
Employee development in order to improve innovation	> = € 50 Mio	3.96[a]	0.021	0.011
	< € 50 Mio	3.40[a]		
Employee retention as a measure to improve innovation	> = € 50 Mio	4.5[a]	0.004	0.002
	< € 50 Mio	3.95[a]		
Employee selection (hiring) as a measure to improve innovation	> = € 50 Mio	4.21[a]	0.099	0.050
	< € 50 Mio	3.84[a]		

[a]The values represent choices on a Likert-scale: "1 = Not true at all" and "5 = Very true"

Table 10 t-Test and H7, proximity

	FoC Size	Mean	p (Sig. 2-sided)	p/2 (Sig. 1-sided)
Employees abroad	> = € 50 Mio	3.96[a]	0.00	0.00
	< € 50 Mio	1.26[a]		
Customers abroad	> = € 50 Mio	4.86[a]	0.00	0.00
	< € 50 Mio	1.77[a]		
Quota employees and customers abroad	> = € 50 Mio	0.89	0.34	0.17
	< € 50 Mio	0.51		
Emotional connection to people in home region	> = € 50 Mio	3.82[b]	0.48	0.24
	< € 50 Mio	4.01[b]		
CSR due to emotional connection to customers	> = € 50 Mio	3.46[b]	0.85	0.43
	< € 50 Mio	3.41[b]		

[a]The values represent intervals on a %-scale (1 ~ "<10%", 2 ~ "10–19%", 3 ~ "20–29%", etc.)
[b]The values represent choices on a Likert-scale: "1 = Not true at all" and "5 = Very true"

more innovative and less dependent on close customer-relationships than small FoCs. To be economically successful under those circumstances, large FoCs are especially depending on the intercultural competences and innovative capabilities of their employees. For this reasons, large FoCs use employee-specific CSR to invest in human and social capital.

The expected differences of small and large FoCs in terms of their degree of internationalization and the driving factors of their competitiveness (close customer relationships vs. ability to innovate) were almost completely confirmed. However, these differences do not in all cases affect CSR. This is especially true for intrinsic CSR. While non-financial motives play an important role for CSR in general, they could not explain why FoCs change the focus of their CSR from customers to employees when their business activities grow. Extrinsic CSR motives seem to be the driving force behind this phenomenon. FoCs are aware of the value of their resources given their specific company environment. Large FoCs also have understood that they can invest in their most valuable resources (i.e. human and social capital) and thus create a competitive advantage by using employee-focused CSR. Small FoCs on the other hand do not seem to use CSR strategically in their markets to leverage their competitive position and to attract customers. We can only speculate why this is the case: either small FoCs are not aware that CSR can actually be used this way or their CSR is driven by factors not considered in this study.

4 Implications, Limitations of the Study and Perspectives for Future Research

This study has important implications for both researchers and practitioners in FoCs. By combining stakeholder theory with the resource based view (VRIN-approach and dynamic capabilities approach) we demonstrated that CSR can be characterized as an investment in intangible human and social capital. The decision to undertake a CSR-investment can be driven both by extrinsic and intrinsic motives. Depending on the underlying motives, we identified multiple determinants, on which stakeholder groups a FoC focuses its CSR: (1) demands for CSR by stakeholders and their relative negotiation position, (2) the value of VRIN resources depending on the specific company environment, (3) the value of dynamic capabilities as determined by the volatility of the company's operating environment and (4) the degree of moral intensity of stakeholder relationships.

The impact of FoC size, its degree of internationalization and competitiveness on these determinants was discussed and then tested empirically. We found that extrinsic motives best explain differences in stakeholder specific CSR activities and that they can be characterized as investments into a company's strategic resources and dynamic capabilities. This is very relevant to the 'holy grail' of CSR research, the link between CSR and financial performance (see Devinney 2009, p. 45), as it substantiates the notion that large FoCs use CSR strategically and do not just treat it as art for art's sake. These considerations might also be of interest for FoC decision makers since we describe under which circumstances, FoCs should address the appropriate CSR activity to their target or critical stakeholder group in order to gain competitive advantages.

We would like to point out that there are a number of limitations to our research findings. First of all, when designing our questionnaires we could not rely on established items. The validity of the questions we used is therefore not clear. Also, results are probably influenced by a non-response bias since FoC owners/managers that are less engaged in CSR are also less likely to answer a questionnaire about CSR (see Bortz and Döring 2006, p. 256). Out of the FoC owners/manager who did participate in our survey, we do now know how many delegated the questionnaire to a third party in the company, e.g. their assistants. Last but not least we find a correlation between firm-size and industry. A disproportionately high amount of large FoCs belongs to the mechanical engineering sector while a disproportionately high amount of small FoCs are active in the service industry. This bias might impact our empirical results, especially those related to innovation.

A promising path for future research based on our findings is the relationship between CSR, innovative capacities and financing of FoCs. Moritz et al. (2015, pp. 7–10) present empirical evidence for Europe that FoCs' demand for finance is influenced by certain different factors like firm size and age, growth and ownership. Existing studies did not take into account that a relationship between the performance of CSR, its issues a firm is striving for and the finance channels. On the other hand an increasing number of empirical research has found evidence for correlations between financing conditions (e.g. cost of capital, capital rationing etc.) and firms' CSR performance (see Schäfer 2012). As a deeper insight into the evidence of such causalities are missing for FoCs it makes sense to carry out further research on that matter.

References

Acs, Z., & Audretsch, D. (1987). Innovation, market structure, and firm size. *The Review of Economics and Statistics, 69*(4), 567–574.

Amram, M., & Kulatilaka, N. (1999). *Real opetions: Managing strategic investment in an uncertain world*. Boston: Oxford University Press.

Arora, P., & Petrova, M. (2010). Corporate social performance, resource dependence and firm performance. *Zeitschrift für Betriebswirtschaft, 80*(Special issue 1), 1–22.

Astrachan, J., Klein, S., & Smyrnios, K. (2002). The F-PEC scale of family influence: A proposal for solving the family business definition problem. *Family Business Review, 15*(1), 45–58.

Augier, M., & Teece, D. (2009). Resources, capabilities, and Penrose effects. In D. Teece (Ed.), *Dynamic capabilities and strategic management – Organizing for innovation and growth* (pp. 113–135). Oxford: Oxford University Press.

Barney, J. B. (1991). Firm resources and sustained competitive advantage. *Journal of Management, 17*(1), 99–120.

BDI/Deutsche Bank. (2016). *Die größten Familienunternehmen in Deutschland. Ergebnisse der Frühjahrsbefragung 2016 – Chartbook I: Innovation und Investition*. Bonn: Durchgeführt vom Institut für Mittelstandsforschung (IfM).

Björk, J., & Magnusson, M. (2009). Where do good innovation ideas come from? Exploring the influence of network connectivity on innovation idea quality. *Journal of Product Innovation Management, 26*(6), 662–670.

Bortz, J., & Döring, N. (2006). *Forschungsmethoden und Evaluation für Human- und Sozialwissenschaftler*. Heidelberg: Springer.

Bowman, E. H., & Hurry, D. (1993). Strategy through the option lens: An integrated view of resource investments and the incremental-choice process. *Academy of Management Review, 18*(4), 760–782.

Branco, M., & Rodrigues, L. (2006). Corporate social responsibility and resource-based perspectives. *Journal of Business Ethics, 69*(2), 111–132.

Brown, J. S., & Duguid, P. (1991). Organizational learning and communities of practice. *Organisation Science, 2*(1), 40–57.

Carroll, A. (1999). Corporate social responsibility – Evolution of a definition construct. *Business & Society, 38*(3), 268–295.

Carter, C. (2005). Purchasing social responsibility and firm performance – The key mediating roles of organizational learning and supplier performance. *International Journal of Physical Distribution & Logistics Management, 35*(3), 177–194.

Chrisman, J., Chua, J., & Sharma, P. (2005). Trends and directions in the development of a strategic management theory of the family firm. *Entrepreneurship Theory and Practice, 29*(5), 555–576.

Chua, J., Chrisman, J., & Sharma, P. (1999). Defining the family business by behavior. *Entrepreneurship Theory and Practice, 23*(4), 19–39.

Commission of the European Union (Ed.). (2002). *European SMEs and social and environmental responsibility*. Observatory of European SMEs, 2002, No. 4, Bruxelles.

Commission of the European Union (Ed.). (2009). *Overview of family-business-related issues – Research, networks, policy measures and existing studies*. Final report of the expert group. Commission of the European Union, Bruxelles.

Conner, K., & Prahalad, C. (1996). A resource-based theory of the firm: Knowledge versus opportunism. *Organization Science, 7*(5), 477–501.

Cornell, B., & Shapiro, A. C. (1987). Corporate stakeholders and corporate finance. *Financial Management, 16*(1), 5–14.

Cramer, J. (2005). Company learning about corporate social responsibility. *Business Strategy and the Environment, 14*(4), 255–266.

Dahlsrud, A. (2008). How corporate social responsibility is defined – An analysis of 37 definitions. *Corporate Social Responsibility and Environmental Management, 15*, 1–13.

Delmastro, M. (2002). The determinants of the management hierarchy – Evidence from Italian plants. *International Journal of Industrial Organization, 20*(1), 119–137.

Devinney, T. (2009). Is the socially responsible corporation a myth – The good, the bad, and the ugly of corporate social responsibility. *Academy of Management Perspectives, 23*(2), 44–56.

Dixit, A. K., & Pindyck, R. S. (1994). *Investment under uncertainty*. Princeton: Princeton University Press.

Dyer, G., Jr., & Whetten, D. (2006). Family firms and social responsibility – Preliminary evidence from the S&P 500. *Entrepreneurship Theory and Practice, 30*(6), 785–802.

Eells, R., & Walton, C. (1974). *Conceptual foundations of business*. Homewood: Kluwer Academic.

Elahuge, E. (2005). Corporate managers' operational discretion to sacrifice profits in the public interest. In B. Hay, R. Stavins, & R. Vietor (Eds.), *Environmental protection and the social responsibility of firms – Perspectives from law, economics, and business* (pp. 13–76). Washington, DC: Resources for the Future.

Eweje, G., & Bentley, T. (2006). *CSR and staff retention in New Zealand companies – A literature review* (Working paper series 2006, no. 6). Auckland: Department of Management and International Business Research, Massey University.

Flören, R. H. (2002). *Crown princes in the clay – An empirical study on the tackling of succession challenges in Dutch family farms*. Dissertation, University of Nyenrode.

Foss, N. J. (1998). *Firms and the coordination of knowledge: Some Austrian insights* (DRUID working papers 98–19). DRUID, Copenhagen Business School, Department of Industrial Economics and Strategy/Aalborg University.

Frank, R. H., Gilovich, T. D., & Regan, D. T. (1996). Do economists make bad citizens? *Journal of Economic Perspectives, 10*(1), 187–192.

Freeman, E. (1984). *Strategic management – A stakeholder approach*. Boston: Cambridge University Press.

Gächter, S., Krogh, G., & von Haefliger, S. (2010). Initiating private collective innovation: The fragility of knowledge sharing. *Research Policy, 39*(7), 893–906.

Galbreath, J. (2005). Which resources matter the most to firm success – An exploratory study of resource-based theory. *Technovation, 25*(9), 979–987.

Graafland, J., Van de Ven, B., & Stoffele, N. (2003). Strategies and instruments for organizing CSR by small and large businesses in the Netherlands. *Journal of Business Ethics, 47*(1), 45–60.

Gunningham, N., Kagan, R., & Thornton, D. (2004). Social license and environmental protection – Why businesses go beyond compliance. *Law & Social Inquiry, 29*(2), 307–341.

Habbershon, T., & Williams, M. (1999). A resource-based framework for assessing the strategic advantages of family firms. *Family Business Review, 12*(1), 1–26.

Halal, W. E. (2001). The collaborative enterprise: A stakeholder model uniting profitability and responsibility. *Journal of Corporate Citizenship, 2*, 27–42.

Hankinson, A., Bartlett, D., & Ducheneaut, B. (1997). The key factors in the small profiles of small-medium enterprise owner-managers that influence business performance – The UK (Rennes) SME survey 1995–1997. *International Journal of Entrepreneurial Behaviour & Research, 3*(3), 168–175.

Howells, J. (2002). Tacit knowledge, innovation and economic geography. *Urban Studies, 39*(5–6), 871–884.

IfM Bonn (Hrsg.). (2007). *Die Bedeutung der außenwirtschaftlichen Aktivitäten für den deutschen Mittelstand, Studie*. IfM-Materialien No. 171. Bonn: Institut für Mittelstandsforschung.

Jácome, R., Lisboa, J., & Yasin, M. (2002). Time-based differentiation: An old strategic hat or an effective strategic choice – An empirical investigation. *European Business Review, 14*(3), 184–193.

Jenkins, H. (2004). A critique of conventional CSR theory – An SME perspective. *Journal of General Management, 29*(4), 37–57.

Jones, T. (1991). Ethical decision making by individuals in organizations – An issue-contingent model. *Academy of Management Review, 16*(2), 366–395.

Keese, D., Tänzler, J.-K., & Hauer, A. (2010). Die Wahrnehmung gesellschaftlicher Verantwortung in Familien- und Nicht-Familienunternehmen. *Zeitschrift für KMU und Entrepreneurship, 58*(3), 197–225.

KfW Bankengruppe (Ed.). (2006). *Die Globalisierung des Mittelstands – Chancen und Risiken, Studie*. Frankfurt a.M: Kreditinstitut für Wiederaufbau.

Klein, S. (2008). Internationale Familienunternehmen – Definition und Selbstbild. In: C. Rödel (Hrsg.), *Internationale Familienunternehmen – Recht, Steuern, Bilanzierung, Finanzierung, Nachfolge, Strategien* (pp. 1–14) München.

Klein, S. (2010). Familienunternehmen – Theoretische und empirische Grundlagen, 3. Aufl., Lohmar.

Knox, S., & Maklan, S. (2004). Corporate social responsibility – Moving beyond investment towards measuring outcomes. *European Management Journal, 22*(5), 508–516.

Kogut, B., & Kulatilaka, N. (2001). Capabilities as real options. *Organization Science, 12*(6), 744–758.

Kogut, B., & Zander, U. (1993). Knowledge of the firm and the evolutionary theory of the multinational corporation. *Journal of International Business Studies, 24*(4), 625–645.

Kyläheiko, K., & Sandström, J. (2007). Strategic options-based framework for management of dynamic capabilities in manufacturing firms. *Journal of Manufacturing Technology Management, 18*(8), 966–984.

Kyläheiko, K., Sandström, J., & Virkkunen, V. (2002). Dynamic capability view in terms of real options. *International Journal of Production Economics, 80*(1), 65–83.

Lattemann, C., Schneider, A., Kupke, P. P. (2007). *Corporate social responsibility and the capabilities based view: A case study of a multinational enterprise (MNE)*. Working paper, Chair for Corporate Governance and E-Commerce, University of Potsdam.

Le Breton-Miller, I., & Miller, D. (2006). Why do some family businesses out-compete-governance, long-term orientations, and sustainable capability. *Entrepreneurship Theory and Practice, 30*(6), 731–746.

Lepoutre, J., & Heene, A. (2006). Investigating the impact of firm size on small business social responsibility – A critical review. *Journal of Business Ethics, 67*(3), 257–273.

Levinthal, D. (2003). Imprinting and the evolution of firm capabilities. In C. Helfat (Ed.), *The SMS Blackwell handbook of organizational capabilities*. Oxford: Wiley.

Litz, R. (1995). The family business – Toward definitional clarity. *Family Business Review, 8*(2), 71–81.

Luo, Y. (2000). Dynamic capabilities in international expansion. *Journal of World Business, 35*(4), 355–378.

Mahoney, J. T., & Pandian, J. R. (1992). The resource-based view within the conversation of strategic management. *Strategic Management Journal, 13*(5), 363–380.

McMahon, J., & Harvey, R. (2006). An analysis of the factor structure of Jones' moral intensity construct. *Journal of Business Ethics, 64*(4), 381–404.

McWilliams, A., & Siegel, D. (2001). Corporate social responsibility – A theory of the firm perspective. *Academy of Management Review, 26*(1), 117–127.

McWilliams, A., Siegel, D., & Wright, P. (2006). Corporate social responsibility – Strategic implications. *Journal of Management Studies, 43*(1), 1–18.

Mittelstaedt, J., Harben, G., & Ward, W. (2003). How small is too small – Firm size as a barrier to exporting from the United States. *Journal of Small Business Management, 41*(1), 68–84.

Moritz, A., Block, J. H., & Heinz, A. (2015). *Financing patterns of European SMEs: An empirical taxonomy* (Working paper 2015/30). Luxembourg: EIF Research & Market Analysis.

Mueller, D., & Tilton, J. (1969). Research and development costs as a barrier to entry. *The Canadian Journal of Economics, 2*(4), 570–579.

Nahapiet, J., & Ghoshal, S. (1998). Social capital, intellectual capital, and the organizational advantage. *Academy of Management Review, 23*(2), 242–266.

Naldi, L., Nordqvist, M., Sjöberg, K., & Wiklund, J. (2007). Entrepreneurial orientation, risk taking, and performance in family firms. *Family Business Review, 20*(1), 33–47.

Nonaka, I. (1991). The knowledge-creating company. *Harvard Business Review, 85*(7/8), 162–171.

Özsomer, A., Calantone, R., & Bonetto, A. (1997). What makes firms more innovative – A look at organizational and environmental factors. *Journal of Business & Industrial Marketing, 12*(6), 400–416.

Penrose, E. (1959). *The theory of the growth of the firm*. Oxford: Oxford University Press.

Perrini, F. (2006). SMEs and CSR theory – Evidence and implications from an Italian perspective. *Journal of Business Ethics, 67*(3), 305–316.

Perrini, F., & Minoja, M. (2008). Strategizing corporate social responsibility – Evidence from an Italian medium-sized, family-owned company. *Business Ethics – A European Review, 17*(1), 47–63.

Pfeffer, J., & Salancik, G. (1978). *The external control of organizations – A resource dependence perspective*. Palo Alto: Stanford University Press.

Prajogo, D., & Sohal, A. (2003). The relationship between TQM practices, quality performance, and innovation performance – An empirical examination. *International Journal of Quality & Reliability Management, 20*(8), 901–918.

Russo, M., & Fouts, P. (1997). A resource-based perspective on corporate environmental performance and profitability. *The Academy of Management Journal, 40*(3), 534–559.

Schäfer, H. (2009). Corporate social responsibility rating. In G. Aras & D. Crowther (Eds.), *A handbook of corporate governance and corporate social responsibility* (pp. 449–465). Surrey: Gower Publishing.

Schäfer, H. (2010). *Gesellschaftliches Engagement des deutschen Mittelstands – Eine empirische Vergleichsanalyse* (Working paper). Betriebswirtschaftliches Institut Abteilung III – Finanzwirtschaft, University of Stuttgart.

Schäfer, H. (2012). *Sustainable finance. A conceptual outline*. (Working paper series, No. 03/2012). University of Stuttgart, BWI/Abt. III. http://ssrn.com/author=432903

Schäfer, H., & Baumann, S. (2013). Managing behavioural risks in logistics-based networks – A project finance approach. *The IUP Journal of Supply Chain Management, XI*(1), 18–35. https://doi.org/10.2139/ssrn.2352812.

Schäfer, H., & Goldschmidt, R. (2010). Corporate social responsibility of large family-owned companies in Germany. *International Journal of Entrepreneurship and Small Business, 11*(3), 285–307.

Sharma, S., & Vredenburg, H. (1998). Proactive corporate environmental strategy and the development of competitively valuable organizational capabilities. *Strategic Management Journal, 19*(8), 729–753.

Sharma, S., Sharma, J., & Devi, A. (2009). Corporate social responsibility – The key role of human resources management. *Business Intelligence Journal, 2*(1), 205–214.

Siegfried, J., & Evans, L. (1994). Empirical studies of entry and exit – A survey of the evidence. *Review of Industrial Organization, 9*(2), 121–155.

Spence, L. (2007). CSR and small business in a European policy context – The five "C"s of CSR and small business research agenda 2007. *Business & Society Review, 112*(4), 533–552.

Subramaniam, M., & Youndt, M. (2005). The influence of intellectual capital on the types of innovative capabilities. *Academy of Management Journal, 48*(3), 450–463.

Teece, D. (2009). Dynamic capabilities and the essence of the multinational enterprise. In D. Teece (Ed.), *Dynamic capabilities and strategic management – Organizing for innovation and growth* (pp. 137–181). Oxford: Oxford University Press.

Teece, D., Pisano, G., & Shuen, A. (1997). Dynamic capabilities and strategic management. *Strategic Management Journal, 18*(7), 509–533.

Thieme, R. J., Song, M., & Shin, G.-C. (2003). Project management characteristics and new product survival. *Journal of Product Innovation Management, 20*(2), 104–119.

Thiessen, M., Hendriks, P., & Essers, C. (2007). Research and development knowledge transfer across national cultures. In D. Pauleen (Ed.), *Cross-cultural perspectives on knowledge management* (pp. 219–244). Westport: Greenwood Publishing.

Trigeorgis, L. (1996). *Real options: Managerial flexibility and strategy in resource allocation*. Cambridge: MIT Press.

Tsai, W., & Ghoshal, S. (1998). Social capital and value creation: The role of intrafirm networks. *Academy of Management Journal, 41*(4), 464–476.

Uhlaner, L., Van Goor-Balk, H., & Masurel, E. (2004). Family business and corporate social responsibility in a sample of Dutch firms. *Journal of Small Business and Enterprise Development, 11*(2), 186–194.

Utterback, J., & Abernathy, W. (1975). A dynamic model of process and product innovation. *Omega, 3*(6), 639–656.

Van de Ven, A. (1986). Central problems in the management of innovation. *Management Science, 32*(5), 590–607.

Van de Ven, B., & Graafland, J. (2006). Strategic and moral motivation for corporate social responsibility. *Journal of Corporate Citizenship, 22*, 111–123.

Vives, A. (2010). Responsible practices in small and medium enterprises. In G. Aras & D. Crowther (Eds.), *A handbook of corporate governance and social responsibility* (pp. 107–130). Farnham: Gower Publishing.

Vlachos, P., Tsamakos, A., Vrechopoulos, A., & Avramidis, P. (2008). Corporate social responsibility – Attributions, loyalty, and the mediating role of trust. *Journal of the Academy of Marketing, 37*, 170–180.

Waddock, S., & Graves, S. (1997). The corporate social performance – Financial performance link. *Strategic Management Journal, 18*(4), 303–319.

WCED – World Commission on Environment and Development. (1987). *Our common future*. Oxford: Oxford University Press.

Weber, M. (2008). The business case for corporate social responsibility – A company-level measurement approach for CSR. *European Management Journal, 26*(4), 247–226.

Weick, K. E., & Roberts, K. H. (1993). Collective mind in organizations: Heedful interrelating on flight decks. *Administrative Science Quarterly, 38*(3), 357–381.

Wheeler, D., & Sillanpää, M. (1997). The stakeholder corporation, London.

Wölfer, K. (2010). Are family-owned businesses better innovators. In A. Gerybadze, H. Ulrich, H. Reiners, & D. Thomaschewski (Eds.), *Innovation and international corporate growth* (pp. 293–415). Berlin: Springer.

Zoeteman, K. (2012). *Sustainable development drivers – The role of leadership in government, business and NGO performance*. Cheltenham: Edward Elgar.

Henry Schäfer is holder of the Chair for Business Administration and Financial Economics at the University of Stuttgart. He is also founding partner of start ups in sustainable finance. Among others he founded EccoWorks, an advisory firm for the integration of sustainability issues in investments and business development strategies. Prior to his academic career he worked as a financial consultant in an international M&A boutique and headed the marketing division of a German medium-sized bank which later on became a subsidiary of SEB.

Prof. Schäfer started his research in SRI and CSR already 20 years ago. He is one of the leading research capacities in these fields with special focus on financial intermediaries, financial markets and family-owned companies. He has published several text books in finance and contributed many articles to academic journals. As a researcher with a practical background and understanding, he deals successfully with the consulting of banks, institutional investors and foundations in SRI and CSR related areas.

Friedrich Völker is leading the competence center "Digital Products" at Alfred Kärcher GmbH & Co KG since October 2015. He is responsible for managing Kärcher's IoT solutions in the B2B sector.

After publishing his dissertation on family business CSR he started his career as the management assistant to the CEO of Kärcher in 2013.

Dr. Völker studied business administration with a focus on family owned enterprises and business law.

He has graduated with both a Master of Science degree (European Business School, Germany) and an MBA (Pepperdine University, USA).

Supplier Engagement in the Sustainable Innovation Process: A Qualitative Analysis of Austrian SMEs

Christine Bachner

1 Introduction and Objectives

Over the last decades, the complexities of doing business have increased many times over and today companies face countless challenges such as resource issues, climate change and pollution. In order to make the most of the opportunities offered by changing global and market dynamics and structures, companies need to innovate constantly and sustainably. In this context, stakeholder engagement can be viewed as a valuable—yet underexplored—source of innovation. The concepts of stakeholder engagement and innovation have so far remained rather isolated from each other, despite the fact that opening up the innovation process to include stakeholders can increase companies' innovation performance and reputation (Gould 2012).

It is generally and inherently accepted that managing stakeholders appropriately can also help reduce risk and improve companies' social responsibility (Jenkins 2006). Proponents of sustainability and Corporate Social Responsibility (CSR) like Stigson (2002) argue that "*it is clear that society expects much more from companies than simply a well-made product or a reliable service at the right price*" (p. 24). By pursuing CSR initiatives, some companies have developed very innovative products and services that simultaneously are beneficial to the company's profitability (Asongu 2007).

Research shows that although typically faced with substantial resource restrictions, Small and Medium-Sized Enterprises (SMEs) are successful innovators, due to their small and versatile structure and entrepreneurial posture (Altenburger and Gaissberger 2014). As little is known about how stakeholder engagement is used in the innovation process of SMEs, this paper is focused on analysing the importance

C. Bachner (✉)
Department of Business, IMC University of Applied Sciences Krems, Krems, Austria
e-mail: christine.bachner@fh-krems.ac.at

© Springer International Publishing AG, part of Springer Nature 2018   335
R. Altenburger (ed.), *Innovation Management and Corporate Social Responsibility*,
CSR, Sustainability, Ethics & Governance,
https://doi.org/10.1007/978-3-319-93629-1_19

of stakeholder engagement for creating sustainable innovations by using in-depth case studies on Austrian SMEs. A special focus will be put on suppliers as partners in the innovation process of sustainable products and processes. Hence, the following research question is to be answered in the course of this study: What role do suppliers play for creating sustainable innovations? Specifically, we seek to identify how and in which of the individual phases of the innovation process suppliers are engaged in. This will also allow us to both build a common frame of understanding issues of sustainability in SMEs and identify future directions of stakeholder engagement research in SMEs, thereby also contributing to the understanding of the sustainable innovation process and its links to stakeholder engagement.

The paper starts with reviewing key concepts on stakeholder engagement and innovation as well as discussing their relationship. After an overview of the methods used in the empirical research, the main results of the study are presented. Good practice case studies of ten Austrian SMEs provide valuable information on the significance and importance of partnering with suppliers for sustainable innovations. A summary of the findings as well as an outlook on future challenges and research activities concludes the paper.[1]

2   Literature Review

Various actors are influenced by and can influence companies' opportunities to achieve a competitive advantage—highlighting the relevance of stakeholders and, thus, the importance of stakeholder management and engagement (Perrini 2006). Freeman (1984) was the first to emphasize the importance of the role of stakeholders in relationship to the organization as he argued that multiple players interact with the organization. For Freeman (1984) a stakeholder is *"any group or individual who can affect or is affected by the achievement of the firm's objectives"* (p. 46). By suggesting that the needs of shareholders cannot be met without satisfying the needs of other stakeholders, stakeholder theory offered a new form of managerial understanding and action in a strategic planning context and a new way to organize thinking about firms' responsibilities (Jonker and Foster 2002; Foster and Jonker 2005). A key principle of stakeholder theory is that the interests of all legitimate stakeholders have value and it is critical that firms' decisions consider these interests (Donaldson and Preston 1995).

Today, the idea that companies have stakeholders is common in management literature. However, the definition of a stakeholder, the purpose and the character of the organization as well as the role of managers still remain somewhat unclear and are contested in literature. Even Freeman changed his definition of stakeholders over

---

[1]This paper is a revised version of a paper entitled 'Supplier Partnerships for Creating Sustainable Innovations—The Case of Austrian SMEs', presented at the RENT XXX Conference, November 16–18, 2016, Antwerp, Belgium.

time, as today Freeman et al. (2010) view stakeholders as *"those groups without whose support, the business would cease to be viable"*. They assume, however, that *"the debate over finding one the true definition of stakeholder is not likely to end"* (p. 26).

Freeman et al. (2010) also stress the importance of stakeholder engagement by arguing that in order *"to successfully create, trade, and sustain value, a business must engage its stakeholders"* (p. 282). Greenwood (2007) views stakeholder engagement as those *"practices that the organization undertakes to involve stakeholders, in a positive manner in organizational activities"* (p. 317). She argues, however, that engagement with stakeholders is not necessarily equal to responsible business behaviour and, thus, that the argument that stakeholder engagement equals responsible treatment of stakeholders is simplistic.

The concepts of stakeholder engagement and innovation can be linked by highlighting that companies increasingly create new sources of value by cooperating with stakeholders (Freeman et al. (2010). Innovation is recognized to play a central role in creating value and sustaining a competitive advantage (Baregheh et al. 2009). Despite the fact that the academic interest in innovation has started more than a 100 years ago, the concept of innovation is still intricate in nature and many ambiguities exist as to its definition (Adams et al. 2006). According to Schumpeter (1934), innovation consists of any one of the following phenomena: (1) introduction of a new good, (2) introduction of a new method of production, (3) opening of a new market, (4) conquest of a new source of supply or raw materials or half-manufactured goods, and (5) implementation of a new form of organization. For Christopher Freeman (1974), *"technical innovation or simply innovation is used to describe the introduction and spread of new and improved products and processes in the economy and 'technological innovation' to describe advances in knowledge"* (p. 18). The Organization for Economic Co-operation and Development (OECD) defines innovation as the *"implementation of a new or significantly improved product (good or service), or process, a new marketing method, or a new organizational method in business practices, workplace organization or external relations"* (OECD and Eurostat 2005, p. 46).

For Klewitz and Hansen (2014) sustainable innovation involves successfully competing in changing markets and environments while at the same time contributing to sustainable development. This kind of innovation demands the unforeseen identification of more efficient methods of doing business or new types of products or services that may not have occurred without CSR initiatives in the first place (Asongu 2007).

Increasingly though, it is being recognized that a single organization cannot innovate in isolation, but that it has to engage with different types of stakeholders to acquire ideas and resources from the external environment to stay abreast of competition. This is what Chesbrough (2003) calls openness or open innovation. Chesbrough's rather broad definition of openness highlights that valuable ideas emerge and can be commercialized from inside or outside the firm.

Research findings highlight the role of the stakeholder as an opportunity for facilitating innovation (Hart and Sharma 2004; Kanter 1999). Gould (2012) argues that open innovation and stakeholder engagement even describe similar organizational processes, as in both cases, the focal organization reaches outside its boundaries making an explicit effort to access essential information. These two concepts, however, have remained isolated from each other and research linking open innovation and stakeholder engagement remains scarce (Gould 2012). Benefits for firms interacting comprehensively with their stakeholders in the innovation process are diverse, including considerable amount of external knowledge exploration and exploitation (Vanhaverbeke et al. 2008), cost reduction, shorter time-to-market as well as gaining stronger credibility and access to partners' network (Martovoy 2014). Sharma (2005) also finds that stakeholder engagement allows the organization to access information from its stakeholders which can then be used help understand and respond to emerging social and environmental issues.

Nevertheless, innovations created together with stakeholders can be challenging as they require the assimilation of knowledge and expectation management at the same time (Magnusson et al. 2003). Boesso and Kumar (2009) highlight the practical difficulties and limitations involved in identifying and soliciting the views of all stakeholders affected by an organization's activities, and in how management may prioritize stakeholder relationships. Collins and Kearins (2007) also mention that there is a risk for the businesses involved in not achieving their preferred outcomes. Compromises may be necessary and financial burdens may not be borne equally by all participants. Organizational stakeholders' interest, time, and resources can vary enormously. Finally, there is the risk at a societal level that the outcomes may not be optimal for sustainability (Collins and Kearins 2007).

Stakeholder research in the past has tended to focus on large companies (Jenkins 2006), despite the fact that small and medium enterprises' combined achievements have the potential of generating a high impact for the global economy and society. Due to their flexibility, SMEs are able to respond quickly to changing circumstances; SMEs can, therefore, rapidly take advantage of new niche markets for sustainable products and services that incorporate social and/or environmental benefits (Jenkins 2004). Hockerts et al. (2007) also find that sustainable innovations, resulting from engaging with stakeholders, do indeed often emerge in SMEs.

With regard to which stakeholder groups are best suited for increasing a firm's innovative performance, the relationship between a company and its suppliers is becoming an important issue (Feng et al. 2010). It is even argued, that supplier involvement has a key role in new product development and that involving suppliers is a way to leverage their knowledge to reduce costs and lead times. Issues such as the level of adaptation between companies' needs and the products or services provided by suppliers, the suppliers' response capacity to firms' demands or the ability to establish routines between both agents to achieve a better performance must be considered in supplier relationships. In this context, communication, networking and collaboration among firms and their suppliers play an important role for the development of innovative activities and knowledge creation (Delgado-Verde et al. 2014).

3   Research Design and Methods

The multiple method case-study analysis (Yin 2003) provides insights into supplier engagement in innovation processes of SMEs. A case study involves the study of an example—a case—of the phenomenon being researched and the use of a variety of types of data and data analysis can be said to be a key feature of the case study method (Yin 2003). We deem the case study methodology an appropriate approach to our research, as there is limited theoretical knowledge of stakeholder engagement in innovation processes and its application in practice (Siggelkow 2007).

For the case studies, we identified ten Austrian SMEs standing out and awarded for their innovative behaviour and sustainability engagement in the past. Data was gathered by conducting 29 semi-structured interviews and on-site visits. All the interviews were recorded, transcribed and analysed based on Mayring's qualitative content analysis approach (Mayring 2010). On top of that, secondary information, such as sustainability reports, journal articles etc., was integrated into the case studies. The data gathered from the interviews, information already available was contrasted with relevant literature, and an analysis was conducted for evaluating the importance of engaging suppliers in the innovation process. Table 1 provides an overview of the case study-companies.

Table 1 Overview of case study companies

Company	Core products/ services	Size (number of employees)	Number of interviews
A	Hospitality	32	3 (Owner, Head of Quality Management, Front Office Manager)
B	Printing	51	3 (Owner, Head of Sales and CSR, Project Manager)
C	Chocolate	160	3 (Owner, Head of Marketing, Head of Production
D	Mineral water Near water drinks	181	3 (Managing Director, Head of Marketing, Head of R&D)
E	Enamel tableware and industry products	121	3 (Owner, Head of Production and R&D, Head of Sales)
F	Beer, soft drinks	180	3 (Managing Director, Environmental Officer, Chief Engineer)
G	Electronic sanitary systems	50	3 (Owner, Operating Manager, Innovation Manager)
H	Tea, coffee, spices, gift items	176	3 (Head of Quality Management, Head of Product Management and Development, Head of Marketing)
I	Paper manufacturing	138	2 (Managing Director, Technical Director)
J	Sportswear manufacturer	200	3 (Managing Director, Head of CSR and Sales Head of Product Management)

The research was being conducted as part of the project "CSR and Innovation" which is funded by the Austrian Research Promotion Agency (FFG) (2013–2018). While the case study research presented in this paper is part of the first phase, the overall aim of this project is to analyse practices used in a CSR-driven innovation process and to develop a model for CSR-driven innovation in SMEs.

4   Research Findings

The SMEs we studied are characterized by individual personalities and personal relationships, limited (financial) resources as well as by the existence of only selected formal internal systems and processes. Values play a crucial role in all SMEs when it comes to sustainability and the company's responsibility. Sustainability implies fairness to stakeholders, such as business partners, employees and the environment. Respectfully dealing with partners and suppliers around the world and enabling a sustainable quality of life through fair pay and long-term relationships is prioritized. These relationships are not to be burdened by ongoing price negotiations. Regionality also plays a central role, as many raw materials and primary products are sourced regionally to preserve Austrian jobs and keep transport routes short and efficient. Company A states that potential future partners must share the SME's attitude towards the environment and quality and provide supplies that meet environmental requirements. This compliance is evaluated regularly.

Overall, the case study-companies demonstrate a strong stakeholder orientation. Stakeholder relations are actively cultivated and seen from a value creation perspective. The relationships SMEs have formed in the past are long-lasting, built on mutual trust and borne by appreciation and respect. Most SMEs have had a long-lasting attachment and devotion to their region of operation, often for generations.

For analysing the innovation process we assumed an innovation process that is linear spanning across several stages—idea generation, idea evaluation, prototyping and market launch. We found that in almost all cases innovations are not created 'behind closed doors', but in close interaction with various stakeholders. As partners for innovation purposes, companies choose those that share their attitude towards and understanding of quality, sustainability and responsibility. Stakeholders offer access to knowledge and expertise for creating sustainable solutions. Table 2 shows the stakeholders SMEs involve in the innovation process, especially when it comes to generating ideas for potential innovations. It becomes clear that, beside employees as the most obvious and easily reachable stakeholder group, suppliers are the most important stakeholder group that SMEs engage with for creating sustainable innovations. Indeed, suppliers are involved in nearly all phases of the innovation process. They often deliver impetus for new ideas and serve as project partners in developing innovations. In the companies we studied supplier involvement may range from giving minor design suggestions to being responsible for the complete new product development, design and engineering of a specific part of assembly. Company G highlights that the trust between the company and their suppliers is more important

Table 2 Different stakeholder groups generating ideas for the company

Company	Industry	Suppliers	Customers	Employees	CSR-networks	Industry-networks	Universities
A	Hospitality	x	x	x	x	x	x
B	Printing	x	x	x			
C	Chocolate	x		x			
D	Mineral water	x	x		x	x	x
E	Enamel tableware	x	x	x	x	x	x
F	Beer, soft drinks	x		x	x	x	x
G	Electronic sanitary systems	x	x	x			x
H	Tea, coffee, spices		x	x			
I	Paper manufacturing	x	x	x		x	x
J	Sportswear manufacturer	x	x	x		x	x
Sum		9/10	8/10	9/10	4/10	6/10	7/10

x denotes stakeholder group is important for the company

than contracts and confidentiality agreements. The company's owner refers to suppliers as partners and stresses that problems can only be solved jointly. According to him this partnership approach is a specific characteristic of medium-sized companies and what makes them stand out.

Submitted ideas are collected in a pool and evaluated. At this stage, the involvement of suppliers is limited.

In the event of an innovation being created in collaboration with an external partner, the mode of cooperation has to be defined. The partners may be supported with investments, if necessary, and their workforce trained appropriately. The feasibility of a sustainable innovation needs to be synchronized with many partners within and outside the company. To do so, companies sometimes have to assume a moderating function. On top of that, demands by different internal stakeholders have to be balanced and compromises found. These aspects along with the fear of suppliers sharing company secrets can be identified as setbacks of working together closely with external partners.

In many of the companies we studied, prototypes are being created, which are assessed and improved if necessary. For prototyping, companies often partner with suppliers, especially when it comes to innovations which require new machines or an adjustment of existing equipment. Employees are encouraged to test the prototypes and give feedback to the project team, which may also include suppliers. Employees, thus, work jointly with suppliers to develop sustainable solutions.

In conclusion, we could identify a positive and valuable influence of suppliers on sustainable product and process innovation. Enduring relationships with suppliers are essential for cooperation in the innovation process. This collaboration is particularly fruitful as it may accelerate the process of innovation activities. We argue that partnerships with suppliers enable these companies to work more effectively with a number of crucial suppliers willing to share responsibility for the success of an innovation project. Identifying and engaging with key suppliers, thus, becomes increasingly critical to long-term corporate viability.

5   Discussion

The last pages emphasized the significance of supplier engagement for creating sustainable innovations in SMEs. Technological, social and environmental innovation is necessary to bring about sustainability, and a supplier partnership approach is critical for achieving results. Suppliers are involved from the front end of the innovation process, starting with generating and evaluating ideas right through to jointly creating and reworking prototypes and launching a new product, service or process. Collaborating with suppliers is a means to exchange experiences and information on best practices and offers benefits such as operational cost savings, further partnership opportunities along with a better approach to emerging consumer concerns. This is especially important for SMEs, which often have to deal with the finiteness of resources. Collaboration with suppliers in dynamic environments can also be related to higher efficiency as it enables the adaptation to rapid changes and facilitates the acquisition and exchange of valuable information and knowledge. Apart from providing highly valuable knowledge for creating sustainable innovations, social and relational capital is produced that can have interesting effects on innovation. Thus, suppliers play a stimulating role for SMEs and are increasingly becoming an indispensable partner in their innovation process. For the researched SMEs, this partnership approach is a specific characteristic of SMEs and what makes them stand out. The closeness and trust, which characterizes many supplier-relationships, turns them into supplier-partnerships implying a more intense and long-term commitment.

There are a number of limitations of our study. First and foremost, we only reviewed a small segment of businesses assuming a general applicability on companies in other countries, sizes and branches. Additionally, we did not focus enough on the risks and disadvantages of collaborating with suppliers for creating sustainable innovations. On top of that, the differences between sustainable innovations and "normal" innovations need to be focused on and worked out more thoroughly. As the case studies unveiled that suppliers are the most important stakeholder group that SMEs engage with for creating sustainable innovations, further studies need to focus on the risks associated with disclosing company secrets to partners and how to deal with these risks.

Ultimately, our research is aimed at helping SMEs gain an innovative edge through the integration of different stakeholder groups and the timely consideration of trends as well as the needs of society. This also allows us to build a common frame of understanding issues of sustainability in SMEs as well as identify future directions of stakeholder engagement research in SMEs.

References

Adams, R., Bessant, J., & Phelps, R. (2006). Innovation management measurement: A review. *International Journal of Management Reviews, 8*, 21–47. https://doi.org/10.1111/j.1468-2370. 2006.00119.x

Altenburger, R., & Gaissberger, C. (2014). Sustainable innovation: Evidence from Austrian SMEs. In *The XXV ISPIM Conference – innovation for sustainable economy & society*. Presented at the ISPIM Conference, Dublin.

Asongu, J. J. (2007). Innovation as an argument for corporate social responsibility. *Journal of Business and Public Policy, 1*(3), 1–21.

Baregheh, A., Rowley, J., & Sambrook, S. (2009). Towards a multidisciplinary definition of innovation. *Management Decision, 47*, 1323–1339. https://doi.org/10.1108/00251740910984578

Boesso, G., & Kumar, K. (2009). An investigation of stakeholder prioritization and engagement: Who or what really counts. *Journal of Accounting and Organizational Change, 5*, 62–80. https://doi.org/10.1108/18325910910932214

Chesbrough, H. W. (2003). *Open innovation: The new imperative for creating and profiting from technology*. Boston, MA: Harvard Business School Press.

Collins, E., & Kearins, K. (2007). Exposing students to the potential and risks of stakeholder engagement when teaching sustainability: A classroom exercise. *Journal of Management Education, 31*, 521–540. https://doi.org/10.1177/1052562906291307

Delgado-Verde, M., Martín de Castro, G., Navas-López, J. E., & Amores-Salvadó, J. (2014). Vertical relationships, complementarity and product innovation: An intellectual capital-based view. *Knowledge Management Research and Practice, 12*, 226–235. https://doi.org/10.1057/kmrp.2012.59

Donaldson, T., & Preston, L. E. (1995). The stakeholder theory of the corporation: Concepts, evidence, and implications. *Academy of Management Review, 20*, 65–91. https://doi.org/10.5465/AMR.1995.9503271992

Feng, T., Sun, L., & Zhang, Y. (2010). The effects of customer and supplier involvement on competitive advantage: An empirical study in China. *Industrial Marketing Management, 39*, 1384–1394. https://doi.org/10.1016/j.indmarman.2010.04.006

Foster, D., & Jonker, J. (2005). Stakeholder relationships: The dialogue of engagement. *Corporate Governance: The International Journal of Business in Society, 5*, 51–57. https://doi.org/10.1108/14720700510630059

Freeman, C. (1974). *The economics of industrial innovation*. Harmondsworth: Penguin Books.

Freeman, R. E. (1984). *Strategic management: A stakeholder approach, Pitman series in business and public policy*. Boston: Pitman.

Freeman, R. E., Harrison, J. S., Wicks, A. S., Parmar, B. L., & de Colle, S. (2010). *Stakeholder theory: The state of the art*. Cambridge: Cambridge University Press.

Greenwood, M. (2007). Stakeholder engagement: Beyond the myth of corporate responsibility. *Journal of Business Ethics, 74*, 315–327. https://doi.org/10.1007/s10551-007-9509-y

Hart, S. L., & Sharma, S. (2004). Engaging fringe stakeholders for competitive imagination. *IEEE Engineering Management Review, 32*, 28–28. https://doi.org/10.1109/EMR.2004.25105

Hockerts, K., Copenhagen Business School, CBS Center for Corporate Social Responsibility, & Copenhagen Business School. (2007). *Managerial perceptions of the business case for corporate social responsibility*. Frederiksberg: CBS Center for Corporate Social Responsibility.

Jenkins, H. (2004). A critique of conventional CSR theory: An SME perspective. *Journal of General Management, 29*, 37–57.

Jenkins, H. (2006). Small business champions for corporate social responsibility. *Journal of Business Ethics, 67*, 241–256. https://doi.org/10.1007/s10551-006-9182-6

Jonker, J., & Foster, D. (2002). Stakeholder excellence: Framing the evolution and complexity of a stakeholder perspective of the firm. *Corporate Social Responsibility and Environmental Management, 9*, 187–195.

Kanter, R. M. (1999). From spare change to real change: The social sector as beta site for business innovation. *Harvard Business Review, 77*, 122–132.

Klewitz, J., & Hansen, E. G. (2014). Sustainability-oriented innovation of SMEs: A systematic review. *Journal of Cleaner Production, 65*, 57–75. https://doi.org/10.1016/j.jclepro.2013.07.017

Magnusson, P. R., Matthing, J., & Kristensson, P. (2003). Managing user involvement in service innovation: Experiments with innovating end users. *Journal of Service Research, 6*, 111–124. https://doi.org/10.1177/1094670503257028

Martovoy, A. (2014). Advantages and disadvantages of open innovation: Evidence from financial services. In A.-L. Mention & M. Torkkeli (Eds.), *Innovation in financial services a dual ambiguity* (pp. 259–294). Newcastle upon Tyne: Cambridge Scholars Publishing.

Mayring, P. (2010). Qualitative Inhaltsanalyse. In G. Mey & K. Mruck (Eds.), *Handbuch Qualitative Forschung in der Psychologie* (pp. 601–613). Wiesbaden: VS Verlag für Sozialwissenschaften.

OECD, Eurostat. (2005). *Oslo manual, the measurement of scientific and technological activities*. Paris: OECD.

Perrini, F. (2006). SMEs and CSR theory: Evidence and implications from an Italian perspective. *Journal of Business Ethics, 67*, 305–316. https://doi.org/10.1007/s10551-006-9186-2

Schumpeter, J. A. (1934). *The theory of economic development: An inquiry into profits, capital, credit, interest, and the business cycle*. Cambridge: Harvard University Press.

Sharma, S. (2005). Through the lens of managerial interpretations: Stakeholder engagement, organizational knowledge and innovation. In S. Sharma & J. A. Aragón Correa (Eds.), *Corporate environmental strategy and competitive advantage, new perspectives in research on corporate sustainability* (pp. 49–70). Cheltenham, UK: Edward Elgar.

Siggelkow, N. (2007). Persuasion with case studies. *Academy of Management Journal, 50*, 20–24. https://doi.org/10.5465/AMJ.2007.24160882

Stigson, B. (2002). Pillars of change: Business is finally learning that taking care of the environment and meeting social responsibilities makes good business sense. *Forum for Applied Research and Public Policy, 16*, 23.

Vanhaverbeke, W., Van de Vrande, V., & Chesbrough, H. (2008). Understanding the advantages of open innovation practices in corporate venturing in terms of real options. *Creativity and Innovation Management, 17*, 251–258. https://doi.org/10.1111/j.1467-8691.2008.00499.x

Wayne Gould, R. (2012). Open innovation and stakeholder engagement. *Journal of Technology Management & Innovation, 7*, 1–11. https://doi.org/10.4067/S0718-27242012000300001

Yin, R. K. (2003). *Case study research: Design and methods* (3rd ed.). Los Angeles: SAGE.

Christine Bachner is working as a researcher at the IMC University of Applied Sciences in Krems and holds a master's degree in International Business (University of Applied Sciences Kufstein) as well as a master's degree in Sociology (Johannes Kepler University Linz). She is currently writing her dissertation at the Research Institute for Family Institute (Vienna University of Economics and Business), focusing on CSR-driven innovation in family businesses. Her research interests and publications focus on (1) Corporate Social Responsibility, (2) Innovation and (3) Family Businesses. Her core competencies include qualitative and quantitative research methods.

Corporate Social Responsibility: Australian Case Study Innovation Capabilities: Not for Profit: Transforming Families and Children

Tricia Murray and Dianna Vitasovic

1 Introduction

Many of the societal issues facing families and children face are considered complex or wicked problems (Rittel and Webber 1973). To solve complex, wicked problems, leaders learn to focus and manage polarities and value the deliberate practice of reframing and re-solutioning dilemmas through sustainable innovation. Strengthening the enterprise innovation capability improves business sustainability and results in sustainable innovation and evolutionary fitness (Caradonna 2014; Teece 2007). Enterprises that continuously implement radical innovation have an appetite to manage risk and thereby continuously adapt and shape business success (Teece 2007).

The questions being considered are: how does society create sustainable social change for families and children? How does an organisation target strategic business activities to address long term social change that make a difference to families and children? Innovation performance is considered through the lens of leadership performance. Leadership in an organisational setting is achieving outcomes, by looking through the wider social lens of their CSR, their environmental, social and economic action and outcomes. However, not all sectors of society use the same terms to describe their innovation actions. How does society provide social change through organisational efforts when the sector language and new concepts used are different?

T. Murray (✉)
Wanslea, Scarborough, WA, Australia
e-mail: tmurray@wanslea.asn.au

D. Vitasovic
Innovation Culture, City Beach, WA, Australia
e-mail: diannav@innovationculture.com.au

© Springer International Publishing AG, part of Springer Nature 2018
R. Altenburger (ed.), *Innovation Management and Corporate Social Responsibility*,
CSR, Sustainability, Ethics & Governance,
https://doi.org/10.1007/978-3-319-93629-1_20

This case study examines how Wanslea, a NfP community organisation, targeted strategic sustainable practices, that resulted in social value for families and children. Outlined are the four major strategic and emerging practices that resulted in building Wanslea's current scaffold that enables adaptation to the environment and delivers social innovation. Initially, Board strategy and board selection; Foundational strategies for aligning research, evaluation and leadership education and professional practices; Brand awareness and reputation; and Partnership collaborations resulted in growth, professional expertise and expanding reach families and children, for resulting in value creation and positively influencing business and wider society.

Outlined are four targeted strategic practices that have resulted in strong community family focus in delivering community and social outcomes, and sustainable business. However, one should be cognisant of the fact that there are a multitude of intervening and interwoven elements that act as subsidiary factors impacting on the success of Wanslea.

Examined are historical examples of how of Wanslea, operating within mission, values and strategy, drives leadership actions for social change. The NfP industry context and its paradoxical tensions, are relevant to Wanslea's strategic landscape, operating within the external context of dynamic political challenges to deliver value creation. Wanslea delivered sustainable business and social outcomes by being proactive in extending its family programs and innovative service design within local and regional communities. The Wanslea brand and reputation are critical for it to strategically fit within an ambiguous political landscape with changing needs of families and children in risk adverse political and business conditions.

2 Social Impact and Paradoxical Tensions

Social impact and value are being measured by effectiveness, program impact, sustainable practices, and return on investment (OECD Development Assistance Committee 2010, p. 21). In Australia, organisations that support very complex social issues are expected to support their approaches with evidence. We see political pressure in the sector calling for quantitatively measured interventions for social programs to demonstrate a return on investment. When the social impact lens places the focus on measurement for organisations we see innovation becoming internally driven through systems, processes, and design of programs.

Tensions and paradoxes (see Fig. 1) are recognised as an inherent part of organisational life (Smith and Lewis 2011). The concept of value in social settings is thwart by multiple perspectives and the solutions depend upon who is asking the question, or if the right question is being asked. The question that needs to be considered is: How does society create sustainable social change for families and children? Increasingly, government sets policies, however, NfP organisations identify gaps in society and, being mission driven, make inroads to address societal issues. Many of these institutions are small and underfunded so they engage across

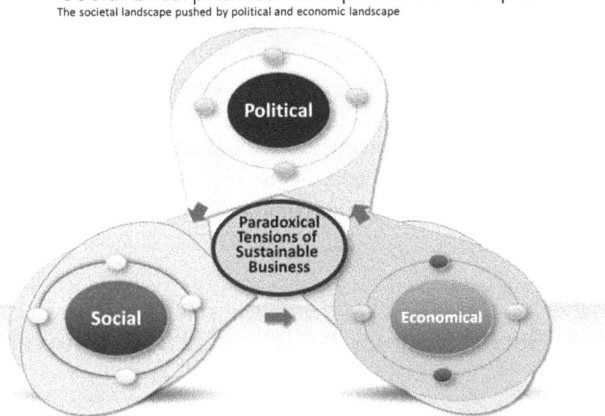

Fig. 1 Paradoxical tensions: Social enterprise landscape and social impact (Vitasovic 2017)

the sector and implement through partnerships, often depending on government grants or other donations.

Depending on the maturity of the NfP sector, leadership really comes from those with organisational capability for creating social innovation that sustains social value for families and children. These NfPs as outlined in Fig. 1, Paradoxical Tensions, operate within a dynamic and complex political, economic and social environment, where policy is dynamic and funding variations dictate the direction of service delivery. Organisations provide social services amongst growing community expectations. They operate in a social and political minefield. To sustain their enterprise, they engage in sector-wide collaborations.

The goal of innovation is to identify the value proposition for customers, and where possible, a scalable market within the context of the social landscape. In the context of delivering social value to address complex needs of society, a NfP determines its mission as its desire for creating social value, and its contribution towards social change. Social Innovation is driven by creating value and scale with social impact. For a NfP to survive economically, a paradoxical tension is automatically created. The NfP needs to identify and maintain a sustainable economic business model and obtain sustainable funding. The paradoxical tensions exist while delivering both social and economic value for wicked social issues, whilst being pushed and pulled between policy reporting and evaluation within an organisational landscape of stretched budgets and limited resourcing. Political funding cycles are short term and social programs require long term thinking. All this operates within paradoxical tensions spinning off a multitude of partnerships, between multiple economic and political stakeholders with different agendas for short term thinking about delivery mechanisms. There is a great deal of interest in the language around '*evidence based practices*' to indicate that innovation continues to take place with little evidence from longitudinal research.

Innovation Management theory is moving from the domain of Research and Development into the future of leadership mainstream practices. The design of social services and programs is dependent upon leaders understanding the context of the innovation, the depth, scope and scale, within the complexity of social relationships.

Social Innovation in the literature has recently grown within a wider context of CSR and Social Impact. From a practical viewpoint, the NfP sector has paid little attention to known systematic concepts of Innovation Management, however their programs have clearly demonstrated innovation management practices.

3   Corporate Social Responsibility and Social Innovation

Western Australia has a long history in the education, mining and agriculture sectors with global influence, practices and growth in these sectors. It is through these global connections that NfPs find the influence of the CSR terms prevailing different industries. It is therefore, from an Australian context, important to define the terms we apply.

Determining working definitions for terms such as CSR, innovation management and social innovation are best understood in an Australian context. Even with the European emergence of an integrative view on corporate sustainability, the literature on corporate sustainability and innovation (social) is growing at a rapid pace with diversity among scholars in multiple scientific and social fields.

The Commonwealth of Australia commissioned research in 2016 with a view to understand the need and make up of community partnerships. The Department of Social Services, together with other university and consulting companies, presented those findings and definitions in a report, Giving Australia project, in December 2016.

CSR is determined by the citizen or business as the licence to operate in that community. With the decrease in resources of the public sector, we increasingly see Australian businesses being asked to contribute to solving wicked social issues, such as families, aged care, Aboriginal health, mental health, youth unemployment and homelessness. In the case of complex family social structures and children, there is an increased awareness that dysfunction in families and children, left unresolved, increases social complications. Organisational responsibility has extended to wider stakeholders of business community and social issues (Googins 2013).

Sustainability was a term applied to the environment and economic responsibility in the mining, oil and gas industries as meaning, "licence to operate". There has been a shift of meaning amongst the global community with a widened focus that includes society and those in need in the community. In the research, people will talk about CSR differently, and the terms blend across sectors, inclusive of the wider stake-holders, meaning society (Aluchna and Idowu 2017; Dahlsrud 2008).

There are many terms that appear to blend across sectors, such as:

Sustainability: focused on the meaning of deliberate practices to politically, eco-nomically, socially, emotionally satisfy the governance and social stakeholders to simultaneously innovate that creates multiple value. It does not necessarily focus on the ecological issues.

Multiple Value Creation: creation of more than one value: children, families, the community, society and economic value. It includes the value of staff, the Board, and the value of being emotionally connected to a higher meaning and purpose that may include volunteerism.

Corporate Social Responsibility: is the way in which business consistently creates shared value in society through economic development, good governance, stakeholder responsiveness and environmental improvement. It implies an integrated, systematic approach by business that builds value (Visser 2011). The term CSR extends to incorporate a corporate responsibility for collective societal wicked problem(s) that impacts on that business and its community. The evaluation of social outcomes reflects in a long term, strategic connection with the way the organisation organises its resources to deliver higher value to society. A wicked social need is addressed with a NfP in collaboration with other agencies, a corporate and or government agency. In general, NfP mission is to look out for the interests of society.

The CSR motivation of a NfP is not the same as that for a corporation. It is likely that corporations use the term CSR, and the NfP uses terms such as social impact and is driven by social mission. Corporations are also looking at shifting their focus from environmental CSR to socially engaging with community philanthropic projects. This adds further dimensions of tension for example, where larger corporations inadvertently add pressure on NfP partnerships when they genuinely look for the good news stories to share their successes, build public goodwill and community recognition.

4 Context of the NFP Industry

Australia faces challenging times in the NfP sector amidst the growing social needs of family and children and funding restrictions.

The Australian Bureau of Statistics (*Australia demographic statistics, March Quarter 2015*-2011—2016) reports that Australia has approximately 24 million people living in 10 million dwellings and people's life expectancy is 82 years. Western Australia has a population of 2.617 million people. Like most developed countries, Australia's population is ageing and the proportion of Australia's population aged 15–64 years remains stable. The proportion aged under 15 years decreased from 21.4% to 18.8% during this period. Children aged 0–14 years increased by 1.3% to 58,900 people. Australian population projections indicate that the number of people aged 65 years and over is projected to exceed the number of children aged 0–14 years by 2030.

In the Giving Australia Report (Scaife et al. 2016), the Reform in Funding Report identified:

> With more than 600,000 NfP organisations in Australia, accounting for more than 3.8% of GDP in 2012–2013, the NfP sector employs almost 10% of Australia's workforce. It also provides an increasing number of services on behalf of government—including those in areas identified as 'sectoral hotspots: aged care, education, medical research and others.

In 2017–2019, policies in contracting arrangements by state and federal governments will alter in unprecedented ways. This is against the background of five changes in the Prime Minister and State Ministers and elections at both levels of government to occur during this period. The NfP sector was reviewed by government in 2008–2009 to identify the sector, growing demand numbers and its viability to deliver sustainable business and social outcomes. The government push is on introducing evidence based Social Impact quantifiable measures to deliver economic and social value. These changes will challenge the existence of smaller NfP organisations.

It is a challenging time for community based NfP organisations as both Australian State and Federal Governments focus on tightening their expenditure and service offering to cope with budget deficits against tensions of increasing demand. This environment also presents opportunities as Governments continue to seek out community based organisations as partners to drive and deliver social innovation, efficiency and effectiveness across the NfP and other sectors. This will be particularly the case for those groups that are seen to present as future financial burdens.

Global, economic, social and commercial imperatives to address family issues across society are of growing concern. Family services can no longer be solely provided by governments. Yet, NfPs generally report that government funding and sponsorships are difficult to obtain for long term sustainable contracts, with the reality being a greater short-term reporting focus. With a shrinking funding model and growing debt by governments, Australia forecasts slower economic growth and continued funding business model disruptions over the next 3 years. Radical social innovation, that has significant strategic impact, represents a shift of 30% reduction in cost, or 5–10x performance improvement (Connor et al. 2008), may well be the mantra of the NfPs against these funding models and contract challenges. Traditionally NfPs direct their profits back into programs. The capacity to provide resources for innovation is limited as it is not a recognised expenditure item by funders. To remain viable, organisations need to understand the political landscape and potential future agendas, especially areas of measuring social impact and the degree of innovation.

5   Historical Context of Wanslea

Wanslea holds an influential position in the community services sector and is recognised as a strong advocate for family and children.

Wanslea was founded in 1943 by Florence Hummerston who dedicated her life to serving the community and was passionate about the welfare of children. In 1941, Hummerston became the Founding President of the Women's Australian National Service (WANS) War Fund. It was during this time that the need for an organisation that could care for the children of mothers who were ill and whose fathers were at war became apparent. Wanslea was established 2 years later. Wanslea has touched the lives of thousands of West Australians in times of need. The goodwill and respect that has been earned by Wanslea continues to be reinvested in the care of families and children through the many programs now offered over its locations across Western Australia.

Wanslea has increased business turnover sixfold since 2004 and has developed a number of collaborations and partnerships to extend its reach. Increases in services and growth resulted in staff numbers rising from 65 to 300 since 2004. Economically the Board considered revenue of $5m vulnerable because volatile financial shocks or changes to any government policy put the organisation at risk. Turnover in 2004 was $6.5m and in 2016–2017 its revenue is $30 million. Strategically back in 2004, the Board review determined the key assets of the organisation, were not its buildings and properties but its programs and people. It identified a considerable number of programs were innovative, evidence based, and its personnel highly trained, with individual reputations in the sector.

The Mission of Wanslea is to promote community, family and individual development through partnerships and services. Wanslea has established strong partnerships with Commonwealth, State and Local Governments, other not for profit organisations and private companies. These relationships are important to Wanslea as the organisation continues to focus on delivering client centred services that are valued by the community. Strategic areas of focus over the past 3 years were to continue to grow and build on past achievements to remain a sustainable, effective, resilient and compassionate organisation, and to innovate through its research and development program.

Wanslea growth of services expanded to sites across the metropolitan Perth and the Bindjareb, Wheatbelt, Great Southern and Goldfields/Esperance regions, spanning 1.6m square kilometres. In 2010, it expanded into childcare and after school care and this area of operation is now one third of the turnover.

Families and children have been at the heart of Wanslea since the organisation was established and they are still the focus of the services offered in four key areas:

- Family Support
- Out of Home Care
- Community Capacity Building
- Child Care

The following sections discuss in turn each of the four targeted emerging practices adopted in the Wanslea case study that demonstrate sustainability and innovation:

1. Selection of Board and Strategy

NfP organisations are registered under Australian standards and governance principles through the Australian Charities and Not-for-profit Commission(ACNC), with constitutional standards set by Standards Australia's Good Governance Principles (AS 8000-2003) and the Australian Securities and Investment Commission (ASIC).

Wanslea's sixfold growth is attributed to maintaining its focus on Mission, the selection and support provided by the Board and a strong leadership team. Wanslea's Board effectiveness is linked to the selection of a highly qualified and capable team and a focus on strategy and governance. The role of the Board is to approve the strategic direction and monitor, protect and enhance the organisational network to the wider community. Wanslea also sought to strengthen its strategic reach through the professional expertise and connections of the Board members. The Board addresses governance, reputation and financial risk mitigation of current services and growth opportunities to complement the existing portfolio of family and children's services.

Significant to the success of Wanslea is the recognition by Board members of the possible negative impact if members are too risk averse to innovation or partnership opportunities. In this case, there is a history of exploring strategic options and determining different scenarios that mitigate risk. This also meant that, as early adopters of good systems and strategic practices, the leadership team was free to explore and create social value by seeking innovative, evidence based global practices.

The Board of Wanslea is committed to its Mission and the provision of evidence based family programs. The Board recognises a sound culture is foundational to the principles of the mission and values if the organisation is to be sustainable. Wanslea mission and values are:

Mission
Wanslea promotes community, family, and individual development through partnerships and services

Vision
Excellence and leadership in services for the community, families and children

Values

- Respect for staff and those engaged with our services
- Integrity in how we work through honest and fair practices
- Collaboration through evidence based practices that ensure quality service provision

Five strategic outcomes designed to grow Wanslea as a NfP are actively measured and reported through to the Wanslea Board:

1. **Financial sustainability**: sustainable business models
2. **Resilient Culture**: workforce engagement to focus on service delivery quality and innovation management
3. **Corporate Business Excellence**: Australian Business Excellence Framework with prioritised incremental and social innovation outcomes
4. **Quality Workforce**: Leadership across the entire organisation with a strong focus on a learning culture
5. **Wanslea Brand**: Recognised for delivering family and children services valued by the community

Strategy

The strategy adopted from 2004, with the uptake of a new CEO, was to seek out partnerships with major parties that brought a significant balance sheet and organisational reputation. The Board and CEO have an appetite for measured risk that sits within a scaffold of measured, underpinning systems.

Wanslea brought the program excellence and skilled personnel to the partnership. In the early days, under these arrangements, Wanslea was the junior partner. Today over time, Wanslea expanded the strategy to include other specialist service providers to complement these skills and as its balance sheet grew it became an equal partner. In more recent times, the strategy has still been partnerships but it now includes senior executives positioning themselves to be voices of influence, through committees, memberships and presentations at conference and forums. There was the deliberate establishment of senior executives as the "go to" people with government and media in selected areas. In the past year, through direct and indirect feedback, Wanslea is seen as having a significant reputation for excellence.

Board Profile and Skills Mix

As with many other NfPs in Australia, participation in the Board is a volunteer position. Tenure profiling and renewal planning have seen a steady flow of membership change through natural attrition and generally resulted in 3–5 year terms.

In the Australian NfP sector, there is a delicate balance between board effectiveness and organisational performance. The Board recognises the need for a diverse Board and workforce capable of leveraging significant social and commercial value. Wanslea's Board has a mix of expertise including social workers; academics in early learning and social work; public sector employees in social services, probity specialist and mental health; marketing specialist; lawyer; Aboriginal/early learning academic; youth social services specialist; business and financial consultants; and a strategic specialist.

The Board skills matrix and expertise represent a cross section and combination of expertise and is augmented through external advisors. The Wanslea Board demonstrates the recruitment of members with specific talents and/or connections within the marketplace. Selection of Board criteria is specific to roles, qualifications and expertise respected in the community and allocated by strategic portfolio and structure of the contracts divisions of work decisions. It is easy to attract, retain and work with talented individuals. People with passion, commitment and the ability to tap into board relationships for partnerships really matter.

Wanslea's golden rule of "no surprises" between the CEO and the Board has built healthy trust and strong performance outcomes. The relationship between the CEO and Chair continues to be transparent and proactive in support of healthy governance. Due to this, the CEO and the leadership team hold a special relationship. Senior leadership members present initiatives directly to the Board, identifying risk and work driven challenges, and inform the Board of the strategic impacts of their current work. This is quite a unique strategic capability.

2. Wanslea Foundations in Research, Evaluation and Education

Social change is a result of capturing credible metrics to determine what is working and what is not. Research and evaluation are a priority and commitment of Wanslea. The social services sector has a high priority for measuring and evaluating outcomes to secure future grants. There is a need to innovate and evaluate and for Wanslea this has meant the allocation of 1% of funded programs to research and development that has enabled the establishment of a research and evaluation initiative resulting in quality decision making where leaders take a strategic approach.

Long term allocation of resources is an investment decision by the Board that is focussed on long term sustainability and excellence. Staff turnover is low and employee engagement is high. This is often attributed to the guiding principles of how organisational excellence is supported through learning, staff development and academic relationships, collaborations and partnerships.

Wanslea deliberately focused on the culture, its people and the way social outcomes would be achieved through targeting research, evaluation and education. It is necessary to apply rigour with reason. Delivering evidence based family and children service programs takes energy and resolve to position foundational scaffolding.

Wanslea invested over the long term into research and evidence based programs in a number of ways. Wanslea was motivated to ensure that practice was informed by the evidence available and that families would access services that have consistency of practice across all local and regional staff. The principles and practices of competent staff are recognised by the community, funding bodies, partners and employees. The investment in research is resulting in a positive impact on the brand and reputation by the industry, stakeholders and consumers.

The families who receive services can be sure that the services are effective. Wanslea's goal is to strategically position itself as a leader in research and evaluation in the community services sector in Western Australia. To do this it initially focused on working with families using the principles of strength based practice. It recognised that beyond policies and rules, there was a focus on evidence-informed approaches to ensure consistency of performance amongst all professional staff in everyday activities.

Mission and Vision to Research and Evaluation

The goal is to strategically position Wanslea as a leader in research and evaluation in the community services sector in Western Australia.

The vision for Research and Evaluation at Wanslea:

- Wanslea is identified as an organisation that uses, implements, develops and builds evidence-based practice
- Wanslea provides best practice services to the children and families with whom it works, seeking and building evidence where there are gaps
- All Wanslea practitioners are able to access and assess best quality research and evidence to inform their work

Wanslea has a research strategy in place, covering priority areas, current and future projects, research partnerships, and the ethics and principles on which its work is based.

With an increased emphasis and requirement for evidence based practice such that one third of federally funded programs must have an evidence base by 2018, Wanslea now has a track record of achievement in this space. Growing complexity of social needs is driving social sector reforms and examination of service design. In 2006 Wanslea was invited to work on the Healthy Start National Strategy led by a collaboration between the Parenting Research Centre (PRC) and the University of Sydney (Dixon 2016) as part of a research trial for parents with disabilities using two evidence-informed programs whereby they eventually co-produced a practice framework that goes beyond this initial program. Dixon (2016) describes the type of evidence based practices necessary and how the staff need to adopt a conscious level of professional practice to achieve social outcomes. The following is a series of key models and thinking underpinning the Wanslea frameworks that are evolving over time.

> Wanslea developed its practice with vulnerable families by linking research evidence to established promising practice. Its investment in research is resulting in recognition in the academic and conference arenas, with a number of staff presenting, peer reviewed journal articles published, and new partnerships to evaluate areas of work, particularly in the early years and grandcarer services. Grants from the Australian Association of Social Workers and Edith Cowan University will focus on documenting practice aspects that can be shared with broader audiences. A successful submission to the Expert Panel for the Family Fun and Learning Program confirmed that our work is evidence-informed and transferrable to other target groups. Data collection has also been introduced to guide and inform decision making at the clinical and organisational level. (Dixon 2016, pp. 6–7).

Learning Organisation and University Links

Learning from peers, universities, and global networks has been inbuilt into the working life of the CEO, the strategic leadership team and employees.

A learning organisation is an aspirational opportunity to develop mastery through peer sharing and for subject experts to create a triple loop learning where the clients receive or society achieves better outcomes (Senge 1990).

Implementation Stages

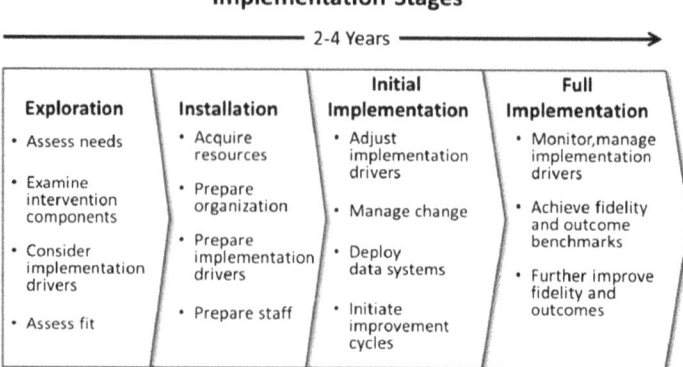

Fig. 2 Implementation science framework. Source: Betram et al. (2013, p. 8)

Leaders in the organisation, are provided with sabbatical time to rejuvenate, time to network with peers locally, and at conferences nationally and globally. Wanslea values its links with academics and university relationships incorporating research and practice development. Strategically there is a funding allocation demonstrating a strong commitment to research and evaluation, to ensure families and children receive services that are effective.

Wanslea's everyday activities supported by Dixon (2016, p. 25) include:

Evidence of effectiveness helps in the selection of what to implement for whom, however,

Evidence of outcomes does not help implement the programme or practice.

Implementation Science is defined as a specified set of activities designed to put into practice an activity or program of known dimensions. Implementation applies four phases (Fixsen et al. 2005), updated, 2013 (see Fig. 2).

Wanslea sought an evidence based approach to ensure consistency of practice and understanding of the social outcomes. Through collaboration with a University, a decision was made to adopt an implementation science approach to developing and embedding the practice framework and the introduction of an Implementation Framework (Fig. 2). The practice evolved from the National Implementation Research Network's Implementation Framework (Fixsen et al. 2005). Implementation stages, used four phases, *Exploration, Installation, Initial Implementation and Full Implementation*. Wanslea practices and programs underwent a structured approach in setting up and evaluating social programs to determine social impact and reporting to regulatory bodies.

Complex social issues and programs are supported at Wanslea by people with dynamic leadership capabilities. To ensure there is a consistency and measurement of social outcomes, leadership addressed adaptive challenges by applying the triangle approach framework in Fig. 3 Improved Outcomes. To improve the outcomes of programs, *organisational drivers* are supported by leadership teams, *leadership*

Fig. 3 Improved outcomes for consistency in program implementation. Source: Betram et al. (2013, p. 11)

drivers are developed to understand complexity science and decision making. These leaders are additionally supported through *competency drivers* (selection, training and coaching).

Utilising the above-mentioned frameworks, Wanslea has implemented and adopted the following guiding principles that Dixon (2016), addressed in their paper, 'Guiding Principles for Wanslea's Practice Framework' for practioners:

- Adult and child voices and choices are heard
- Building on natural supports
- Collaboration across Wanslea
- Working within the community
- Promoting culturally competent practice
- Providing individualised support: services and supports are customised to the unique needs of assessment and goal setting
- Behaviour change
- Skills development
- Working with strengths

Corporate Business Excellence and Innovation

To achieve one of the five strategic outcomes, the Australian Business Excellence Framework (Fig. 4), with prioritised incremental and social innovation outcomes, was included as part of the strategy.

Wanslea introduced the Australian Business Excellence Framework (ABEF) to improve business excellence and sustainable performance of leaders and outcomes,

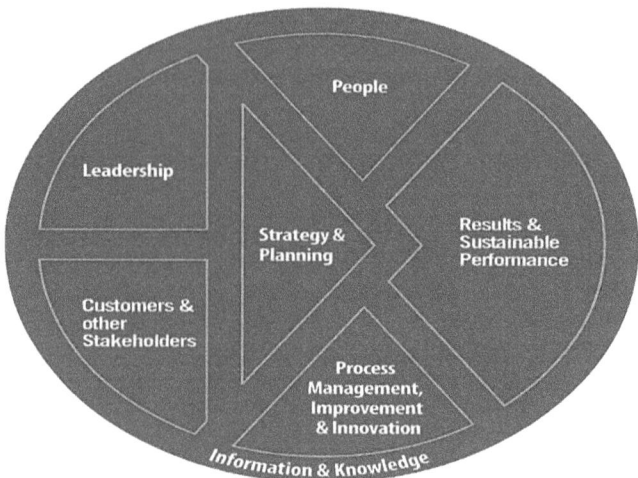

Fig. 4 Australian business excellence framework. Source: https://www.saiglobal.com/ABEF

systems and processes, applying the seven categories and 16 sub-categories of the model. The sixth and seventh category of the ABEF are Innovation and Quality Improvement Results and Sustainable Performance.

The ABEF is an integrated leadership and management system, designed to incorporate essential service and program governance for contracts, and ensure a high level of performance.

It was recognised that to qualify for large scale, long term projects with ongoing funding, embedding a quality framework in all aspects of projects and reporting, improves business performance and brand in the market. A quality framework contributes to sustainable economic mechanisms and increases the sphere of influence in the industry.

Innovation management theory is constantly transforming. As a term, often Social Innovation means a business model or service innovation for a social group. The different types of Innovation such as product, process, or market are different depending upon the scope of change especially if it is incremental, radical or disruptive innovation.

Social Innovation and Innovation Management were a means of 'thinking differently' and partnering with other NfPs by capturing the value and strengths approach of sharing expertise and innovative capability, directly relates to societal value. To be a successful organisation, adopting the ambiguity and paradox lens, directly applies to simultaneously exploring small incremental innovation whilst exploiting new and larger opportunities (O'Reilly and Tushman 2008).

Innovation is anchored in data and evidence. The community sector looks to establish sustainable social value with families and children amongst a dynamic and complex environment. Social Impact is a measure of the interventions and societal outcomes. Innovation at Wanslea can be identified by focusing on a range of internal and external innovation management systems:

1. **Purpose and Mission:** family and children who are vulnerable
2. **Process and practices:**

   a. **Corporate Business Excellence** establishing an internal system for the purpose, design and development of innovation;
   b. **Evidence Based Measures:** design and delivery;
   c. **Implementation Framework**: addressing both a consistency of practice to determine what is successful and what is working; and
   d. **Leadership development:** focus and training on the internal systems to look for innovations and consistency of leadership practices across the divisions and geographical locations (mission and branding).

3. **Performance:** Measure and evaluate for program and service design outcomes
4. **Design of Programs:** Social Impact measurement with the intention to look for innovations
5. **Research & Development, Evaluation & Global Networks:** Infrastructure to support the identification of innovation
6. **Business Model:**

   a. Partnership & Collaborations;
   b. Service Design; and
   c. Continuous Improvement (Australian Business Excellence Framework).

3. Brand Awareness and Reputation

Building a brand and reputation has been a strategic decision and is a catalyst for growth. Government and sector agencies recognise Wanslea's strength in advocacy for family and children. Wanslea is invited as having expertise, by partners for projects and research.

Quality programs have strong brand awareness. Clarity of the brand and reputation in the sector have been a result of sticking to the mission and values and ensuring that the leadership decisions align to the strategic goals. The growth of family programs and services is a result of organisational cohesion, employee engagement, attracting exciting innovative projects and being recognised by partners as having a strategic fit. Funding opportunities through invitation, as opposed to tendering, is a result of being recognised as an organisation focused on outcomes and having clarity around sustainable business, economic, financial and social practices.

As a result of Wanslea's engagement with family and children, there is retention of engaged families, increased volume of volunteers, recognition as advocacy partners in the community and professional sectors. This has been supported by two independent surveys across staff and stakeholders in the past 3 years.

4. Partnerships and Collaborations

Complex society problems are not always easily digested or solved by passionate individuals or organisations. No single organisation alone can adequately deal with the growing level of family complexities. As society witnesses a growing divide between those 'that have and those that don't', complex family structures require different forms of family and children's services.

Organisations engage in cross industry partnerships to fill gaps in service or extend their reach. Strong quality partnerships provide access across the sector and networks, build connections, grow expertise and provide opportunities to foster new thinking with peers. In public and private NfP partnerships, it can increase the reach of the supply chain to broaden the horizons for the future.

Wanslea deliberately sustains growth through targeting values driven partnerships. Today Wanslea has more than 11 agreements that improve this social reach. Partnerships and collaboration have been critical to growth and now reputation. A cornerstone of its work is a belief in relationships with local communities and partnerships with other providers. Through a healthy governance structure, partnerships are 'fit for context and capacity' and are appropriately resourced.

Relationships, trust building and leadership are critical to successful collaboration. Time is required to build trusting relationships with contact partners. The focus continues to be to explore options and agree on important social principles. Clear scope, expectation and accountability agreements are documented. Leaders who value the collective impact recognise long term opportunities, build innovative solutions and organisational capacity that are sustainable in the long term. Since 2004, Wanslea identified three types of partnerships that increase innovation and sustainability business model to deliver social change: (1) partnerships that supported joint funding and fee for services; (2) partnerships where additional *expertise for innovation* enhance social impact and outcomes; and (3) partnerships where enhanced growth of service and social impacts are the objective.

For each project scope, plans and rigorous communication plans were in place. Wanslea worked on the principles of 'strength based leadership' and appreciated the strengths and collective impact each party brought to the partnership. To sustain collaboration and partnerships, Wanslea established a Value Proposition, ROI measures and partnership focus. Identified is the business model for social impact and innovation, with clarity of agreements for updating changes during the life of an innovation or project.

6   Conclusion

In this specific case study, the social and the economic objectives are paradoxically being challenged by the capacity of the industry and, as a result, the sector is challenged to provide ROI measures for its programs in a time of fiscal restraint.

Wanslea is a successful case of an NfP organisation that has successfully pursued the creation of social value for families and children. This case outlines an Australian example where, through sound business practices, utilising the triad of political, economic and social dimensions successfully manages the paradoxical tensions to deliver social innovation.

The understanding and use of terms CSR, sustainability and innovation management were not terms that were formally used by Wanslea. However, in its application, this case contributes to our understanding that measuring social impact is evolving, and that we do see social innovation occurring even though different business terms are being used by NfPs. It is clear that the culture of the organisation and good business practices constantly, and strategically adapt, and share issues will enable them to be sustainable in the long term.

The Board and CEO have an appetite for measured risk that sits within a scaffold of measured, underpinning systems. Tensions and paradoxes are an inherent part of organisational life and political funding for addressing wicked social problems. Sustainable innovation is strategically targeted through partnerships to arrive at scale. Multiple tools and frameworks, including the Australian Business Excellence Framework, research and evaluation, and social impact evidence of innovation implementation, will allow the organisation to be well positioned for future challenges.

References

Aluchna, M., & Idowu, S. O. (Eds.). (2017). *The dynamics of corporate social responsibility, CSR, sustainability, ethics & governance, a critical approach to theory and practice*. Heidelberg: Springer.

Australian Bureau of Statistics. (2016). *Australia demographic statistics, March Quarter 2015*. Retrieved December 20, 2016 from http://www.abs.gov.au/ausstats/abs@.nsf/Latestproducts

Betram, R. M., Blasé, K. A., & Fixsen, D. L. (2013a). *Improving programs and outcomes: Implementation frameworks*. Bridging the research & practice gap symposium, Houston, Texas, April 2013, & Review in a special issue of research on social work practice (pp. 7–11).

Caradonna, J. L. (2014). *Sustainability: A History*. Oxford: Oxford University Press.

Connor, G. C., Leifer, R., Paulson, A. S., & Peters, L. S. (2008). *Grabbing lightening: Building a capability for breakthrough innovation*. Hoboken: Jossey-Bass, Wiley.

Dahlsrud, A. (2008). How corporate social responsibility is defined: An analysis of 37 definitions. *Corporate Social Responsibility and Environmental Management, 15*(1), 1–13.

Dixon, P. (2016). *Dirst 1000 days: Early intervention to support vulnerable families and their children*. Family & Relationship Services Australia (FRSA) 2016 Conference e-Journal (pp. 23–23).

Fixsen, D. L., Naoon, S. F., Blase, K. A., Friedman, R. M., & Wallace, F. (2005). *Implementation research: A synthesis of the literature*. Tampa: University of South Florida, Louis de la Parte Florida Mental Health Institute, the National Implementation Research Network.

Googins, B. (2013). Transforming corporate social responsibility: Leading with innovation. In T. Osburg & R. Schmidpeter (Eds.), *Social innovation – solutions for a sustainable future*. Heidelberg: Springer.

O'Reilly, C., & Tushman, M. L. (2008). *Ambidexterity as a dynamic capability: Resolving the innovator's dilemma. Research in Organizational Behavior, 36*, 165–186.

OECD's Development Assistance Committee (DAC). (2010). *Better policies for better lives Report*. Paris: OECD Publishing.

Rittel, H. W., & Webber, M. M. (1973). Dilemmas in a general theory of planning. *Policy Sciences, 4*(2), 155–169.

Scaife, W., McGregor-Lowndes, M., Barraket, J., & Burns, W. (2016). *Giving Australia 2016: Literature review summary report*. Brisbane: The Australian Centre for Philanthropy and Nonprofit Studies, Queensland University of Technology, Centre for Social Impact, Swinburne University of Technology and the Centre for Corporate Public Affairs.

Senge, P. M. (1990). *The fifth discipline: The art and practice of the learning organization.* New York: Doubleday/Currency.

Smith, W. K., & Lewis, M. (2011). Toward a theory of paradox: A dynamic equilibrium model of organizing. *Academy of Management Review, 36*(2), 382–403.

Teece, D. J. (2007). Explicating dynamic capabilities: The nature and microfoundations of (sustainable) enterprise performance. *Strategic Management Journal, 28*, 1319–1350.

Visser, W. (2011). *The ages and stages of CSR: Towards the future with CSR* 2.0. (CSR international paper series, no. 3). First published in Social Space 2011.

Tricia has been the Chief Executive Officer of Wanslea Family Services since 2004. Wanslea provides family support, out of home care, community capacity building and a range of children's services in metropolitan, rural and regional areas of Western Australia.

Tricia has worked in the community not for profit sector in NSW and Western Australia for over thirty years, primarily in management positions, with a focus on child protection, homelessness, family violence and children's services.

Tricia is also the Chair of the Children and Family Welfare Agencies Association; member of the boards of Community Employers of WA and Families Australia; member of the General Council of the Chamber of Commerce and Industry of WA and the Ministerial Advisory Committee for Child Protection. She has a degree in Social Work, Master of Service Administration, graduate of the Australian Institute of Company Directors; is a Justice of the Peace and a Fellow of the Australian Institute of Management.

Dianna Vitasovic is the Director of Innovation Culture since 1988. Innovation Culture provides management consulting services across multiple industry sectors in local and international organisations. Dianna builds thought leadership to future proof organisations in times of dynamic global competition and growing social complexities. Her expertise is in the progressive field of systematic business innovation culture, building leadership capability during transitions.

Dianna's background in business consulting and business education enables her to challenge strategies to foster innovation cultures. She partners with Senior Leadership Teams to challenge their thinking and clarifying their purpose, priorities, performance and values alignment, stimulating sustainable and systematic innovation with quality decisions.

Since 2007 Dianna has been a Principal in the Centre for Innovation in Decision Quality (CIDQ), a Curtin Business School initiative combining research and practice in the area of Innovation in Decision Quality. Experience in design and delivery of Executive Education and lecturing on MBA programs, in Organisational Behaviour, Strategic Human Resources Management and Leading Innovation and Change.

Dianna is a member of ISPIM International Scientific Professional Innovation Management as a Scientific Panel Case Reviewer for Conferences: 2011 Winner of the ISPIM Scientific Panel Award. Committee member for GIO, 2016 ICT public sector forum committee (The Future of Digital Government). Past advisor for Public Sector Innovation over 4 years. Consults with clients for Radical or Breakthrough Innovation projects. Until 2015, Advisory Board Chair for Inspire Foundation, Youth Mental Health and Welfare. Fellow of the Institute for Learning Practioners.

Lightning Source UK Ltd.
Milton Keynes UK
UKHW02n2031270918
329549UK00009B/251/P